Pandemic Diaries

Soumik Gangopadhyay · Soma Sur ·
Rabin Mazumder · Saugat Ghosh
Editors

Pandemic Diaries

A Casebook of Corporate and Society Practices

Springer

Editors
Soumik Gangopadhyay
Institute of Engineering and Management
University of Engineering and Management
Kolkata, West Bengal, India

Soma Sur
Xavier Business School
St. Xavier's University
Kolkata, West Bengal, India

Rabin Mazumder
Institute of Engineering and Management
University of Engineering and Management
Kolkata, West Bengal, India

Saugat Ghosh
Xavier Business School
St. Xavier's University
Kolkata, West Bengal, India

ISBN 978-981-96-5414-7 ISBN 978-981-96-5415-4 (eBook)
https://doi.org/10.1007/978-981-96-5415-4

This Springer imprint is published by the registered company Springer Nature Singapore Pte Ltd.
The registered company address is: 152 Beach Road, #21-01/04 Gateway East, Singapore 189721, Singapore

Preface

COVID-19 pandemic has challenged the fundamental existence of mankind. A microscopic organism posed the greatest question to mankind about its existence and the most evolved animal of the planet earth was flabbergasted. Zero preventive option, unidentified solution, high risk of infection, unexpected mortality have compelled the society hide behind the doors due to lockdown, work from home, and quarantine as a strategic step to escape the infection. Life and livelihood faced the greatest challenge of the century. On the one hand the death tolls were rising and the other was witnessing disastrous fall in commercial value. Many firms collapsed, many are still struggling to survive and even the giants were also perplexed and severely hurt. During turbulence time or crisis, the very necessities of implementing innovative practices were redefined and society and commercial organizations became agile and resilient to cope up with the situation. Sustainability is a pervasive concept applicable across social science, economics, and commerce. Moreover, it is a time-tested approach for stability of growth and prosperity of business. But when the growth of economy regularly reach higher altitudes, a prevailing feeling and aspiration of further growth often leads to untenable steps. But business houses ignore such follies as short-term economic clouds the long-term vision. Innovation of different magnitude became the strategic weapon in this struggle for existence. Applicability of innovative approaches is sector specific. It does not always guarantee success but have profound value as a reference. It can even ensure specific outcome which led way to further pursuit in the process of innovation. Many of them with higher magnitude are well witnessed, documented, and presented in front of the world for further study and improvement. But there are several discrete practices of different scales, which enacted a pivotal role in the process of corporate survival, but are often ignored/unknown due to research and publication in this field. Case-based research analysis provides access to these hidden, ignored but effective strategies. The proposed book will be a ready reference for practitioners and budding managers. This can facilitate the understanding of the depth of a problem from a human—situational perspective and the illustration and the solution of the problem stimulate new practices as it becomes an example for others to follow. The proposed book is aimed to document the human effort to attain sustainability in their business using the case study methodology.

Kolkata, India

Soumik Gangopadhyay
Soma Sur
Rabin Mazumder
Saugat Ghosh

Acknowledgments

At the outset of this acknowledgement, we are thankful to the entire team of Springer Nature Publication for their dedicated efforts and their immense enthusiasm in shaping this creative and unique book, compiling a vast array of research works on a single theme. The hard work and constant partnering done by the editorial board are especially noteworthy.

We are thankful to all the authors of the chapters of this book, who have showcased their research acumen in their work. Their strict adherence to the rules laid by the editorial team and given timeline helped us to compile quality works within the given duration.

Constructive criticism is always good and is particularly very helpful when it is unbiased. We convey our immense gratitude to all reviewers whose valuable inputs and comments have helped us to take this book closer to the perfection. Their knowledge and vast research experience helped us to identify quality research works which helped us in introducing this unique book to the researchers and practitioners of diverse field.

No work of essence is possible without the support and patronage of the organizations where we work and we all are thankful to work in organizations which provide the impetus to pursue such endeavours. We convey our sincere gratitude to Prof (Dr.) Satyajit Chakarbarti, Director, IEM & UEM Group, Kolkata; Rev. Dr. Jhon Felix Raj, S.J., Vice- Chancellor, St. Xavier's University, Kolkata, India; Late Prof. (Dr.) Satyajit Chakarbarti, President, IEM & UEM Group, Kolkata, India, and Professor Banani Chakarbarti, Registrar, IEM & UEM Group, Kolkata, India, for their constant motivation and support during the project.

This book has been planned on the support of many people to whom we are indebted. We gratefully acknowledge the contribution of the internal editorial team members who, through their learned criticism, contributed substantially to improve the content, quality, and structure of the book.

We are also on the note regarding our family and all of us would like to thank our families deeply and sincerely, who have been extremely supportive toward this vital project. We also extend our thanks to our well-wishers and friends for their constant support and encouragement to make this book live.

About This Book

The book, "*Pandemic Diaries: A Casebook of Corporate and Society Practices*", will provide an insight into the struggle of businesses for their existence. The grave issues during pandemic, the threats, and the process of curving opportunities are the main focus of the book. Using the case-based approach, the book aims to cover the recent trends, issues, and practices by focusing on Corporate Governance, Corporate Social Responsibility, Information Technology, Psychological stability, Marketing, human resource, Sustainability during COVID-19 pandemic. It is intended to help the readers in understanding the strategic management practices of business and the application of innovations which took place during pandemic and their effectiveness on sustainability of the business. Sustainable issues in the context of both subjective and objective dimensions have been analyzed by the authors of the chapters. Chapters of the book will explore strategic applications of creative issues and trends in different field of management. Thoughtfully narrated game changing strategic case-based practices with respect to marketing management, human resource management, financial management; supply chain management as applicable for the corporates and societal aspects for sustenance has been narrated in this book is an eye opener for the budding managers, practitioners, and researchers.

Contents

Editors and Contributors

Md. Zahir Uddin Arif Department of Marketing, Faculty of Business Studies, Jagannath University, Dhaka, Bangladesh

Hanzala Awais Department of Commerce, St. Xavier's College (Autonomous), Kolkata, West Bengal, India

Darshna Bhatter St. Xavier's College (Autonomous), Kolkata, West Bengal, India

Soma Bose Biswas Associate Professor, Department of Management Science, Techno International New Town (Affiliated to MAKAUT), Kolkata, India

Indranil Bose School of Management, MIT University of Meghalaya, Shillong, India

Aman Chadha Department of Management Studies, Indian Institute of Information Technology, Allahabad, India

Debashree Chakravarty School of Management, KIIT University, Bhubaneswar, India

Arundhati Das Chatterjee St. Xavier's University, Kolkata, India

Subrata Chattopadhyay Institute of Engineering & Management, School of University of Engineering and Management, Kolkata, West Bengal, India

Manoj Kumar Chaudhary Amity Business School, Amity University, Kolkata, India

Udit Chawla Institute of Engineering & Management, School of University of Engineering and Management, Kolkata, West Bengal, India

Peu Das Department of Business Administration, Institute of Engineering and Management, University of Engineering & Management, Kolkata, India

Sanjay Das Department of Management and Marketing, West Bengal State University, Barasat, West Bengal, India

Siddharth Das Institute of Management, Nirma University, Ahmedabad, India

Kamalika Dasgupta Department of Basic Science and Humanities, Institute of Engineering & Management, University of Engineering and Management Kolkata, Kolkata, India

Sohini Datta Institute of Engineering & Management, School of University of Engineering and Management, Kolkata, West Bengal, India

Sanchita De Department of Basic Science and Humanities, Institute of Engineering & Management, University of Engineering and Management Kolkata, Kolkata, India

Arijita Dutta Department of Economics, University of Calcutta, Kolkata, India

Soumik Gangopadhyay Department of Business Administration, Institute of Engineering &Management, UEM-Kolkata, Kolkata, India

Avradeep Ganguly Department of Management, School of Business, Adamas University, Kolkata, India

Debopriya Ghatak Department of Psychology, Adamas University, Barasat, India

Saugat Ghosh Xavier Business School, St. Xavier's University, Kolkata, India

Pritha Ghosh University of Engineering & Management, Kolkata, India

Jayendra S. Gokhale Embry Riddle Aeronautical University, Daytona Beach, FL, USA

Akriti Gupta Department of Management Studies, Indian Institute of Information Technology, Allahabad, India

Sananda Halder Department of Business Administration, Institute of Engineering &Management, UEM-Kolkata, Kolkata, India

Shweta Kishore Department of Business Administration, Institute of Engineering & Management, University of Engineering and Management, Kolkata, West Bengal, India

Sabat Kumar Digal Department of Commerce, Rama Devi Women's University, Odisha, India

Shreosee Kushari Institute of Management, Nirma University, Ahmedabad, India

Tamal Maity CAPLET INDIA PVT. LTD., Kolkata, West Bengal, India

Amrita Majumdar Amity Business School, Amity University, Kolkata, West Bengal, India

Sudipta Majumdar School of Business, Adamas University, Kolkata, India

Nishi Malhotra Indian Institute of Management, Sambalpur, India

Rabin Mazumder Department of Business Administration, Institute of Engineering & Management, University of Engineering & Management, Kolkata, India

Debolina Mitra Indian Institute of Management, Bangalore, India

Shubham Mondal Kargo Motors, Kolkata, India

Swati Mukherjee College of Management, Institute of Engineering & Management, Kolkata, India

Devika Mulchandani The Calcutta Anglo Gujarati College, Kolkata, West Bengal, India

B. C. M. Patnaik School of Management, KIIT University, Bhubaneswar, India

Gouranga Patra Department of Management, School of Business, Adamas University, Kolkata, India

Ipsita C. Patranabis Associate Professor, Department of Management Science, Techno International New Town (Affiliated to MAKAUT), Kolkata, India

Arna Ray Department of Commerce, School of Business and Economics, Adamas University, Kolkata, India

Rupsha Roy Amity Business School, Amity University, Kolkata, India

Sayani Saha Faculty of Commerce and Management, St. Xavier's University, Kolkata, India

Swikriti Saha Department of Computer Application, Institute of Engineering & Management, UEM-Kolkata, Kolkata, India

Soumen Santra Department of Master of Computer Application, Techno International New Town, Kolkata, West Bengal, India

Anirban Sarkar Department of Management and Marketing, West Bengal State University, Barasat, West Bengal, India

Shankha Shubhra Sarkar KIIT School of Management, Bhubaneswar, India

Ipseeta Satpathy School of Management, KIIT University, Bhubaneswar, India

Paramita Barman Sen Faculty of Commerce and Management, St. Xavier's University, Kolkata, India

Soumojoy Sen British Telecom, Kolkata, India

Soma Sur Xavier Business School, St. Xavier's University, Kolkata, India

Vijayshri Tewari Department of Management Studies, Indian Institute of Information Technology, Allahabad, India

Jhilli Das Tewary Department of Psychology, St. Xavier's University, Kolkata, India

Part I
COVID-19 and Strategic Management Practices: A Paradigm Shift

Chapter 1
Blended Learning: A Complementary Approach to Traditional Classroom Education in the Post-COVID-19 Era

Sanchita De and Kamalika Dasgupta

Abstract This study discovers the growing preference for blended learning approach in educational system in India. Blended mode of learning has become increasingly popular in India, as well as in the world particularly in the post-COVID-19 period. Before that countries like India did not have the mass awareness about such a parallel mode of teaching–learning, though some of the first world nations were already implementing them for a long time. Blended learning integrates online and face-to-face methods, resulting in flexible and personalized learning experience. The current research investigates the determinants of growing popularity of blended learning among the Indian students in the post-COVID-19 era. To understand what makes them choose hybrid approach over traditional in-person classes a pilot survey was conducted with participants aged 18–35 years. The regression analysis finds out that technological proficiency, age and income have positive impact on seamless learning by the students in a blended environment, while students living in rural areas face difficulties in hybrid learning system because of having limited access to stable internet connection, the urban and sub-urban areas with stable internet connection and other facilities do not even want to go back to the traditional system. This present study altogether provides some valuable insights into the factors which initiated the adoption of blended learning as complementary approach to traditional classroom learning system in India in the post-COVID-19 period and paves the way for future research in this field.

Keywords Online learning · Hybrid learning · e-learning · Blended mode · e-content · COVID-19 · Pandemic · Paradigm shift

S. De (✉) · K. Dasgupta
Department of Basic Science and Humanities, Institute of Engineering & Management, University of Engineering and Management Kolkata, Kolkata, India
e-mail: sanchita.de@uem.edu.in

S. Gangopadhyay et al. (eds.), *Pandemic Diaries*,
https://doi.org/10.1007/978-981-96-5415-4_1

1.1 Introduction

Blended learning is a learning method which merges traditional in-person learning techniques with virtual learning method, generating a dynamic, flexible and engaging learning environment. This newly emerging educational method integrated the advantages of face-to-face interaction such direct communication and hands-on learning with the flexibility of digital learning. This hybrid model offers several assistances to the students by increasing their access to larger domain of educational resources. It also brings about opportunities for the students so that their digital literacy skills will get developed. In recent times, this innovative learning is becoming popular in different educational set-ups as technology gets upgraded. The blended learning tool is a strong alternative to conventional learning methods due to the advancement of technology and increased awareness of the limitations of traditional teaching methods.

1.1.1 The Evolution of Blended Learning

With the introduction of digital technology in education system, blended learning originated its root in late twentieth century. At the beginning of this transformation phase, the use of technology in offline classroom was restricted to few supplementary study materials like PowerPoint presentation and the use of other digital resources. As technology has evolved, Learning Management Systems (LMS) and other digital tools have been developed to enable more comprehensive blending of instructional methods for educators.

The rising access to stable internet and the explosion of digital devices accelerated the adoption of hybrid learning further. Online learning inputs such as e-textbook, virtual expert talks became much more accessible for the educators as well as learners. Furthermore, the COVID-19 pandemic in 2020 forced all the educational institutions worldwide to transform the teaching–learning method from a relatively rigid physical classroom learning approach to a comparatively flexible technology-driven distance learning approach bringing about significant dependence on blended learning. In the present time, blended learning has got wide recognition as a valuable learning instrument to foster student participation, improvement in learning outcome and to enhance lifelong learning skills.

1.1.2 Key Components of Blended Learning

Blended learning incorporates a range of instructional formats, including the following core components:

(a) **Face-to-Face Instruction:** The traditional teaching–learning facilitates direct communication, instantaneous feedback and hands-on learning like experiments, presentation, demonstration and discussion in a physical classroom setting to enhance a sense of community. This is the conventional teaching–learning method with which the students are mostly used to. Group work, discussions and hands-on projects leverage peer learning in such a set up. This component of blended learning also makes it possible for the educators to provide individual care and attention to the students and cater to their individual learning needs.
(b) **Online Learning:** Online learning offers digital study contents and resources through the internet to allow students to access those materials at their own pace and as per their own schedule. Laptops, tablets and smartphones are used to access online learning materials and to participate in online classes. Discussion forums, video conferencing (through Zoom, G meet, etc.) and online collaboration platforms makes student engagement possible in this set up. But it is impossible for the educators to attend to the individual needs of the students.
(c) **Learning Management Systems (LMS):** Learning Management Systems (LMS) refer to the digital platforms used for accessing course materials, submission of coursework, assessments and communication which helps the educators to deliver content, track student performance and simplify collaboration. These are the most important factors behind the successful and smooth implementation of blended learning.
(d) **Synchronous and Asynchronous Learning:** Synchronous learning incorporates live interactive sessions with all participants like remote learning, webinars, etc. whereas asynchronous learning permits the students to independently engage with study materials as per their own pace and schedule. Synchronous learning is possible both in online and face-to-face mode. But asynchronous learning is only possible in online mode. Synchronous learning involves real-time interaction, instant doubt clearing and immediate feedback, which creates a sense of community among the students. However, in asynchronous learning doesn't give a chance to connect with the peers and hence the students might feel isolated sometimes.
(e) **Digital Collaboration Tools:** Digital collaboration tools like discussion forum enable students to work together by sharing idea and providing feedback. The critical thinking and problem-solving skills of the students can be developed when this active learning system gets adopted.

1.1.3 Benefits of Blended Learning

In modern education, blended learning is a popular choice for the educators as well as students for offering the following advantages:

(a) **Flexibility and Accessibility:** Blended learning gives the students the accessibility to the course material at any time at any place. This flexibility in learning style makes the education easily accessible for a larger audience, including remote and international students.

(b) **Personalized Learning:** More personalized learning can be promoted through blended education approach as it allows the students for their own improvement at their own pace and as per their own need and schedule.

(c) **Enhanced Student Engagement:** Blended learning creates more engaging learning environment for the students by efficiently utilising multimedia resources related to studies such as video which easily attracts students' attention. For the clear understanding of the complex theories by the students, this dynamic education method is employed frequently by many educational institutes.

(d) **Improved Learning Outcomes:** Blended learning may lead to improvement in learning outcomes for the students as it promotes critical thinking skills, problem-solving capacity and self-dependent learning of the students which are very much required to achieve success in twenty-first century.

(e) **Cost-Effectiveness:** Blended learning is the most cost-effective education method as it curtails the requirement for physical infrastructure for delivering lecture in classroom or in seminar. Digital tools can easily be reused and updated, cutting down the cost of learning. Moreover, blended learning reduces the travelling cost and accommodation cost for the students as well as educators.

1.1.4 Challenges of Blended Learning

Despite its many advantages, blended learning also presents several challenges that must be addressed to make it more effective:

(a) **Digital Divide:** Students belonging to low-income group or the students living in rural regions may not have easy access to technology and stable internet. Those students remain deprived of enjoying the advantages of blended learning. Equal access to technology and internet must be ensured to provide the benefit of hybrid learning to the entire student community. This is the problem of "inconsistent implementation" which has become especially prominent in developing countries like India. While blended learning has emerged as the "new normal" in the urban areas, the rural areas are still lagging behind in this aspect.

(b) **Technical Skills and Support:** If the teachers-learners do not possess adequate technical skills to deal with digital tools, blended learning will not be effective. Proper workshop and training programme to improve the pedagogical skills must be organized, offering technical knowledge to educators and students. Blended learning is time consuming and resource intensive, therefore constant professional development in this regard is very essential for its smooth operation.

(c) **Student Motivation and Self-Discipline:** Blended learning needs high level self-motivation and discipline for the students while participating in asynchronous activities, otherwise the students will suffer from the struggle of time management, resulting in lower rates of participation and completion. Sometime it is hard to keep the students motivated in the online aspects of blended learning, especially when they are not self-motivated towards learning, they get distracted or feel isolated in online environment. In order to avoid such demotivation and distraction student-engaging activities (in terms of games and competitive quizzes) and constant interaction are very much needed.

(d) **Quality Assurance:** All institutions adopting blended learning should establish proper assessment system to monitor the effectiveness of blended learning programme to assure quality education and maintain high educational standards. The courses taught should integrate online and face-to-face components in a balanced manner. Learning outcomes should be clear to both the educators and the learners. Also blended learning should always mean active learning, where students would be engaged in interactive discussions and group activities should be integrated into both environments. This makes both the environments equally interesting.

1.1.5 Blended Learning: Personalized Learning Pathways Powered by AI Algorithms

The incorporation of artificial intelligence (AI) into blended learning at higher institutions of learning has had dramatic impacts on the nature of training as well as the learning habits of the students. AI-powered tools greatly assist in generating not only a more personalized learning experience, but in maximizing the use of available resources and delivering learning in a more interactive and better focused manner. AI is radically enhancing blended learning by integrating the e-learning system with adaptive learning technology. The system continuously pays attention to the child's way of the learning, the child's skills and difficulties, and customizes the rate and content of the syllabus in an entirely individualistic manner. For instance, AI programmes embedded in learning management systems such as Coursera and EdX suggest videos, articles and quizzes as additional training materials based on the learner's achievement and targets. In simple terms, the system makes sure that learners only get the right help they require while at the same time keeping them motivated by not overloading them with overly complex or too basic materials. Such customized paths are necessary in higher learning where students are bound to have a wide range of knowledge as students already have different educational backgrounds.

AI also strengthens assessment and feedback systems in mixed-mode teaching. The machine learning possessed automated grading systems are able to mark pieces of work, quizzes and even essays with great effectiveness and within the shortest time possible. With these, educationists can concentrate on other more complicated qualitative aspects of teaching and provide students with feedback on their actions immediately. Tools that are powered by artificial intelligence such as GradeMark from Turnitin or Grammarly are concerned with the grade but they are equally interested in the overall work of the student in relation to the content that is being submitted.

Besides, these systems also can help to detect persistent mistakes and/or forms of misapprehension, allowing the teachers to be more preventative and to tackle particular problems. For example, in programming or in mathematics, AI systems like Codio or ALEKS are capable of explaining why each of a student's answers was wrong helping the student arrive at a more correct understanding. Such speed and precision of feedback provision mechanisms are especially appreciated in the course of university education because it is regarded as twinned with appraisal of deep thinking and dealing with practical issues.

Blended learning takes on a whole new dimension with AI as another core attribute abandoning the aspect of communication to the virtual worlds. Chatbots and virtual advisors, including IBM's Watson or Microsoft's Cortana, can be helpful in interclass communication as students and teachers can ask or answer basic questions and, in some instances, exchange academic advice. These tools are especially handy in large-scale blended learning models where otherwise teachers may struggle to interact with students in a smaller time span. For example, in a course at Georgia Tech, an AI programme named "Jill Watson" has on many occasions replaced human beings in answering questions posed by students so that these people can attend to situations where discussions have more depth and periphery. A sort of virtual advisor, AI can as well skim through verbal and nonverbal communication data on forums where group projects are posted and highlight students whose activity levels appear to be low or activity levels alongside effective collaboration are below par. In such a case the teachers come up with the appropriate actions to help stem the challenges and ensure that all students take part in learning activities. Above all, let's not forget that AI plays a significant role in creating and enabling learners engaging in experiences within the blended learning models. Blended learning in augmented and virtual reality-powered environments aids students in deepening their understanding of advanced knowledge.

AR tools enhance the physical world by overlaying information on it, thereby exposing the students to new ways of engaging with their environment. In areas such as archaeology or environmental science, AR components can simulate outdoor activities in the classroom, allowing students to visit ancient or ecological places in the virtual world. Technologies such as these increase the interest of students, while enabling them to acquire skills that are relevant to their future employment.

Student achievement prediction, the use of AI in blended learning, is changing for the better the way in which higher education institutions assess and assist students in their development. Nariakis: "Many of these artificial intelligences are imagining courses as a student lens. They set out to study the management system of learning processes (LMS) which data has been archived for a long time. Distance learning achievements, forum attendance, task execution, and quizzes complete were these requirements for assessment to offer preventive measures". For example, Blackboard and Canvas use AI metrics to produce reports on students who might require additional attention and make such updates available to educators. This kind of strategy allows institutions to have all the necessary resources to provide students with timely assistance using study interventions, for instance emergency help lessons or working out a separate study plan. Predictive analytics may also identify which topics are more challenging for the students which educators design curriculum, accordingly teaching strategies may be refined.

AI-driven tools reduce the administrative burden on educators in higher education because tasks such as scheduling, enrolment management and resource distribution may be automated employing AI. As a result, efficient blended learning environment can be created by allowing the educators to focus mainly on teaching. AI algorithms in blended learning provide maximum learning experience for the students. It may lead to more collaborative works in between students and teachers. Inclusivity and accessibility in higher education may get promoted through AI-powered tools in blended learning. Students with disabilities may be included in higher studies in blended learning system by offering customizable learning interfaces and voice recognition software. Otter.ai or Microsoft's Immersive Reader transcribes lecture and make the content easily accessible to hearing-impaired students. AI-driven language translation tools remove language restrictions, enabling foreign students to participate in blended learning programme overcoming their linguistic challenges. As AI technologies evolve, they would be capable of addressing complex educational needs. Advancement in AI system makes blended learning more dynamic, interactive and effective, helping students to be more equipped with skill and knowledge. As technology will get more advanced, the significance of AI in blended learning will be higher. As a consequence, new possibilities for innovative and excellence in education would be offered to make blended learning system much more powerful and inclusive in nature.

1.1.6 Objective of the Study

The goal of blended mode of learning is to leverage the strengths of both face-to-face and online methods to create a more flexible, personalized, and effective learning experience. The idea of blended learning became popular in the post-COVID-19 period, particularly in India. Now that we are back to normal life, blended learning has emerged as a "new normal" in the field of education. Students across the country who have access to online tools are now preferring online courses and study materials to keep up their pace of learning. This was a practice in the western world, but the concept of hybrid learning is particularly new in India. So it is actually worth trying to find out why so many students are preferring hybrid mode over offline classes? Is it because hybrid mode of learning has positive aspects of both online and offline learning which makes the entire learning process smooth? Or is it something else? There is dearth in the literature highlighting these issues particularly in Indian context. The present study will fill up this gap and contribute to the existing literature in this direction.

1.2 Survey of Literature

Blended learning brings a transformation in education by combining traditional offline learning method with remote learning, targeting better teaching style and student outcome [1, 2]. This method is being adopted by many educational institutions for achieving high-quality learning atmosphere [3–5]. Hybrid learning brings about revolution in twenty-first century through providing diverse learning experience for the students by blending virtual classroom-based study resources [3, 4, 6, 7]. A personalized learning ambience can be created for the students through the combination of in-person interaction and online educational content in blended learning method which would ultimately result in large benefit for the learners [5]. The primary objective of blended learning is shift from traditional lecture-based teaching to more productive learner-centred approach which is more engaging for the students [8]. Blended learning can be viewed as an evolution in educational method to fulfil students' learning needs [9]. This new education system involves a strategic combination of online and classroom teaching–learning to optimize student outcomes [10]. Kazu and Demirkol [11] have described blended learning as the mixture of face-to-face instruction and distance learning. They have also termed this new educational method as hybrid learning and mixed learning. The article by Oliver and Trigwell [12] argues against the concept of "blended learning". It suggests that this new teaching–learning process is based on unclear ideas. According to this article, this new teaching–learning system does not focus on actual learning. This research proposes a new way to think about it through variation theory.

1.3 Data and Methodology

We did a pilot study using the convenience sampling method. The target group of respondents are aged between 18–35 years. This age group justifies the research objectives as they usually have the access to all the necessary resources required for blended learning. The questionnaire consisted of questions related to the respondents "demographics, their technological knowhow and their seamless learning experience using hybrid mode". The demographic information consisted responses on their age, location (rural, urban, semi-urban, etc.), number of heads in the family, monthly income and the level of education. Majority of the respondents (almost 60 percent) belong to the age of 19–25 years which states that the data mostly captures the point of view of graduate (42.1 percent) and post-graduate (12.1 per cent) students. Nearly 52 percent of the respondents belong to a nuclear family consisting of 1–3 members and almost 68 per cent are residing in urban areas. This indicates that our surveyed population have easy access to both online and offline tools of learning.

In order to have preference for blended learning over traditional mode of learning one needs to possess technological expertise and must also have access to necessary tools and techniques. After demographics the next section of the questionnaire asks the respondents questions related to their technological knowhow. The names and the descriptions of the technological knowhow components are listed in Table 1.1.

The above-mentioned components are clubbed together to form a "Technological Knowhow Index" as per the standard indexing procedure using PCA (see following section for the detailed results), for the students which later serves as one of the independent variables in our analysis. The other independent variables are the basic demographic ones, i.e. age, education level, monthly income, family size and residing area. As per the hypothesis of the research, the ease of learning is considered as the dependent variable. However, how seamless the hybrid learning experience is for the students is not possible to capture with just one component. Taking insights from the available literature and our understanding we have decided to gather information on total of eighteen components that will perfectly capture the seamless learning experience using hybrid mode. The names and the descriptions of these components are listed in Table 1.2.

The above-mentioned components are again clubbed together to form the "Seamless Learning Index" (see following section for detailed description) which is used as the dependent variable in our analysis. We have employed the Multiple Linear Regression Model (MLRM) and find out the Ordinary Least Square (OLS) estimates to identify the factors which are significantly responsible to provide a hassle-free learning experience to the students using hybrid mode. The equation of the MLRM is defined as follows:

$$y_i = \alpha + \beta_1 x_{1i} + \beta_2 x_{2i} + \beta_3 x_{3i} + \beta_4 x_{4i} + \beta_5 x_{5i} + \beta_6 x_{6i}$$

where $i = 1(1)133$, $x_1 =$ technological knowhow index, $x_2 =$ location, $x_3 =$ monthly income,

Table 1.1 Description of the components of students' technology background

Component	Description
Internet stability	How stable internet connection do you get in your area?
Mode of learning	Which mode of learning do you mostly use?
Familiarity	Are you familiar with hybrid mode of teaching–learning?
Time of use	When did you start using blended mode of learning?
Preference	What do you prefer in the blended course?
Typing proficiency	The student is proficient with typing and keyboarding
MS word proficiency	The student is proficient with MS Word
PowerPoint proficiency	The student is proficient with MS PowerPoint
Excel proficiency	The student is proficient with MS Excel
FrontPage proficiency	The student is proficient with MS FrontPage
Internet proficiency	The student is proficient with WWW and web search
E-mail proficiency	The student is proficient in using E-mail
Comfort to use technology	Overall comfort level with technology

Table 1.2 Description of the components of seamless learning experience

Component	Description
Login	How difficult it is to login to the online learning system?
Instruction	How difficult it is to understand the instructions?
Exercise	How difficult it is to complete the exercises?
Grade checking	How difficult it is to check grades?
Technical support	Is technical assistance easily available?
Place of use	Where do you usually use the computer?
Mode of access	How do you access Internet?
Frequency of use	How often do you use the computer?
Pace of learning	Can you control the pace of your learning in hybrid mode?
Understanding material	Are the online course materials were difficult to follow?
Understanding assignments	Are online assignments are helpful in understanding the course content?
Blended learning is better	Time is spent better way in blended learning than in just offline learning
Time management	Is time management difficult in blended mode?
Idea sharing	Is it difficult to share ideas with classmates in blended mode?
Parallel method	Can you do online and offline classes simultaneously?
Recommendation	How strongly would you recommend blended learning to a friend?
Complementarity	Are face-to-face learning and online learning complements to each other?
Effectiveness	

x_4 = level of education, x_5 = age, x_6 = family size. $\beta_1, \beta_2, \beta_3, \beta_4, \beta_5, \beta_6$ represent respective estimated regression coefficients and y_i stands for the dependent variable "seamless learning index".

1.4 Results and Discussion

The aforementioned two indexes are constructed after the Principal Component Analysis (PCA) which helped us to understand the weightage of each of the component. Not all the components carry equal weightage. Some are more important than others. It also helps in reducing the dimensionality of data while preserving as much variability as possible. Indexes often involve multiple variables or indicators. PCA reduces the number of these variables by transforming them into a smaller set of uncorrelated variables called principal components. This makes the data easier to analyse and interpret. By concentrating on the principal components with the highest variance, we effectively reduce the impact of less informative or

Table 1.3 PCA of the components of "Technological Knowhow Index"

Components	Eigen values
Internet stability	4.14
Mode of learning	1.42
Familiarity	1.32
Time of use	1.08
Preference	0.86
Typing proficiency	0.79
MS word proficiency	0.76
PowerPoint proficiency	0.60
Excel proficiency	0.58
FrontPage proficiency	0.45
Internet proficiency	0.36
E-mail proficiency	0.34
Comfort to use technology	0.24

Source Authors' own calculation

noisy features from the index. The Eigen values from the PCA results are considered as weights while preparing the indexes. Table 1.3 shows the PCA results of the components of "Technological Knowhow Index".

From the result above it is clearly understood that internet stability is the most important factor when it comes to the students' technological background. A stable internet connection is the prerequisite to have access over online learning resources. It is followed by the mode of learning (majority of the respondents prefer hybrid mode), familiarity with hybrid mode of education (majority said yes), and time of use (majority said post-COVID-19). Thus the weightage found from the PCA results are perfectly consistent with the survey and the literature.

Next we present the PCA results of the components of "Seamless Learning Experience Index". Going by the similar understanding here also we find that the component "login" holds most of the importance when it comes to a hassle-free blended learning experience. Along with classroom, face-to-face teaching if a student finds it difficult to login to the online portal then her blended learning experience might not seem so seamless. This is followed by ease of instruction following, ability to solve exercise, checking grades, having technical assistance and the place (institute or home) where students use the computer, etc.

Table 1.4 represents the weightage of the respective components in computing the index of dependent variable (y_i).

The regression results presented in Table 1.5 indicate that the Technological Knowhow Index is a significant determinant of the Seamless Learning Index. It has a positive impact on seamless learning, suggesting that a student who is technologically proficient and well-equipped will perceive the entire learning system as more seamless when blended learning is implemented. Such students are likely to

Table 1.4 PCA of the components of "Seamless Learning Experience" Index

Components	Eigen values
Login	5.41
Instruction	1.95
Exercise	1.58
Grade checking	1.25
Technical support	1.11
Place of use	1.002
Mode of access	0.81
Frequency of use	0.76
Pace of learning	0.68
Understanding material	0.646
Pace of learning	0.68
Understanding material	0.646
Understanding assignments	0.640
Blended learning is better	0.57
Time management	0.48
Idea sharing	0.39
Parallel method	0.35
Recommendation	0.30
Complementarity	0.27
Effectiveness	0.21

Source Authors' own calculation

have a more favourable view of blended learning compared to those with relatively limited access to technology.

Age is another factor that significantly and positively influences seamless learning, meaning that the learning process becomes more seamless for a student as they get older. It is quite logical that as the students age, they are likely to pursue their higher education in which the benefits of blended learning will be realized. Monthly income also has a positive effect on seamless learning for students. This is because of

Table 1.5 OLS regression results

Variables	Coefficient	Std. error	P-value
Technological knowhow index	**0.82**	0.12	**0.000*** **
Location	-0.001	0.92	0.988
Monthly income	5.33	1.57	0.734
Age	**0.03**	0.017	**0.032** **

Source Authors' own calculation
Note ***, **: Significant at 1% & 5% level respectively

the fact that higher monthly income makes students more technologically enriched, which may lead to seamless learning experience for them by participating in blended learning.

Finally, the location of a student's residence has a negative impact on seamless learning. This present study has surveyed students from both rural and urban areas, using a binary variable with "0" assigned to students living in rural areas and "1" to those living in urban areas. It has been observed that there exists a negative impact of location of the residence on seamless learning experience by students living in rural regions. Participating in blended learning system is very challenging for rural students. They do not have easy access to stable internet connectivity, resulting in incomplete learning experience for them.

1.5 Conclusion

Blended learning represents a transition of the education system from traditional approach to combination of offline and remote learning approach. By combining good elements of traditional and online learning, this blended learning approach offers more flexible, engaging and personalized learning experience, becoming potential learning for the modern learners. The hybrid learning will become the integral part of future education system as technology evolves.

The present study explores the growing preference for blended learning among students in India, particularly during post-COVID-19 pandemic. This research fills a gap in the existing literature by examining the students' perception regarding blended learning. Through a pilot survey, this study has surveyed respondents aged 18–35 years. The regression analysis shows that technological proficiency, age and income positively influence seamless learning by students in a blended environment, while rural students face difficulties due to limited access to internet connection. The study provides valuable visions into the factors which may lead to much higher participation of students in blended learning system in India.

The future of blended learning is very promising with the continuous advancement in technology. Almost all educational institutes want to explore new technique of teaching for the betterment of the students, as a result of it, blended learning system will receive much more acceptability to large section of educators and learners. Emerging technologies like artificial intelligence will make the blended learning more flexible, interactive and personalized not only for the better learning experience for students but also for adult learners for their professional development.

The National Education Policy (NEP) 2020 in India highly recommends blended learning methods to be adopted in higher education. According to the NEP 2020, blended learning technology is, therefore, a revolutionary approach to educating and learning in higher education. Blended learning is a combination of traditional face-to-face classroom teaching coupled with online digital tools. Blended learning, with its flexibility and inclusive character, enhances the quality of education, establishes lifelong learning, critical thinking, and problem-solving skills among students.

This innovative approach to learning fosters engagement, collaboration, and self-paced learning, thus enhancing the total learning experience. Blended learning, by employing technology, ensures equity and easy access to education and individualized learning paths, thus catering to the diverse student needs. It is aligning the twenty-first-century skills that students must learn in a fast-changing digital world. Nevertheless, they have to be executed with utmost care since they involve huge investment, planning and hard work on it. It includes building technology infrastructure supporting the effective implementation of the systems, like digital classrooms, high-speed internet, and comprehensive learning management systems. Additionally, it is important to provide training and capacity building of the faculty through training programmes. Teachers need to be trained to become competent in designing and delivering blended learning courses, and to integrate some modern pedagogical practices and digital tools. Thus, NEP 2020 envisages blended learning as an impactful approach to higher education, which can live up to the challenge of accessibility, quality and equity. Through well-monitored implementation, this new teaching–learning approach may redefine the Indian educational scenario to become an inclusive, dynamic and future-ready angle.

References

1. Horn, M. B., Gu, A., & Evans, M. (2014, September). Knocking down barriers: How California superintendents are implementing blended learning. Christensen Institute. Retrieved from www.christenseninstitute.org
2. Moran, G. (2010, September 20). The rise of the virtual classroom. Entrepreneur, 38(10), 112. Retrieved from www.entrepreneur.com/magazine/entrepreneur
3. Bonk, C. J., & Graham, C. R. (2007). *The handbook of blended learning environments: Global perspectives, local designs*. Jossey-Bass/Pfeiffer.
4. Horn, M. B., & Staker, H. (2015). Blended: Using disruptive innovation to improve schools. Jossey-Bass.
5. Watson, J. (2008). Blended learning: The convergence of online and face-to-face education. North American Council for Online Learning.
6. Horn, M. B., & Staker, H. (2015). Shaping culture for blended learning. American Association of School Administrators.
7. Tucker, R. C., Wycoff, T., & Green, T. J. (2017). Blended learning in action: A practical guide toward sustainable change. Corwin.
8. Buckley, D. P. (2002). In pursuit of the learning paradigm. Coupling faculty transformation and institutional change. EDUCAUSE Review, 37(1), 29–38.
9. Thorne, K. (2003). *Blended learning: How to integrate online and traditional learning*. Kogan Page Limited.
10. Garrison, D. R., & Kanuka, H. (2004). Blended learning: Uncovering its transformative potential in higher education. *The Internet and Higher Education, 7*(2), 95–105.
11. Kazu, I. Y., & Demirkol, M. (2014). Effect of blended learning environment model on high school students' academic achievement. *Turkish Online Journal Of Educational Technology - TOJET, 13*(1), 78–87.
12. Oliver, M., & Trigwell, K. (2005). Can 'blended learning' be redeemed? *E-learning and Digital Media, 2*, 17–26.

Sanchita De is an Associate Professor in Department of Basic Science & Humanities, Institute of Engineering & Management, Kolkata, University of Engineering and Management Kolkata. She has 12 years teaching experience. She teaches courses on Economics, Finance and Entrepreneurship. She was awarded Ph.D (Economics) Degree from Jadavpur University in the year 2022. Her research interest includes Development Economics, Health Economics, Environmental Economics, Gender Economics and Industrial Organisation. She has authored two books and multiple book chapters and research articles in various journals and books published by international and national publishers, few of which are indexed in Scopus and UGC Care. The orcid ID of the author is https://orcid.org/0009-0004-5891-8567

Kamalika Dasgupta is an Assistant Professor at the Department of Business Administration of Institute of Engineering & Management, Salt Lake, Kolkata under University of Engineering and Management, Kolkata. She has 10 years of work experience in academics. She takes courses on Economics, Statistics, Research Methodology and Business Analytics. She is nearing the submission of her Ph.D. at the department of Economics, Jadavpur University. Her research interests include Microeconomics and Behavioural Economics. She has authored multiple papers and book chapters in reputed national and international journals some of which are indexed in Scopus, Crossref, UGC care etc. The orcid ID of the author is https://orcid.org/0000-0003-2045-7529.

Chapter 2
Shifting Trend of Customer Purchase Decision Within the Omni-channel Capability During the Pandemic

Sananda Halder, Soumik Gangopadhyay, Sabat Kumar Digal, and Swikriti Saha

Abstract Customer purchase decision is a widely discussed agenda across industries. It is more relevant with respect to the highly competitive FMCG market. In this era of consumerism, it matters a lot to the marketers. During the Covid-19 pandemic, it was evident that the FMCG market had changed from demand-driven scenario to supply-driven orientation. Time of delivery and price were no more the dominating factor for each occasion of such purchase. New purchase habit has emerged and the current study is an effort toward disclosing it. Omni-channel retailing has emerged as one of the popular strategies in the emerging era of digitization. This study has been conducted on the consumer in an attempt to analyze the change in the mode of purchase during the Covid-19 pandemic and its impact within Dakshina Kannada, Udupi and Manipal districts. One of the major goals of the study is to gain deeper insights in the consumer purchase decision making during the unexpected scenario of pandemics from retailers. The study has revealed that the choice of distribution channel is highly contextual and not conceptual. Reaching consumers in the unexpected scenario of pandemic was not easy for every retailer. It has also shown the customer intention while purchasing in an omni-channel capabilities. Emphasizing on the factors determining those purchase intentions during the hard phase of

S. Halder · S. Gangopadhyay (✉)
Department of Business Administration, Institute of Engineering & Management, UEM-Kolkata, Kolkata, India
e-mail: soumik.gangopadhyay@iem.edu.in

S. Halder
e-mail: sananda.halder@iem.edu.in

S. Kumar Digal
Department of Commerce, Rama Devi Women's University, Odisha, India
e-mail: sabatkumardigal@rdwu.ac.in

S. Saha
Department of Computer Application, Institute of Engineering & Management, UEM-Kolkata, Kolkata, India
e-mail: saha.swikriti07@gmail.com

S. Gangopadhyay et al. (eds.), *Pandemic Diaries*,
https://doi.org/10.1007/978-981-96-5415-4_2

pandemic. Therefore, traditional retailers should increase their online presence in an emerging omni-channel setup to cope up with the latest trend.

Keywords Omni-channel · Customer · Purchase decision · Decision-making

2.1 Introduction

There are 1.38 billion consumers in the Indian market who view shopping as their personal journey and are selective in their habits in order to obtain information and indicate their preferences conveniently across several platforms [1]. In India, the revolutionary shift of the Indian market to virtual or online channels has changed the way things are done [2]. Indian retailers have always focused on ensuring product availability when the customer needs it. Moreover, advanced technology has changed buyer behavior. Therefore, the omni-channel approach toward customer intent will be impactful when retailers use digital strategies like accessibility and tracking of orders; and promotion based on location [3]. The retail conditions are however changing in India and new channels are emerging to meet shifting demands. [1]. Most Indian companies use omni-channel to assert that the company strengthens consumer values toward the corporate by possessing the capacity to stay constantly engaged by connecting consumers to various avenues simultaneously [4]. Consumer decides when and how to buy [5]. Earlier studies found that it is not just procuring goods or services that receives attention in consumer behavior but, the process that begins with factors such as price, place, product, promotion, psychology and physiology that profoundly impact consumer behavior [6]. There are multiple variables to measure the online shopping behavior such as timing, intentions, demographic and socio-cultural factors of the buyers, and it could also be the factor in influencing consumer purchase decision [7]. Researchers mentioned that factors such as time, availability of products, affordable price and promotions that e-retailers provide and ease in payment methods encourage the young minds to shop more online [8]. It has been revealed that demographic factors of the consumers, i.e., gender, age, marital status, family size and their income leave a notable effect on the online shopping [2]. Buyers across cities, towns even rural areas in India are updated on different products available for sale. Even, culture also plays an elemental role in a person's need which influences customer's purchase decision makings. For example, trends of the western countries have an influence on the purchase behavior among Indian consumers [9]. When it comes to the 4Ps of a product the buying intentions and pattern of the consumer changes; consumers mostly follow the market and fashion trend which inorganically influence the buying pattern of the consumer [10]. To actively manage consumers' channel choices, corporates need an understanding of the factors that affect consumers' choice of supply channel across stages of the buying process [11].

The process of retailing is undergoing revolutionary transformation propelled by the new process of digitalization [12]. The recent changes made in the retailing industry can be termed as a radical conversion into 'E-Tailing' which is a fusion of digital technology advancement with traditional retail practices [13]. Along with these revolutionary changes emerging in the industry "Omni-Channel Retailing" has sprung up as a new prudent attitude to offer satisfactory experience to customers in the process of change to match the customer expectations [14]. An effective omni-channel strategy can be designed if the collaboration and the development of goals are totally in line [15]. Therefore, the retail customer experience has been deliberately growing in complexity to consistently perform among distinct touch points [16]. It is found that large number of respondents purchased consumer products from private retail shop followed by super bazaar and consumer co-operative store and housewife dominated to play a vital role in taking purchase decision [17].

Ever since the Covid-19 pandemic has surfaced across different countries, there has been a shift in the physical and mental lifestyle of people [18]. There has also been a disruption in the economy and human lives; no sector has been spared and has been affected adversely because of coronavirus [19]. The present situation has caused retailers to enter into newer businesses due to the constantly evolving market situation [20]. Many retailers have reconstructed their trade during the pandemic and have adapted various methods to reach out to consumers. This has opened several avenues and challenges for both modern and traditional retailers. Thereafter, the lockdown period saw customers' decision of buying was moved to what was needed in that hour rather than buying what they needed under certain conditions [21]. As a consequence, traditional retailers have been hit by competition coming from online retailers due to the present lockdown situation [22]. Study conducted by Hwang et al. [23] has established the pain of many retailers during the harsh reality of the pandemic by closing their stores temporarily and moving to online retailing. It was also found that the customer demand is the catalyst for the transition from physical stores to online stores. Interestingly, each year, customer visits to retail stores are found to be decreasing significantly because of the emergence of omni-channels [23]. Earlier studies also established that retailers are experiencing stiff challenges to adopt omni-channel retailing methods to sustain in the business [24]. This has created a need and opened up a spectrum of opportunities for retailers to change their traditional ways of conducting business and move into new omni-channel retailing methods to expand their consumer outreach [25]. Retailers are required to adopt an efficacious omni-channel plan of action in order to meet industry trends [26–28]. But, data regarding pattern of customer purchases can guide the retailers working in supply chain for run time analysis [29]. This can help them identify their online customers and extend their outreach in the process of adapting to online retail [24].

This study is an attempt to understand the decision-making process of omni-channel customers and how they access various retail mode of operations to purchase convenience goods during the pandemic. Objectives of the study were to gain insight into decision making of omni-channel users with respect to convenience goods, to analyze the change in the mode of purchase during the pandemic and its impact within Dakshina Kannada, Udupi and Manipal districts and to recommend traditional retailers to enhance their online presence in an emerging omni-commerce setup.

2.2 Literature Review

2.2.1 Customer Decision Making

To understand and explore the possible impact on the buyer's experience due to the presence of omni-channel retailing, previous researchers attempted to integrate channels which are evolutionary from multichannel including traditional retailing [30]. In order to deliver a similar retailing experience of customers both online and in stores, retailers have to make great effort by providing a unified customer-oriented technology [31, 32]. As a result, integration of multiple channels was found to be providing a seamless and unconventional shopping experience to customers [33]. Earlier research initiatives have documented that the omni-channels allow customers to shop from anywhere at any time. Omni-channel strategy as explored is found to be benefiting consumers with a unique experience that removes barriers between multi-channels [34]. Along with integration of channels, customer service also drives buyer's experience, the digital, local and mobile consumers of the US perceptions expressed that in order to get product reviews, there needs to be a purchase as well, a post-purchase phase which can build customer's trust and loyalty. It will further prompt consumers to make more purchases [35].

Study conducted by Holmes et al. [36] on the application of mobile channels for the decision-making process reported differences in the utilization of online, offline and other mediums. This further facilitates the identification of consumer decision making at the time of purchasing across multiple channels and the factors influencing this decision making during the pandemic. Studies reaffirmed with their robust initiative that the consumers of the United Kingdom (UK) and the United States (US) say that convenience, uniformity and transparency of channels positively influence customer decision making [37]. The globalization has led to competition among the retailers and has led to the development of the logistics and distribution system that has consequently led to the rapid change in consumer behaviors, needs and expectations [38]. Furthermore, if an organization wants to introduce omni-channel in its system, it has to constantly focus on the customer's experience. The brands who can understand the need of their customers are champions. It is not just about the price, product, promotion and place anymore; rather, it has become the need of the hour to seriously consider customers' experience, engagement, purchase experience and re-purchase intentions. Any experience of shopping through any channel should help the customer to recognize the brand. The great omni-channel strategy begins with the collaboration of goals and business development going together in a line [15, 39].

2.2.2 *Omni-Channel Retail*

Rapid advancement in the field of technology is bringing more players to the market and as a result it is preventing the creation of competitive advantage [36, 40]. To accomplish effortless and uniform experience consumer move across digital and physical channels. This technology enhancement has been done to develop a seamless experience for the customer via the process of omni-channel [41]. This has influenced the less digitalized retailers to adopt digital touch points in their services to inhibit the increasing competition [42]. The multichannel retailing experience of breaking down all the odd barriers and effecting change in customer behavior and expectation is significant. The shift of retail industry toward a customer-centric concierge model is creating new opportunities for retailers and supply chain partners [31]. In the epoch of expeditious technology development retail practices are going through dramatic alteration. Customers are also having wide access to a variety of products and services extending over various multichannel. Therefore, retailers have strategies for the effective use of omni-channel to get rid of the entanglement of digitized retail business environment [43, 44]. A significant impact on customer buying behavior has been developed due to the evolving technology and growth of digital channels in business. To respond to these changes the concept of omni-channel retailing has been advocated by researchers and practitioners in recent years. It can be termed as a strategic approach to providing cohesive experience to customers [45]. Omni-channel can be defined as the synchronized operating model in which a customer will get all the channels aligned together. The ever-changing digital eco-system is creating need for a seamless and personalized customer-driven multichannel approach replacing old customer-centric approaches [46].

The 2020 pandemic is an unpredictable disaster of this era. It has led to a dramatic loss of life worldwide along with turning tables toward major thriving industries, one of which is also the Food Processing Industry. Thanks to this development consumers are moving from traditionally used channels to newer modes of Purchases. We know that consumers are moving toward omni-channel because its very service pattern offers convenience, affordability and accessibility to customers. They enjoy a better experience. This shift has forced a lot of retailers to try new modes of purchases and also affected traditional retailers. The research has studied this shift toward different modes of purchases during the pandemic to understand if there has been a significant change and adaptation by omni-channel users. Traditional retailers continue to sell using old techniques and maintain their customer base through goodwill and loyalty. In order to upscale and reduce the impact of the pandemic on these retailers, it is needed that we measure this shift to better decide the most likely adapted mode of purchase and reasons that influence the same.

2.3 Research Methodology

Focusing on the omni-channel retailer consumers we designed an online survey. We decided to have our target consumers from Dakshina Kannada and Udupi district through the questionnaire designed for them. The study needed Omni-channel consumers who at least use one channel from the same retailer when they do their purchases. For this purpose, our respondents were screened before circulating the questionnaire to them to ensure that the selected respondents fit in with our definition of omni-channel consumer. In all, 103 responses were collected. These responses indicated consumer behavior with regard to their average purchases during the pandemic (2020).

The study was carried out by selecting convenience goods for several reasons. First, convenience goods are one among the most necessary and essential parts of one's livelihood. During the pandemic, the opportunity for convenience goods increased significantly. Additionally, such goods began to follow omni-channel strategies where customers are allowed to combine different online channels (the company website, social media and the mobile app) with the offline channels throughout their shopping journey. Information regarding the product can be searched using the Convenience Goods App to buy the product on the website that allows pickup or delivery of the product from the physical store. However, the most important reason for choosing convenience goods is to study the factors influencing omni-channel customers' shift in the decision making before and during the pandemic. This research wants to determine the different drivers to use different channels and/or technologies to buy convenience goods during the pandemic.

There are two parts in the questionnaire. The first part consists of statements about the most used channel of purchase and if there was a shift in the mode of purchase. Based on their shopping experience, respondents rate their agreement with various factors that influenced their decisions which resulted in a shift. The questionnaire follows a close-ended format of questions.

2.3.1 Data Collection

Primary data were collected from consumers of omni-channel in the form of responses for the questionnaire. Secondary data were collected from other sources like books, articles and research papers.

2.4 Analysis and Findings

Out of the 103 responses, it is found that 67% belonged to the age category of 18–25 and most of them were from Dakshina Kannada and Udupi district. The respondents belonged to different occupations like self-employed, employed, unemployed and students where 41.7% had an income below 10,000, 19.4% had an income between 10,000 and 25,000, 21.4% had an income of 25,000–50,000 and the rest had an income above 50,000 (Refer Table 2.1).

It is also found that 86.4% of the respondents stayed with their family, 9.7% of them were with their friends and the rest stayed alone during the pandemic. It is observed that consumers living with their family have a higher shift in their mode of purchase. Before the pandemic 75.7% of the respondents used Offline stores/vendors as their mode of purchase; 14.6% of them were into website/app.

Based on mode, 8.7% used on-call mode and 1% of the respondents used social media as their mode (Refer to Table 2.2).

Table 2.1 Description of sample

Indicator	Universe	Sample%
Age	18–25	67
	26–35	21
	36–50	5
	50 >	7
Location	Dakshina Kannada	40
	Udupi	32
	Manipal	8
	Other	20
Occupation	Self-Employed	9
	Employed	42
	Unemployed	9
	Student	41
Income	< 10,000	41.7
	10,000–25,000	19.4
	25,000–50,000	21.4
	50,000 >	17.5

Table 2.2 Demography-based analysis

		Sample %
Home environment	Family	86.4
	Friends	9.7
	Alone	3.9
Mode of purchase before Pandemic	Offline store/vendors	75.7
	Website/App-based (online groceries)	14.6
	On-call (direct purchases through a phone call)	8.7
	Social Media	1

Pandemic has brought lots of changes in everyone's live. It has also changed our ways of purchasing convenience goods because of various factors. Among our respondents 59 of them have shifted their mode of purchase from their previous mode during the pandemic; the remaining 44 respondents have not shifted their mode of purchase.

To understand the shift in the decision making of consumers five reasons were selected and given to the respondents. The first reason was that they had no other choice in the mode of purchase where 23.7% shifted from their previous mode for this reason and 6.8% of the respondents chose this option for not changing their mode of purchase. The second reason was concerned with the safety of the mode of purchase; 22% chose a different mode of purchase because of this reason and 11.4% of the respondents did not change their mode for the safety reason. Third reason was accessibility to the mode of purchase, 42% of which is the highest among our respondents who shifted for this reason and 52.3% did not shift because other modes were not accessible. The fourth reason was that the respondent was not involved in the decision making for the mode of purchase. A total of 16.5% responded that they were not involved. And the rest responded saying that there was no particular reason.

Different factors affect our decision making. This paper mainly focuses on identifying how different factors like economic, marketing, social and cultural issues influence customer decision making. The first factor analyzed in this paper is economic factor under which there is affordability, finance, necessity and availability, whereas 40.7% of the respondents chose necessity for goods as a factor which affected them the most to change their mode of purchase, and 40.9% chose the same factor for not shifting to any other mode. The second economic factor which affected the shift was availability of mode of purchase. Here 37.3% selected this factor for their change in shift and 36.4% selected it as a factor for not changing the mode. There are two other economic factors that affected the decision making. They are affordability and finance. A total of 22.1% selected these factors as a reason for their shift in decision making and 22.8% selected it as a reason for not shifting to any other mode of purchase.

The second factor under marketing: Product and price was the most important factor leading to the change/no change in the decision making regarding the mode of purchase. It is observed that product was the factor for the shift for 42.6% respondents and for 41.4% it was the factor that led them not to change their mode of purchase. For 40.2% of the respondents, price was the marketing reason for them to shift and for 43.6% it was a factor not leading them to change. Other factors like place and promotion did not affect their decision making much.

Third factor is culture. This factor did not affect the decision making as all other factors did. The most powerful cultural factor was Food Habits. 22% told it was the reason for them to change their mode of purchase and 38.6% told they didn't change to any other mode because of their food habit and religion. Religion was the least reason which led only 3.4% to change the mode and it was not selected by anyone as a reason for them to not change their mode of purchase.

The fourth factor is social factor. Among the different social factors like family, friends, social status and reference group, family is the factor which has affected the most. 37.3% chose this as a reason for them to shift their mode of purchase and 36.4% chose this reason for them not to shift to any other mode of purchase (Refer Table 2.3).

People think, feel and act in different ways, qualitative interpretation is used to gain information on what people think, feel and do. The data were collected through online surveys. The description and analysis of data present the information collected in a narrative form. It is observed that 42.7% have said that there has been no shift from their old mode of purchase while 57.3 have had shifted. Out of the 59 respondents who said yes, 44.1% of the users have moved toward online mode of purchase during the pandemic, while offline stores come second in preference with 30.5%. We asked our respondents how likely they are to select different modes, and the majority of the users have opted for offline and website. Social media has secured a lower vote for convenience goods, whereas on call comes before opting for social media. On a side note, we did observe that respondents under 25 had a more preference toward social media retailing. We stated factors such as quality, customer service and user journey for retaining a specific mode of purchase, while quality and customer service took

Table 2.3 Responses of the consumers

Change in the shift	Scale of response	Yes % 59 Responses	No % 44 Responses
Reasons for change/No change	I had no other choice.	23.7	6.8
	I do not feel safe with the old mode of purchase	22	11.4
	It was convenient and accessible for me to use.	42.4	52.3
	I was never involved in the decision related to purchases	5.1	11.4
	No reason	6.8	18.2
Due to economic factors	Affordability	15.3	11.4
	Finance	6.8	11.4
	Necessity	40.7	40.9
	Availability	37.3	36.4
Due to marketing factors	Product	42.6	41.4
	Price	40.2	43.6
	Place	10.4	12.8
	Promotion	6.8	2.2
Due to cultural factors	Religion	3.4	0
	Tradition (Habitual Practices)	3.4	13.6
	Food Habits	22	38.6
	Language	3.4	4.5
	Moral Values	5.1	6.8
	None of the Above	62.7	36.4
Due to social factors	Family	37.3	36.4
	Friends	11.9	4.5
	Social Status	6.8	2.3
	Reference Groups	3.4	6.8
	None of the above	40.7	50

over 90% of the decision to buy. User journey from start to end did not matter as much as the above stated with 60%. With this we can confirm that there has been a change in the mode of purchase due to various reasons due to which retailers have not been able to fully retain users through old channels during the pandemic.

Customers' choice usually changes at every phase of their decision-making process and the choice of channel influences this decision [13, 39]. Discoveries from a research by Mason et al. [11] propose that the Covid-19 pandemic has modified customers' needs, shopping practices, buying practices just as their post-purchase satisfaction levels [11]. Findings also suggest that it might now be harder for advertisers to acquire customer loyalty since buyers' satisfaction levels have fallen. During the study, we have analyzed and noted that consumer purchase behavior depends on two kinds of intentions, one is subjective and another is objective. Before Covid-19 pandemic, it was observed that availability, accessibility, convenience, perceived risk, affordability, image and loyalty were the factors influencing consumer's objective and subjective intentions for doing purchase. Changing food habit, necessity and availability of items, price of the product and emergency like compulsion drives people to select the mode of channel to purchase goods. During Covid-19 as the scenario got changed so we have observed that now emergency purchasing context has dominated the consumer purchase decision making (Fig. 2.1).

The proposed model shows that while purchasing a product from omni-channel the consumer is having two types of intentions—one subjective and the other objective in nature. The objective intentions are determined by factors such as availability, accessibility, convenience and perceived usefulness while in the other part of the coin subjective intentions are determined by affordability, image, loyalty and perceived risk. During the pandemic, it was observed that Emergency Purchasing Context or

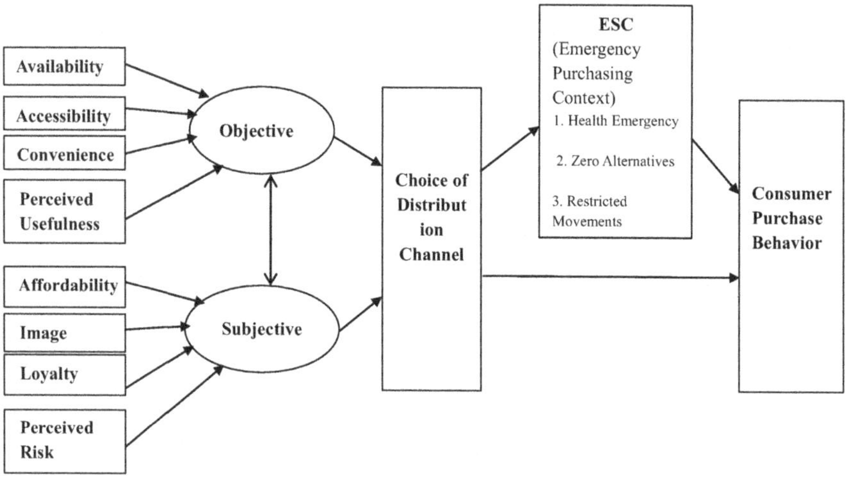

Fig. 2.1 Hypothetical model of opting distribution channel by consumers. *Source* Author's proposed model

ESC was a factor in the decision making. The variables which created the ESC were health emergency, zero alternatives and restricted movements. This model can be checked in future to know the forces affective consumer decision making while purchasing in omni-channel which will be beneficial for the businesses. It will also help the business for proper cost estimation of distribution channel and to focus more on the relevant channel as per the consumer preferences and business growth economically.

2.5 Conclusion

The rise of omni-channel business creates a new path for researchers and practitioners, providing some distinctive challenges as well [41]. The increasing expansion and implementation of technological advancements are transforming retail practices and consumer buying patterns, creating several research areas in the context of expeditious multichannel and multidimensional retail business environment [41]. The study revealed that convenience and accessibility to mode of purchase are very important, due to which traditional retailers selling convenience goods, have to adopt omni-channel retailing and be present online and offline, develop websites, create apps and be present on call, in order to reach consumers. Necessity and availability of goods is an essential economical factor, traditional and omni-channel retailers should identify the need of different products by the consumers and make it available through different modes of retailing. Price and product were the two things depending on which most of the consumers made their decisions, there being a high competition in the market providing goods to the consumers with a good price, having quality goods and giving customers the best service experience is essential. Traditional retailers can make accessible products at the same price on all platforms. Our findings tell us that consumers are purchasing different items based on their habits. Traditional retailers should keep a track of consumer purchases by using different information systems so that they can monitor and identify the habits of their consumers and deliver items readily. It is found that living with the family during the pandemic has led to the shift in mode of purchase due to the change in the decision making of the consumers. This type of change cannot be predicted and retailers have to be ready to face such changes by adopting the new technology and new ways of omni-channel retailing. Traditional retailers presence not only establishes an omni-user base but also improves their user journey across various platforms. This not only caters to need but also to accessibility and reliability.

The focus of the study is to understand customer decision making of omni-channel users. Across various modes of retailing, users have access to different modes to purchase convenience goods. During the pandemic, there was a shift in this mode due to external and internal factors. This study helped us analyze this change and helped us recommend measures that traditional retailers can adapt so that their users are met at the end of every channel. The study summed up the different factors affecting customer decision making, various points of preference at the time of purchase as well as studied the level of shift which will help the retailers to deliver better services and run a profitable business.

This study mainly focuses on the shift to omni-channel by adopting technology in traditional retail. This study has shown how much Covid-19 as a pandemic has influenced consumers to change their mode of purchase and also emphasized on the internal and external factors other than pandemic for that shift. As the research was mostly focused on convenience goods which is having a very competitive market all over the world and surviving on that market was always very challenging for retailers. As this study was conducted during the period of Covid-19, the businesses had to face unexpected situations. The situation was not in anyone's control but through the shift into technology, retailers have survived this situation. In the era of digitization, adoption of technology is very much needed for every business, still so many businesses haven't adopt edit yet. This study and the author's created model can help those businesses to understand the need to shift in technology and adoption of omni-channel to their businesses. This will not only help them to satisfy the need of their consumer but also help them to increase the sale and profit maximization.

References

1. Banerjee, M. (2019). Development of omni-channel in India: Retail landscape, drivers, and challenges. In W. Piotrowicz & R. Cuthbertson (Eds.), *Exploring omni-channel retailing* (pp. 93–108). Springer. https://doi.org/10.1007/978-3-319-98273-1_6
2. Richa, N. (2012). Impact of demographic factors of consumers on online shopping behavior: A study of consumers in India. *International Journal of Engineering and Management Science, 3*(1), 43–52.
3. Rajan, S., Swaminathan, M., & Pavithra, K. (2017). Key drivers of purchase intent by Indian consumers in omni-channel shopping. *Indian Journal of Marketing, 7*(5), 19–25.
4. Solomon, M. (2015). The OC customer experience: Expert systems, 360-degree views, and AI. *Forbes*. Retrieved from https://www.forbes.com
5. Rastogi, A. K. (2010). A study of Indian online consumers & their buying behavior. *International Research Journal, 1*(10), 80–82.
6. Vibhuti, S., Tyagi, A., & Pandey, A. (2014). A case study on consumer buying behavior towards selected FMCG products. *International Journal of Scientific Research and Management (IJSRM), 2*(4), 1–8. https://doi.org/10.18535/ijsrm/v2i4.04
7. Prasad, A., & Gudimetla, S. (2019). A study of digital shopping behavior of women with respect to beauty and personal care products. In *Proceedings of the 10th International Conference on Digital Strategies for Organizational Success.*
8. Bhatt, S., & Bhatt, A. (2012). Factors influencing online shopping: An empirical study in Ahmedabad. *The IUP Journal of Marketing Management, 9*(4), 51–65.
9. Dey, P. (2017). A study on changing buying behavior of Indian customers. *Global Journal of Marketing Management and Research, 7*(1), 1–4.
10. Anjana, S. S. (2018). A study on factors influencing cosmetic buying behavior of consumers. *International Journal of Pure and Applied Mathematics, 118*(9), 453–459. https://www.ijpam.eu/contents/2018-118-9
11. Mason, A., Narcum, J., & Mason, K. (2020). Changes in consumer decision-making resulting from the COVID-19 pandemic. *Journal of Customer Behavior, 19*(4), 299–321. https://doi.org/10.1002/cb.1840

12. Hickman, E., Kharouf, H., & Sekhon, H. (2019). An omni-channel approach to retailing: Demystifying and identifying the factors influencing an omni-channel experience. *International Review of Retail, Distribution, and Consumer Research, 30*(3), 266–288. https://doi.org/10.1080/09593969.2019.1645952
13. Verhoef, P. C., Kannan, P. K., & Inman, J. J. (2015). From multi-channel retailing to omni-channel retailing: Introduction to the special issue on multi-channel retailing. *Journal of Retailing, 91*(2), 174–181. https://doi.org/10.1016/j.jretai.2015.02.005
14. Poranki, K. R. (2015). Consumer attitudes and perceptions on personal care products and cosmetics at Visakhapatnam, India. *Research Journal of Social Science & Management, 5*, 220–225.
15. Garberson, N. (2015). Marketing: An introduction to omni-channel marketing. *Corridor Business Journal, 12*, 14–15.
16. Pantano, E., & Viassone, M. (2015). Engaging customers in a new integrated multichannel retail setting: Challenges for retailers. *Journal of Retailing and Consumer Services, 25*, 1–10. https://doi.org/10.1016/j.jretconser.2015.04.003
17. Sube, S., Ramesh, K., Rohit, P., & Tiwari, K. M. (2020). Impact of COVID-19 on logistics systems and disruptions in the food supply chain. *International Journal of Production Research*. https://doi.org/10.1080/00207543.2020.1792000
18. Ramalingam, S., Govindan, K., & Broumi, S. (2021). Analysis of COVID-19 via fuzzy cognitive maps and neutrosophic cognitive maps. *Neutrosophic Sets and Systems, 42*, 102–116. https://doi.org/10.5281/zenodo.4424985
19. Desai, K. (2014). A study on consumer buying behavior of cosmetic products in Kolhapur. *Reviews of Literature, 1*(10), 2347–2723.
20. Adalarasu. (2020). Retailing and its impact during Covid-19 in Coimbatore City. *Aut Research Journal, 11*(12). http://www.autrj.com/gallery/13-aut-dec2020.pdf
21. Alexander, L. (2011). All's fair in India's cosmetics market. *Beauty Packaging. Journal of Pakistan Association of Dermatologists, 17*, 100–104.
22. Karekar, S. (2020). How is the retail sector adapting to disruptions through business transformations? *Psychology and Education, 57*(9), 2394–2402.
23. Hwang, E. H., Nageswaran, L., & Cho, S. H. (2020). Impact of COVID-19 on omni-channel retail: Drivers of online sales during the pandemic. *SSRN*. https://ssrn.com/abstract=3609945
24. Sramek, J., Ishfaq, D., Gibson, B. J., & Defee, C. (2020). Examining retail business model transformation: A longitudinal study of the transition to omni-channel order fulfillment. *International Journal of Physical Distribution & Logistics Management, 50*(5), 557–576. https://doi.org/10.1108/IJPDLM-02-2019-0055
25. Roggeveen, A. L., & Sethuraman, R. (2020). How the COVID-19 pandemic may change the world of retailing. *Journal of Retailing, 96*(2), 169–171. https://doi.org/10.1016/j.jretai.2020.04.002
26. Du, Y. (2021). An overview of omni-channel inventory strategy under uncertain demand. *American Journal of Industrial and Business Management, 11*, 525–535. https://doi.org/10.4236/ajibm.2021.114036
27. Jocevski, A., Miragliotta, G., & Mangiaracina, R. (2019). Transitions towards omni-channel retailing strategies: A business model perspective. *International Journal of Retail & Distribution Management, 47*(2), 78–93. https://doi.org/10.1108/IJRDM-06-2018-0166
28. Srivastava, S., & Yadav, V. (2021). Retail post-COVID growth strategy: Is omni-channel the way? *Emerald Emerging Markets Case Studies, 11*(1), 1–10. https://doi.org/10.1108/EEMCS-04-2021-0273
29. Bradlow, E. T., & Kopalle, P. K. (2017). The role of big data and predictive analytics in retailing. *Journal of Retailing, 93*(1), 79–95. https://doi.org/10.1016/j.jretai.2016.12.002
30. Verhoef, P. C., Neslin, S. A., & Vroomen, B. (2007). Multichannel customer management: Understanding the research-shopper phenomenon. *International Journal of Research in Marketing, 24*(2), 129–148. https://doi.org/10.1016/j.ijresmar.2007.01.002
31. Bilgihan, K., & Zhang, X. (2016). Towards a unified customer experience in online shopping environments: Antecedents and outcomes. *International Journal of Quality and Service Sciences, 8*(1), 102–119. https://doi.org/10.1108/IJQSS-06-2015-0052

32. Savastano, M., Bellini, P., D'Ascenzo, D., & De Marco, M. (2019). Technology adoption for the integration of online–offline purchasing: Omni-channel strategies in the retail environment. *International Journal of Retail & Distribution Management, 47*(5), 474–492. https://doi.org/10.1108/IJRDM-10-2018-0222
33. Karanko, K. (2020). Non-monetary incentives in online purchase decisions: An exploratory research on Finnish consumers' purchase behavior in the 2020s (Master's thesis, University of Turku).
34. Shi, S., Wang, Y., Chen, X., & Zhang, Q. (2020). Conceptualization of omni-channel customer experience and its impact on shopping intention: A mixed-method approach. *International Journal of Information Management, 50*(February), 325–336. https://doi.org/10.1016/j.ijinfomgt.2019.05.003
35. Kang, M. (2019). What drives omni-channel shopping behaviors? Fashion lifestyle of social-local-mobile consumers. *Journal of Fashion Marketing and Management: An International Journal.* https://doi.org/10.1108/JFMM-10-2018-0149
36. Holmes, G., Byrne, A., & Rowley, J. (2014). Mobile shopping behavior: Insights into attitudes, shopping process involvement, and location. *International Journal of Retail & Distribution Management, 42*(1), 25–39. https://doi.org/10.1108/IJRDM-02-2013-0032
37. Xun, X., & Jackson, C. (2019). Examining customer channel selection intention in the omni-channel retail environment. *International Journal of Production Economics, 208*(February), 434–445. https://doi.org/10.1016/j.ijpe.2018.11.013
38. De Sousa, M. D., Barbosa, L., Oliveira, L., et al. (2021). Challenges, opportunities, and lessons learned: Sustainability in Brazilian omni-channel retail. *Sustainability, 13*(2), 666. https://doi.org/10.3390/su13020666. This article belongs to the Special Issue *Green City Logistics*. https://www.mdpi.com/2071-1050/13/2/666/htm
39. Gensler, S., Verhoef, P. C., & Böhm, M. (2012). Understanding consumers' multichannel choices across the different stages of the buying process. *Marketing Letters, 23*(4), 987–1003. https://doi.org/10.1007/s11002-012-9172-x
40. Hanaysha, J., Shaikh, A., & Alzoubi, H. (2021). Importance of marketing mix elements in determining consumer purchase decisions in the retail market. *International Journal of Service Science, Management, Engineering and Technology (IJSSMET), 12*(6), 56–72. https://doi.org/10.4018/IJSSMET.20210701.oa8
41. Lazaris, C., & Vrechopoulos, A. (2014). Human-computer vs. consumer-store interaction in a multichannel retail environment: Some multidisciplinary research directions. In *Proceedings of the 16th International Conference on Human-Computer Interaction* (pp. 22–27). Creta Maris, Heraklion, Crete, Greece.
42. Bhatt, K., & Sankhla, P. (2017). A study on consumer buying behavior towards cosmetic products. *International Journal of Engineering Technology Science and Research, 4*(12), 2394–3386.
43. www.nemertes.library.upatras.gr/jspui/bitstream/10889/11570/1/Dissertation_Final.pdf
44. www.timesofindia.indiatimes.com/readersblog/midweekread/impact-of-covid-19-on-indian-economy26770/
45. Khan, S. (2012). Factors affecting buying behavior of females for purchase of cosmetics. *International Review of Business and Social Sciences, 1*(9), 68–76.
46. Chandra, P., & Sastry, N. S. (2004). Competitiveness of Indian manufacturing: Findings of the 2001 national manufacturing survey. *Technology, Operations, and Management.* Forthcoming.
47. Ailawadi, K. L., & Farris, P. W. (2017). Managing multi- and omni-channel distribution: Metrics and research directions. *Journal of Retailing, 93*(1), 120–135. https://doi.org/10.1016/j.jretai.2016.07.004
48. Burt, S., & Sparks, L. (2003). E-commerce and the retail process: A review. *Journal of Retailing and Consumer Services, 10*(5), 275–286. https://doi.org/10.1016/S0969-6989(03)00033-9
49. Cai, Y.-J., & Lo, C. K. Y. (2020). Omni-channel management in the new retailing era: A systematic review and future research agenda. *International Journal of Production Economics, 229*(November), 107729. https://doi.org/10.1016/j.ijpe.2020.107729

50. Chen, H., Lee, Y., Tu, Y., & Chao, Y. (2015). Consumer purchase intention for skin-care products. *Semantic Scholar*. Retrieved from http://www.semanticscholar.org/paper
51. Dobberstein, K., Neumann, H., & Zils, M. (2005). Logistics in emerging markets. *McKinsey Quarterly, 1*, 15–17.
52. Gupta, S., & Singh, R. (2020). Managing operations by a logistics company for sustainable service quality: An Indian perspective. *Management of Environmental Quality, 31*(5), 1309–1327. https://doi.org/10.1108/MEQ-12-2019-0234
53. Häubl, G., & Trifts, V. (2000). Consumer decision making in online shopping environments: The effects of the decision-making process. *Marketing Science, 19*(1), 4–21. https://doi.org/10.1287/mksc.19.1.4.15181
54. Ishfaq, D., Gibson, B. J., & Raja, U. (2016). Realignment of the physical distribution process in omni-channel fulfillment. *International Journal of Physical Distribution & Logistics Management, 46*(6), 543–561. https://doi.org/10.1108/IJPDLM-06-2015-0159
55. Jadhav, V., & Khanna, M. (2016). Factors influencing online buying behavior of college students: A qualitative analysis. *The Qualitative Report, 21*(1), 1–4. https://doi.org/10.46743/2160-3715/2016.2285
56. Jawahar, J. V., & Tamizhjyothi, K. (2013). Consumer attitude towards cosmetic products. *International Journal of Engineering and Management Research, 3*(6), 2249–2585.
57. Kaspiri, L. (2018). Anywhere, anytime… it's about all customers! The omni-channel approach for an advanced shopping experience. *Retail and Consumer Research, 8*(4), 1–10.
58. Markowicz, F. (2017). Optimizing order-routing decisions: Leveraging omni-channel supply chain fulfillment (Doctoral dissertation, Massachusetts Institute of Technology).
59. Ortis, I. (2010). Unified retailing—Breaking multichannel barriers. *IDC Retail Insights Report*.
60. Poranki, K. R., & Perwej, A. (2014). The buying attitudes of consumers of cosmetic products in Saudi Arabia. *Research Journal of Social Science & Management, 4*, 138–145.
61. Sangvikar, B., Kolte, A., & Pawar, A. (2019). Competitive strategies for unorganized retail business: Understanding structure, operations, and profitability of small mom-and-pop stores in India. *International Journal on Emerging Technologies, 10*(3), 253–259.
62. Sayali, D. (2020). Impact of COVID-19 on Indian economy. *Midweek Read*. Retrieved from https://www.midweekread.com/covid-19-indian-economy
63. Sharma, A., Bhola, S., Malyan, S., & Patni, N. (2013). Impact of brand loyalty on buying behavior of women consumers for beauty care products: Delhi region. *Global Journal of Management and Business Studies, 3*(7), 817–824.
64. Xing, Y., Grant, D. B., McKinnon, A. C., & Fernie, J. (2010). Physical distribution service quality in online retailing. *International Journal of Physical Distribution & Logistics Management, 40*(5), 415–432. https://doi.org/10.1108/09600031011055127
65. www.utupub.fi/bitstream/handle/10024/151451/Karanko_Thomas_opinnayte.pdf?sequence=1

Sananda Halder is an Assistant Professor of business administration of IEM, Kolkata, India. Currently, she is pursuing her doctor of philosophy in Management.

Soumik Gangopadhyay is a Professor, Dept. of BBA, IEM-Kolkata (University of Engineering & Mgmt. Kolkata) India. He has 23+ years of experience combined in academics and industry, effectively conducted training sessions extensively in areas of marketing for senior Indian defense personnel, & corporate executives of CESC, Elillily, Indian Oil Corporation, Reserve Bank of India, Shyamali Paribahan, Caplet Pharma etc. He has worked in the pharmaceutical industry for 8+ years. He has received Excellence in Research of 2022-23 from IEM-UEM Group and Gold medal (2004, 2007) from ZYDUS CADILA Healthcare for best performance. Successfully presented research article in conferences held at IIM-Ahmedabad, IIM-Kozhikode, MAHE, NMIMS, Bangkok (award winner), Indonesia, Colombo (award winner). Published article in

UGC approved, ABDC, SCOPUS indexed journals such as Indian Journal of Marketing, JHM (SAGE), IJCMS, The Indian Economic Journal, IJBIR(Inderscience), Springer Nature etc. He is the reviewer of "Health Education" (Emerald), American Journal of Health Promotion, . He has reviewed 25+ SCOPUS or WOS journal articles. As an author, he has contributed two chapters in an edited book of Prof. Philip Kotler published by Routledge. He has recently published an edited book "Perspectives in Sustainable Management Practices" (ISBN: 9781032440026) with Routledge.

Sabat Kumar Digal is a professor of Rama Devi Women's University, Bhubaneswar, Odisha. He began his career in teaching with the Degree College in Purulia, West Bengal. Thereafter, he joined his own alma mater, the Post Graduate Department of Commerce, Utkal University, Bhubaneswar, India on 25 June, 2004 and continued till 26th July, 2018. Besides research, his areas of teaching interest lie in International Finance, Financial Institutions and Markets and Micro Finance and Entrepreneurial Finance. Currently, he is pursuing research on the sustainable finance and entrepreneurship among the rural people at the Panchayat level by using locally available products.

Swikriti Saha is an Assistant Professor of English of IEM, Kolkata and pursuing her Phd in humanities at UEM, Kolkata. She has worked as Assistant Relationship Manager at Crizac UK after completion of her M.A (English) in 2023.

Chapter 3
Redefining the Teacher's Role in the "Phygital" Learning Era: Bridging Digital and Physical Education Spaces

Amrita Majumdar and Sudipta Majumdar

Abstract If Education 4.0 is to become Education 5.0, technology must be incorporated into the classroom. The COVID-19 epidemic has exacerbated the recent trend of human existence becoming increasingly concentrated on technology. We learnt from the epidemic to use technology in every aspect of our lives. Additionally, even if adaptation was unavoidable, it became more challenging in the realm of education. The National Education Policy 2020 proposed by the Government of India was a step in the right direction for education. This policy places a high importance on education quality, with digital transformation being one of its main areas of focus. This policy places a high importance on education quality, with digital transformation being one of its main areas of focus. Teachers, who have simultaneously changed and adapted to survive in this "Phygital" learning environment—which blends the digital and physical aspects of education—are the primary pillar of this changing cosmos. This study focuses on the adjustments educators made to adjust to the physical classroom setting in higher education. This paper will also outline the different tools and strategies teachers utilize to give pupils a top-notch education. Using the relevant statistical tools, the factors affecting the quality of instruction and education delivery in this digital age will be determined. Through digital transformation, this study will assist in the development of strategies for delivering high-quality education in a variety of sectors. Educational institution managers can fund teachers' ongoing professional development to give them the tools they need to successfully navigate both digital and physical learning settings.

Keywords NEP2020 · Phygital · Digital transformation · Higher education institutions · Education 4.0 · COVID-19 · Education 5.0 · Quality of education

A. Majumdar (✉)
Amity Business School, Amity University, Kolkata, West Bengal, India
e-mail: amritamajumdar14@gmail.com

S. Majumdar
School of Business, Adamas University, Kolkata, India

S. Gangopadhyay et al. (eds.), *Pandemic Diaries*,
https://doi.org/10.1007/978-981-96-5415-4_3

3.1 Introduction

Due to swift advancement of technology from Industry 4.0 to 5.0, organizational competition is escalating in all sectors and industries. In order to differentiate oneself from competitors, anticipate the formation of new markets, build competitive advantages, and improve organizational performance and sustainability, technology competitiveness is consequently considered as a critical component [1]. Every problem has the potential for an invention in today's technologically evolved society. The pandemic has compelled us to investigate alternative teaching strategies to the traditional face-to-face style [22].

The National Education Policy 2020 recognizes both the benefits of using technology and some of its potential drawbacks. It became important to take action to reduce NEP's downsides while it was being put into practice. We must integrate existing digital platforms and ICT-based educational initiatives into all levels of learning environments in order to maximize their effectiveness in meeting the current and future difficulties of providing high-quality education for everyone [3].

The major goal of the NEP is to create a system that is based on Indian ethics and actively contributes to India's transformation into a vibrant, just, and equitable knowledge society by offering top-notch education to everyone. As a result, India emerges as the knowledge superpower of the globe. The curriculum and pedagogy of the institutions are primarily intended to promote a deeper sense of duty and respect for the fundamental duties of the nation [20].

To improve learning outcomes in society, Industry 5.0 has created a need for teachers and students to acquire new skill sets [4]. These abilities, which are necessary for teachers to become skilled members of our workforce in a rapidly changing world, are unavoidable. Adhikari [2] proposed a novel framework for learning called "Phygital" in their study, which combines physical and digital learning. In a classroom setting, this medium creates the ideal fusion of many digital mediums (such as content links, interactive software, various web portals, and online resources). Different important components, such as Knowledge, Interaction, Personalization, Competency, and Career Development, are included in this proposed model.

National Education Policy and "Phygital" Learning Environment

The Government of India's National Education Policy (NEP) 2020 provides a revolutionary framework for updating the educational system to satisfy the needs of a technologically advanced future. This policy highlights the necessity of modifying teaching and learning procedures to allow for the merging of digital and physical learning environments, as embodied in the idea of "phygital" learning. NEP 2020 recognizes that the COVID-19 epidemic has sped up the use of technology, which has drastically changed the way education is provided and called for a move toward blended and hybrid models that strike a balance between traditional and digital pedagogies.

The research goals of examining teacher responsibilities in a physical learning environment are strongly aligned with NEP 2020's primary focus area of digital transformation. The policy promotes the use of technology to close the inequalities in educational quality, equity, and access. It imagines a system in which educators have the knowledge and tools necessary to move between traditional classroom settings and online learning environments with ease. This goal serves as the foundation for the study, which examines how teachers have responded to these demands by using resources including learning management systems (LMSs), online tests, and collaborative platforms to improve learning outcomes and student engagement.

A key component of NEP 2020 is teacher development, which highlights the value of ongoing professional development in preparing teachers to handle the demands of a hybrid learning paradigm. The policy suggests comprehensive teacher preparation programs that include pedagogical innovations, digital literacy, and the application of cutting-edge technology like virtual reality (VR) and artificial intelligence (AI). By looking at how educators use professional development opportunities to improve their teaching methods and adjust to the demands of both digital and physical learning settings, this study explores the efficacy of these programs.

Additionally, NEP 2020 places a strong emphasis on student-centric strategies, encouraging individualized and hands-on learning that aligns with Education 5.0 ideals. To promote critical thinking, creativity, and problem-solving abilities in a physical setting, educators must use collaborative tools and adaptive learning technologies. The study explores the operationalization of these pedagogical changes by examining the resources and strategies educators employ to design inclusive and captivating learning environments that support the goals of the policy.

Recognizing the necessity of strong technological support to facilitate the phygital change, the policy also addresses infrastructure and resource distribution. With an emphasis on topics like classroom technology, internet access, and the availability of digital resources, the study examines how institutional support—as directed by NEP 2020—affects the adoption of phygital learning environments.

This research paper offers insights into how the policy's vision might be turned into practical measures by examining the junction of NEP 2020 and the changing role of teachers. It emphasizes how important it is to make investments in teacher preparation, technology infrastructure, and creative teaching methods in order to meet the objectives of NEP 2020 and create an education system that is prepared for the future.

3.2 Survey of Literature

Although technology has always been used to enhance learning, current advancements in digital technology, particularly those pertaining to social media, have taken learning in significant new and, for some, completely unexpected areas. Businesses that want to stay competitive globally, preserve their institutional knowledge, and thrive in the gig economy must embrace social learning. In conclusion, social learning seems to be positioned as an important component of workplace education in the present, considering that up to 80% of learning is informal. In other words, the company that is successful in the digital sphere will be the one that promotes digital learning [11].

The Evolution from Physical/Digital to "Phygital" Learning Environment

The term "phygital," a blend of physical and digital, encapsulates the hybrid educational model gaining traction in recent years. With the integration of technology into traditional education systems, the need for a balanced approach to physical and digital spaces has become paramount. Studies state that COVID-19 pandemic expedites this transition, forcing education institutions worldwide to adopt technology-driven pedagogies. The shift has underscored the importance of teachers as critical mediators in this transition, blending traditional and modern educational techniques.

The Evolution from Education 4.0 to Education 5.0

When it comes to bridge the gap between digital and physical learning environments, teachers' role becomes very much essential. The existing literature lists several opportunities and difficulties related to this dual function. For example, while preserving interpersonal relationships in traditional classroom settings, educators must adjust to the use of learning management systems (LMSs), online collaborative tools, and gamification techniques [16], teachers' confidence, and teaching results have improved as a result of professional development programs designed to increase their digital literacy [12].

By combining virtual and in-person learning, blended learning enables teachers to employ both synchronous and asynchronous teaching strategies. By improving accessibility and inclusion, this dual modality makes it possible for students from a variety of backgrounds to engage in and gain from their education [9].

In a phygital classroom, students participate in interactive lectures in real-world settings while using digital platforms for assignments, conversations, and additional materials. In order to accommodate different learning styles, this integration offers chances for reflective and cooperative learning [5].

Inclusion of Phygital Innovations in Learning Platform

A paradigm shift in the way teaching and learning are organized is represented by the combination of digital and physical spaces in the classroom. The "phygital" learning environment is characterized by this method, which aims to combine the advantages of conventional, in-person instruction with the adaptability and creativity of digital technology. This convergence ensures inclusion, adaptability, and sustainability in the learning process, which is in line with the objectives of Education 4.0 and beyond [5].

The physical classroom is being transformed by new technologies like augmented reality (AR) and virtual reality (VR). These resources offer immersive experiences that let students investigate ideas and settings outside of the confines of the conventional classroom [12]. For example, AR may superimpose interactive images on real textbooks, while VR can mimic scientific investigations.

Furthermore, by evaluating student achievement and recommending relevant resources, AI-driven systems offer individualized learning pathways [10]. These developments promote a comprehensive learning environment by improving social and cognitive presence.

For the most effective use of the changes and opportunities brought about by a technology mix and its accelerated impact on society, Gobble defines digital transformation as the comprehensive transformation of business operations and organizations, processes, competencies, and models [8].

Cukurova [6] listed many assessment metrics to examine the influence on the teaching–learning metric that will result in the greatest value for both parties. Anecdotal, descriptive, correlational, and causal evidence are the four basic types of evidence stated by the author in this work.

An institution's operations, strategic objectives, and value proposition can all be altered via digital transformation (Dx), which is a set of fundamental and well-coordinated changes in technology, culture, and staff. Technical developments and trends are driving Dx, enabling a new approach to everything from the integration of digital infrastructures to the communication between the IT department and campus administrators. All of these projects are geared toward enhancing student results, developing new research capacities, better teaching and learning strategies, and evolving business models [19].

3.2.1 Different e-Learning Models

The guiding principles for the usage of educational technology are integrated into a number of e-Learning models.

Technological Pedagogical Content Knowledge (TPCK) Model Proposed by Matthew Koehler et al. [13]

One of the most well-known and widely accepted models in the education sector is the technological-pedagogical-content knowledge (TPCK) model, which states that teachers must acquire three types of knowledge: technological, pedagogical, and content knowledge that is appropriate for using information and communication technologies (ICT) in education.

Content knowledge focuses on the discipline or subject being taught, whereas technological knowledge describes the usage of technologies that are part of the curriculum. Pedagogical knowledge is the term for methods and strategies used to encourage learning. New domains of knowledge are created by the intersection of these two existing fields: While pedagogical-content knowledge (PCK) focuses on the strategies a teacher uses to teach a subject, technological-content knowledge (TCK) aims to comprehend how the use of technology might alter the subject matter. However, the existence, components, and potential of different technologies as they relate to the teaching and learning framework are the focus of technological-pedagogical knowledge (TPK) [13, 15].

Unified Theory of Acceptance and Use of Technology (UTAUT) by Viswanath Venkatesh et al. [21]

The Unified Theory of Acceptance and Use of Technology (UTAUT) model is a theoretical framework used to understand and predict user behavior regarding technology acceptance and usage. Developed by Venkatesh et al. [21], UTAUT consolidates various technology acceptance models and identifies key determinants influencing technology adoption. This model can be effectively applied to the research paper titled "Redefining the Teacher's Role in the "Phygital" Learning Era" to investigate how teachers utilize and adjust to technology in the context of hybrid learning. Performance Expectancy (PE), Effort Expectancy (EE), Social Influence (SI), and Facilitating Conditions (FC) are the four main constructs and moderating variables that the model finds as having an impact on behavioral intention and usage behavior. The PE addresses the extent to which a person thinks that utilizing technology would enhance their performance at work. The EE covers the ease of use associated with the technology. The SI involves the extent to which an individual perceives those others (e.g., peers, supervisors) think they should use the technology and the FC covers. The degree to which an individual believes that an organizational and technical infrastructure exists to support the use of the technology. The UTAUT model is particularly valuable in contexts where technology adoption plays a critical role, such as in hybrid or "phygital" learning environments, where teachers must integrate physical and digital tools effectively.

The Community of Enquiry Model by D. Randy Garrison et al. [18]

Computer conferences in education are explained by the Community of Enquiry Model. The paradigm contains the following three crucial components: cognitive presence, social presence, and teaching presence. The interaction between these three components is established by this model, which takes place during a computer conference. The overarching goal of all higher education is critical thinking, which is shown by the cognitive presence. The ability of a participant to engage with their society or community is represented by the second component, social presence. This element serves as a crucial auxiliary part of cognitive presence. The design of the educational experience is covered by the third crucial component, which is teaching presence. The first one includes the design and development of course content along with its assessment and evaluation criteria whereas the another one includes the execution process of the same [18, 7].

The Community of Enquiry Model is an effort to assess different strategies to facilitate the learners in a computer conferencing learning environment.

3.3 Methods

The objectives of the study are as follows:

1. To identify different tools and techniques used by the teachers to impart quality education to the students.
2. To identify the factors influencing quality education in the digital working environment.

The researcher had reviewed several articles which include policy documents, research papers, online materials, etc. This is an exploratory study to identify the different factors influencing the quality education imparted by teachers in the digital working environment. To administer this research work a structured questionnaire has been developed considering the Community of Enquiry Model [18] which is a blend of three major elements cognitive presence, social presence, and teaching presence.

In this research study, the researcher considers the following components under different elements like ensuring that students engage in deep, reflective learning processes using tools like discussion forums, virtual simulations, and physical classroom interactions under cognitive presence. Encouraging collaboration and community building among students through hybrid activities, such as group projects, conducted both online and offline under social presence and adapting instructional strategies to facilitate seamless transitions between digital and physical spaces, designing interactive content, and providing consistent guidance under Teaching Presence.

The researcher had reviewed several articles which include policy documents, research papers, online materials, etc. This study is an exploratory one designed to determine the various variables affecting the caliber of instruction provided by teachers in a digital working environment. A systematic questionnaire that incorporates the Community of Enquiry Model, combining three key components of cognitive presence, social presence, and teaching presence, has been constructed to administer this research project.

A structured questionnaire with multiple questions on the cognitive, social, and teaching elements was used to gather the data. The purpose of convenience sampling is to get input solely from academicians in higher education institutions. 102 of the 112 responders' insightful comments have been taken into consideration for additional analysis. The data was analyzed using SPSS 19.0, a statistical software program.

3.4 Results and Discussion

The study's conclusions shed important light on how higher education instructors have adjusted to the phygital learning environment by utilizing digital tools and resources while preserving crucial components of face-to-face interaction. The findings from the statistical analysis of the data are presented in this section along with their implications for the goals of the study and theoretical frameworks like the Community of Inquiry and UTAUT models. The analysis's main themes, obstacles, and tactics are emphasized, providing practical suggestions for enhancing instruction in the age of hybrid learning. With an emphasis on institutional support networks and professional growth, the conversation also takes into account the larger backdrop of the digital transition in education (Table 3.1).

Table 3.1 Social element

Parameter (social element)		Variable
Emotional expression	The socio-emotional interaction is important to ensure better learning outcome	V1
	The use of emotions or other symbolic displays are the efforts to add the affective component	V2
Open communication	The open communication also carries the misunderstanding among the learners	V3
	Lack of visual cues becomes a challenge to establish social presence	V4
Group cohesion	The collaborative process should be always encouraged and practiced	V5
	In order to create and validate meaning, the collaboration pulls the student into a shared experience	V6
Dependent variable (DV)	The educational experience through the social element is satisfactory	V7

Source Researcher's analysis

The above table depicts the composition of "Social Element" with six independent variables, i.e., V1, V2, V3, V4,V5 & V6.

H0: The social element does not influence the educational experience of teachers of Higher Education Institutes.

H1: The social element influences the educational experience of teachers of Higher Education Institutes (Tables 3.2 and 3.3).

The output from the aforementioned SPSS table shows that the Beta values for V4 and V5, respectively, are 0.566 and -0.164, respectively. This means that the statement "Lack of visual cues becomes a challenge to establish social presence" has the greatest influence on the criterion variable, i.e., satisfaction on the educational experience of teachers through social element. With regard to the criterion variable, "The collaborative process should always be encouraged and practiced" has the least bearing.

The aforementioned ANOVA table also shows the exact ANOVA significant level (p value). This study's level of significance is 5%, or = 0.05. The table shows that

Table 3.2 ANOVA results of social element and educational experience of teachers

Model		Sum of squares	Df	Mean square	F	Sig
1	Regression	90.267	4	22.567	306.354	0.000[a]
	Residual	7.145	97	0.074		
	Total	97.412	101			

a. Predictors: (Constant), V6, V5, V4, V3

b. Dependent Variable: V7

Source SPSS output

Table 3.3 Regression output of social element and educational experience of teachers

Model		Unstandardized coefficients		Standardized coefficients	T	Sig
		B	Std. error	Beta		
1	(Constant)	0.146	0.144		1.010	0.315
	V3	0.412	0.054	0.410	7.675	0.000
	V4	0.544	0.050	0.566	10.876	0.000
	V5	-0.120	0.023	-0.164	-5.299	0.000
	V6	0.135	0.036	0.130	3.782	0.000

a. Dependent variable: V7

Source SPSS output

the value of "p" is less than the value of "0.05" As a result, the alternative hypothesis is accepted and the null hypothesis is not accepted (Table 3.4).

The above table depicts the composition of "Cognitive Element" with eight independent variables, i.e., V1, V2, V3, V4,V5, V6, V7 & V8.

H0: The cognitive element does not influence the educational experience of teachers of Higher Education Institutes.

H1: The cognitive element influences the educational experience of teachers of Higher Education Institutes (Tables 3.5 and 3.6).

Table 3.4 Cognitive element

Parameter (Cognitive element)		Variable
Triggering event	The communication is restricted and encouraged by the medium chosen	V1
	Asynchronous mode of instruction facilitates deep and meaningful learning in higher education	V2
Exploration	My learners are less interactive during this asynchronous mode of instruction	V3
	My learners have developed high critical thinking during asynchronous mode of instruction	V4
Integration	My learners are able to link the facts, ideas, and notion with the learning outcome of the course	V5
	The learners are able to connect the ideas to solutions more in case of face to face learning rather than digital learning	V6
Resolution	I welcome new ideas from the learners	V7
	The development of KSA (Knowledge, Skill, and Attitude) is not at par in "Phygital" working environment	V8
Dependent variable (DV)	The educational experience through the cognitive element is satisfactory	V9

Source Researcher's analysis

Table 3.5 ANOVA results of cognitive element and educational experience of teachers

Model		Sum of squares	df	Mean square	F	Sig
1	Regression	67.199	8	8.400	125.747	0.000[a]
	Residual	6.212	93	0.067		
	Total	73.412	101			
a. Predictors: (Constant), V8, V2, V5, V1, V4, V3, V7, V6						
b. Dependent Variable: V9						

Source SPSS output

Table 3.6 Regression output of cognitive element and educational experience of teachers

Model		Unstandardized coefficients		Standardized coefficients	t	Sig
		B	Std. error	Beta		
1	(Constant)	0.175	0.223		0.783	0.435
	V1	0.332	0.054	0.381	6.122	0.000
	V2	0.045	0.049	0.044	0.912	0.364
	V3	-0.011	0.065	-0.013	-0.168	0.867
	V4	0.528	0.059	0.621	8.931	0.000
	V5	-0.274	0.040	-0.416	-6.829	0.000
	V6	-0.263	0.077	-0.345	-3.397	0.001
	V7	0.524	0.080	0.636	6.587	0.000
	V8	0.107	0.052	0.114	2.046	0.044
a. Dependent Variable: V9						

Source SPSS output

The results from the SPSS table above show that the Beta values for V7 and V3 are 0.636 and -0.013, respectively. This suggests that the statement "I welcome new ideas from the learners" has the greatest influence on the criterion variable, satisfaction on the educational experience of instructors through social aspect. The effect on the criteria variable of "My learners are less interactive during this asynchronous mode of instruction" is the smallest.

The aforementioned ANOVA table also shows the exact ANOVA significant level (p value). This study's level of significance is 5% or = 0.05. The table shows that the value of "p" is less than the value of "0.05". As a result, the alternative hypothesis is accepted and the null hypothesis is not accepted (Table 3.7).

The above table depicts the composition of "Teaching Element" with eight independent variables, i.e., V1, V2, V3, V4,V5, V6, V7, V8 & V9.

H0: The teaching element does not influence the educational experience of teachers of Higher Education Institutes.

Table 3.7 Teaching element

Parameter (Teaching element)		Variable
Instructional management	I am able to establish and sustain my role as a teacher in this "Phygital" working environment	V1
	The instruction through "asynchronous text based communication" is more learner friendly	V2
	Effective Teacher/Tutor presence is possible and acceptable in this "phygital learning environment	V3
Building understanding	Establishment of comfort level becomes challenging in this asynchronous medium of instruction	V4
	Critical thinking can be developed through digital instruction method	V5
	Learner's expectation can be communicated and negotiated in this "Phygital" learning environment	V6
Direct instruction	Due to the asynchronous nature of the medium, learners are provided time to reflect, then contribute to the discussion after they have formulated their thoughts	V7
	The content need to be limited to ensure deep learning among learners	V8
	The topic of the discussion should last for a week/two	V9
Dependent variable (DV)	The educational experience through the teaching element is satisfactory	V10

Source Researcher's analysis

H1: The teaching element influences the educational experience of teachers of Higher Education Institutes (Tables 3.8 and 3.9).

The output from the aforementioned SPSS table shows that the Beta values for V8 and V4 are 0.910 and -0.137, respectively. This means that the statement "The content need to be limited to ensure deep learning among learners" has the greatest impact on the criterion variable, i.e., Teachers' satisfaction with their educational experience through social element and "Establishment of comfort level becomes challenging in this asynchronous medium of instruction" has the least influence on the criterion variable.

Table 3.8 ANOVA results of teaching element and educational experience of teachers

Model		Sum of squares	df	Mean square	F	Sig
1	Regression	79.051	9	8.783	163.298	0.000[a]
	Residual	4.949	92	0.054		
	Total	84.000	101			
a. Predictors: (Constant), V9, V6, V1, V3, V8, V7, V2, V4, V5						
b. Dependent Variable: V10						

Source SPSS output

Table 3.9 Regression output of teaching element and educational experience of teachers

Model		Unstandardized coefficients		Standardized coefficients	t	Sig
		B	Std. error	Beta		
1	(Constant)	1.030	0.146		7.063	0.000
	V1	0.287	0.031	0.392	9.171	0.000
	V2	-0.312	0.051	-0.354	-6.067	0.000
	V3	0.082	0.048	0.106	1.725	0.088
	V4	-0.155	0.098	-0.137	-1.586	0.116
	V5	-0.428	0.153	-0.370	-2.804	0.006
	V6	0.136	0.040	0.169	3.387	0.001
	V7	0.264	0.096	0.249	2.761	0.007
	V8	0.730	0.049	0.910	14.958	0.000
	V9	0.182	0.046	0.181	3.932	0.000

a. Dependent variable: V10

Source SPSS output

The aforementioned ANOVA table also shows the exact ANOVA significant level (p value). This study's level of significance is 5% or = 0.05. The table shows that the value of "p" is less than the value of "0.05" Consequently, the alternative hypothesis is accepted and the null hypothesis is not accepted.

3.5 Conclusion

Emerging technologies are essential for the modernization of management education. The epidemic has shown us that innovation is present in every industry, including education. The use of the Unified Theory of Acceptance and Use of Technology (UTAUT) to comprehend the adoption of various emerging technologies in classrooms was discussed by CC Lewis, 2013 [14] in his study work. When talking about the acceptance of technology, elements like performance expectations and effort expectations were crucial.

The transition to the phygital learning era represents a paradigm shift in education, redefining the role of teachers as pivotal agents of change. This study emphasizes the various adjustments educators have made to close the gap between online and in-person learning environments, utilizing cutting-edge techniques and resources to guarantee high-quality instruction. The results highlight how important it is for educators to have access to strong digital infrastructures, institutional support, and ongoing professional development in order to properly navigate this changing environment.

This study looks into how teachers' educational experiences in the Education 5.0 system are impacted by the social, teaching, and cognitive components [7]. This study also explored the different challenges faced by faculty members in a technologically enhanced context. These comprise, among other things, student interest in the subject, low student response to online learning, maintaining the link between digital and non-digital environments, infrastructure support for the use of digital technologies, and students' preparedness for embracing new technologies. As new technologies enter the teaching and learning process, there is an increasing need to address these issues [17]. We must not forget the crucial role that instructors play in the adoption of cutting-edge technologies in the field of education. In addition to content creation, it also entails delivery of the material and cooperative methods for carrying them out among the students [2]. The institutions, administrators, and policymakers play a crucial role in ensuring that instructors have the required training, that the necessary infrastructure is in place to keep things running smoothly, and that both educators and students are informed of what is going on. The teachers involved with school education can participate in this study. The educational experiences of both students and faculty members can be compared in this study's bilateral form.

Challenges and Opportunities Faced in This "Phygital" Learning Environment

Challenges: For instructors who must navigate both physical and digital teaching spaces, the shift to a "phygital" learning environment poses a number of difficulties. The **digital gap**, which emphasizes disparities in access to technology and the internet, is one of the main issues. It can be challenging to establish an egalitarian learning environment in rural or under-resourced institutions because teachers frequently lack the resources, infrastructure, and training necessary to successfully adopt blended learning.

The **absence of thorough professional growth** presents another significant obstacle. Even while contemporary education regulations place a strong emphasis on adopting technology, many educators feel unprepared to meet the demands of incorporating sophisticated digital tools into their lesson plans. They need advice on using cutting-edge teaching techniques that enhance digital tools in addition to technical training.

Overuse of technology is another urgent issue. To effectively plan, implement, and evaluate physical education, teachers must become proficient in a variety of platforms, apps, and tools. This frequently results in stress and burnout, especially when combined with more conventional duties like curriculum development and classroom management.

Phygital environments present a special difficulty for student engagement. Digital tools can be interactive, but they also run the risk of distracting children or making them disengaged. Teachers need to strike a balance between these factors and use tactics to keep students' attention and involvement steady in both the real and virtual worlds.

Another area that presents difficulties is **assessment.** Innovative approaches are needed to assess student performance in a blended learning environment in order to guarantee validity, reliability, and fairness. It's still difficult to create measurements that accurately reflect the subtleties of learning in both contexts.

Lastly, the efficacy of phygital learning may be impeded by **gaps in policy implementation**. The on-ground reality frequently lacks coherence because of bureaucratic delays, insufficient money, or inconsistent implementation, even as efforts like India's NEP 2020 promote digital transformation.

Opportunities: The phygital learning period has revolutionary possibilities for both teachers and students in spite of these obstacles. The ability to customize learning experiences is one important opportunity. With the use of digital resources, educators may design flexible learning programs that are suited to the needs of each individual student, resulting in more focused instruction and better learning outcomes.

Through interactive platforms, virtual simulations, and real-world applications, the phygital model encourages students to actively engage with content and promotes collaborative and experiential learning. By taking advantage of these chances, educators can enhance the dynamic and relevance of learning.

Another significant potential is professional development. Teachers have opportunities to improve their skills, obtain new certificates, and broaden their knowledge of digital pedagogy through the use of technology in the classroom. Organizations that fund ongoing professional development initiatives enable teachers to remain competitive in a quickly changing environment.

Additionally, the hybrid method encourages resource sharing and worldwide networking. Teachers and students can engage in cross-cultural learning activities, interact with classmates worldwide, and access a wide range of online resources. In addition to expanding perspectives, this also gets students ready for a world that is becoming more interconnected.

The phygital age gives educational institutions the opportunity to make decisions based on facts. Administrators can more efficiently distribute resources and teachers can improve their teaching methods with the use of analytics from digital platforms and Learning Management Systems (LMS).

Last but not least, the phygital learning environment supports Education 5.0's goals, which prioritize a human-centered approach to education. Teachers are given the freedom to combine the best elements of digital and traditional teaching methods, encouraging students' creativity, critical thinking, and problem-solving abilities.

In conclusion, adopting a phygital learning paradigm presents a number of potential to redefine the educational process, even though it also entails overcoming considerable obstacles. Teachers may establish an inclusive, dynamic, and future-ready learning ecosystem by carefully tackling obstacles and utilizing technology's advantages.

By identifying key factors influencing teaching quality and student engagement in phygital environments, this research contributes to the broader discourse on educational transformation in the digital age. As the education sector moves toward the vision of Education 5.0, a concerted effort from policymakers, institutional leaders, and educators is essential to foster an inclusive, adaptive, and technology-integrated learning ecosystem. The insights gleaned from this study can serve as a foundation for developing strategies that not only enhance teaching practices but also ensure equitable access to high-quality education for learners across diverse contexts.

References

1. Acur, N.K.-N. (2010). Exploring the impact of technological competence development on speed and NPD program performance. *Journal of Product Innovation Management, 27*(6), 915–929.
2. Adhikari, M. S. (2019). Digital transformation of education landscape: The 'phygital' learning framework. *Manpower Journal, 53.*
3. Agarwal, K. (2022). Digitalized education and NEP 2020: reinventing the classroom. *Research Journal of Philosophy & Social Sciences, XLVIII*(1), 63–72.
4. Backhouse, J. (2016). *Knowledge management and e-learning.* CRC Press.
5. Bonk, C. J., & Graham, C. R. (2012). *The Handbook of Blended Learning: Global Perspectives.*
6. Cukurova, M. (2018). Measuring the impact of emerging technologies in education: A pragmatic approach. Cham, Switzerland: Second Handbook of Information Technology in Primary and Secondary Education. Springer.
7. Garrison, T. D. (1995). Critical thinking in distance education: Developing critical communities in an audio teleconference context. *Higher Education, 29*(2), 183–199.
8. Gobble, M. M. (2018). Digital strategy and digital transformation. *Research-Technology Management, 61*(5), 66–71.
9. Graham, C. R. (2006). Blended learning systems: Definition, current trends, and future directions. *The Handbook of Blended Learning.*
10. Gros, B. (2016). The design of smart educational environments. *Smart Learning Environments, 3*(1).
11. Hinchcliffe, D. (2017). The Digital Transformation of Learning: Social, Informal, Self Service and Enjoyable. ZD NET.
12. Huang, R., Tlili, A., Chang, T. W., Zhang, X., & Cheng, W. (2020). Digital transformation in education: Frameworks and key considerations. *Educational Technology Research and Development, 68*(3), 1457–1475.
13. Koehler, M. (2009). What is technological pedagogical content knowledge (TPACK)? *Contemporary Issues in Technology and Teacher Education, 9*(1), 60–70.
14. Lewis, C. C. F. (2013). Faculty use of established and emerging technologies in higher education: A unified theory of acceptance and use of technology perspective. *International Journal of Higher Education, 2*(2), 22–34.
15. Mishra, P. (2006). Technological pedagogical content knowledge: A framework for teacher knowledge. *Teachers College Record, 108*(6), 1017–1054.
16. Mishra, L., Gupta, T., & Shree, A. (2020). Online teaching-learning in higher education during lockdown: Challenges and opportunities. *Asian Journal of Distance Education, 15*(2), 45–66.
17. Onyema, E. M. (2019). Integration of emerging technologies in teaching and learning process in Nigeria: The challenges. *Central Asian Journal of Mathematical Theory and Computer Sciences, 01*(01).
18. Randy Garrison, D. T. A. (1999). Critical inquiry in a text-based environment: Computer conferencing in higher education. *The Internet and Higher Education,* 2(2-3), 87-105.
19. Susan Grajek, B. R. (2019, July 8). Getting ready for digital transformation: Change your culture, workforce, and technology. *Educase Review.*
20. Uma, S., & Singh, S. S. (2020, December). Insights and perspectives of NEP in transforming higher education in India. *International Research Journal of Humanities and Interdisciplinary Studies, 1*(2), 30–41.
21. Viswanath, M. G. (September 2003). User acceptance of information technology: Toward a unified view. *MIS Quarterly, 27*(3), 425–478.
22. Vorst, C. V. (July, 2021). Higher education turnaround supporting digital transformation. In Proceedings of Economics and Finance Conferences (No. 12413357), *International Institute of Social and Economic Sciences, 105–122.*

Amrita Majumdar is currently working as an Associate Professor in the Amity School of Business at Amity University, Kolkata. She has more than 15 years of experience in corporate as well as teaching & academic administration in various Institutions and Universities in the domain of Business Administration. She is also involved in writing research papers, book chapters, training students & faculty members in the area of human resource management and teaching pedagogy. Her area of research is in the area of Psychological Contract in Academics. Her area of interest lies in Human Resource Management, Organisation Behaviour, Online Teaching- Learning Pedagogy etc.

Sudipta Majumdar is currently working as an Associate Professor in the School of Business at Adamas University, Kolkata. He is a graduate in Engineering, Post Graduate and Ph. D. in Management. He has corporate exposure in the areas of Sales and Management, Management Consulting. Besides that, he has more than 16 years of experience of teaching in Graduate and Post Graduate Programs. Mr. Majumdar is also actively involved in research in the domains of Technology Management, Consumer behavior and allied areas, conducting workshops, Faculty Development Programs, Management Development Programs, mentoring and consulting.

Chapter 4
Fostering Employee Engagement Through Leadership Strategies in Disruptive Environment: Learning from Experiences

Soma Bose Biswas, Ipsita C. Patranabis, and Indranil Bose

Abstract Leadership has always been a sought after role and a topic of discussion across ages. It is a role which is desirous for many but very few are able to effectively perform as a leader. In the present context, the challenges of leadership are increasing in leaps and bounds than that in the recent past. Pre- and post-pandemic has witnessed a radical change in the work settings and employee expectations. The convergence of the VUCA and BANI environment witnessed a pervasive transformation in the business world triggering a host of challenges and opportunities for organizations across industries (Andrea and G. F., & Jimenez-Hernandez, P. R. in Advances in Human Resources Management and Organizational Development Book Series 20–43, 2024). Leaders had to experience a dramatic role progression to meet the expectations of the organizations, their team members, and external demands. This article delves into the leaders' insight to the changing experiences during turbulent times and gives this study a new dimension to deal with the complex situations. The in-depth interviews conducted with the leaders, of minimum experience of ten years and span of control of at least five people, reveal that in the multicultural hybrid work settings, being a mentor and an empathetic catalyst is immensely appreciated by all stakeholders than that of a regular boss. Employees need support and comfort to perform and they look up to the leader as an advisor than that of an evaluator. The interactions with the leaders encouraged the researchers to recommend a dynamic work environment where the employees experience a psychosocial security which in turn can create an engaging and stable structure.

Keywords Empathy · Hybrid workforce · Engagement · Psychosocial context

---

S. B. Biswas (✉) · I. C. Patranabis
Associate Professor, Department of Management Science, Techno International New Town (Affiliated to MAKAUT), Kolkata, India
e-mail: somabbiswas@gmail.com

I. Bose
School of Management, MIT University of Meghalaya, Shillong, India

S. Gangopadhyay et al. (eds.), *Pandemic Diaries*,
https://doi.org/10.1007/978-981-96-5415-4_4

4.1 Introduction

Work from home, agility, resilience, employees' health and well-being, great resignation: these terms now became very common in the organizational parlance after the deadly pandemic. But business houses have already been baffled over how to react to change for the past 12–15 years due to the rapid technological development of Industry 4.0 and shifts in the national and worldwide industry landscape. Moreover, due to the pandemic, it has become more challenging to maintain a flexible workforce and keep employees motivated to do their best. There has been a dramatic shift in leadership styles across these epochs, with implications for how they are practiced, how goals are communicated, how tasks are initiated, and how leaders interact with their subordinates. Pandemic presented some universal challenges for business leaders, including the increasing difficulty of working remotely while keeping employees motivated, managing a remote workforce, and adjusting to new demands on their leadership skill sets in terms of learning new skills and techniques quickly [1]. Returning to the new normal was not easy as the workforce started experiencing new standards, unfamiliar policies, unknown norms, and exceptional business strategies. The huge disruptive and chaotic effect of this has left the employees as well as leaders perplexed and traumatized. Business environment has always been identified as uncertain and complex but the pandemic has given a new form to the environment as random and inexplicable. Coupled with this organizations started to combat the threat of sustainability in case they lack effective functioning of the business due to skill gap [2]. This approach of the businesses has put the employees under stress as they felt a potential intimidation of job loss in case of failure to rapidly adapt and acquire new skills. Here, role of a leader became pivotal to proactively respond to the environmental challenge by rebuilding the structure and process, preparing the team to recover the loss, as well as reinstituting future possibilities [3]. Effective leadership required the ability to optimize operations in a hybrid setting, make people feel safe and productive, respond quickly and compassionately, be agile and communicate difficult decisions with empathy, and reduce stress among the employees [4]. The purpose of this research is to investigate how leaders adjusted their styles in the wake of the pandemic in order to meet the evolving needs of the people and businesses in a hybrid and boundaryless organization. Thus the paper aims to study:

1. The contemporary challenges faced by organizations in a turbulent work setting.
2. Role of empathetic leadership in a hybrid work environment.
3. A contemporary leadership framework synchronizing the relation between organizational challenges, leadership strategies, and employee engagement in a disruptive business environment.

4.2 Review of Literature

4.2.1 Contextual Factors and Organizational Challenges

The pandemic marked the convergence of the VUCA (volatile, uncertain, complex, and ambiguous) and BANI (Brittle, Anxious, Non-linear, and Incomprehensible) environment which witnessed a pervasive transformation in the business world triggering a host of challenges and opportunities for organizations across industries [5]. Past researchers further suggested that, organizational survival in this disruptive situation is possible by significantly nurturing the softer elements of the leadership capabilities to stabilize the turbulence and continue the journey of growth and development.

4.2.2 Leadership Strategies in Disruptive Work Settings

As the BANI environment is characterized by digitization, advanced decision-making tools, and work from remote locations, an appropriate leadership style is the key to navigate in this turbulent work setting. According to [6], training to upskill employees is a driving factor in this unreliable situation. This indicates concern for employees and a leader's support toward employee development thus strengthening the morale and sense of trust and security of the employees. Leadership and its impact on employee performance are best understood when seen in the larger organizational and societal contexts [7]. Remote workplaces, forced flexibility, technology, and distribution of workload are some of the challenges posed by the pandemic to the leaders [8]. The transient business landscape of the BANI period demanded faster technological adaptability and an optimistic approach toward the new normal [9]. A fresh outlook of business leaders was the need of the hour which could strike a perfect balance between the incomprehensible workplace dynamics and anxious employees. Early researchers even examined the conceptual link between employee needs, emotional competency of leaders, and transformational leadership styles [10]. They further mentioned that engaged workforce makes a significant positive difference to all types of businesses by achieving the financial goals and taking control of the competition. Thus leaders should work toward designing strategies to create higher levels of employee engagement in the emerging workplaces. Researchers have already drawn attention to the fact that leadership behavior can encourage employee engagement to a great extent [11]. Empirical evidences from the past reveal application of emotional intelligence by leaders as a strategy for improving performance at work [12]. This provides discerning insights for this current study to unearth the contemporary leadership approaches toward employees in the post-pandemic complex and unpredictable environment. According to [13], empathy is an integral competency of emotional intelligence and forms the essential building block for effective leadership and relationship management. This particular leadership trait of empathy was the focal point of discussion during pandemic [14]. Evidences from recent research cite that empathetic leaders can effectively respond to crisis situation [15] and successfully build a well-accepted team who can readily contribute and meet the organizational expectations [16]. Following this we tend to explore the power of empathy in stabilizing employee psychosocial state of mind and thus enhancing the level of engagement.

4.2.3 The Influence of Empathy on Employee Engagement

To be successful in today's business world, a leader must be able to successfully navigate a variety of situations and challenges, including working in a multicultural setting, predicting the future, guiding their team, and alleviating the stress of their employees all while maintaining a high moral standard [17]. Empathy is the ability to understand and re-experience another's feelings [18]. Use of empathetic language, and emotional support from the leader affects follower performance, job satisfaction, and innovation [19]. The influence of empathy is found to be moderated by the ethical context and moral awareness [20]. According to [21], evidences from the pandemic indicated that empathetic leadership style enhances organizational effectiveness by creating a congenial work environment which fosters well-being, employee engagement, and retention. This has been clearly depicted in the researcher's recent work through a 'multi-level conceptual framework of the impact of empathy on leadership effectiveness' where the author outlines that empathy in leadership has a three-tier multi-level impact namely, leader, follower, and organization. Drawing reference from [21] research findings, it can be clearly stated that a self-aware leader by virtue of his enhanced ability can empower his or her followers through work appreciation and individual development. Leaders can improve their followers' feeling of psychological safety if they can identify with their experiences, empathize with the mind and provide them with a sense of emotional support and safety [22]. Trust in leader was found to mediate the managers' leadership style and perceived psychological well-being of the followers [23]. Organizational norms and principles are generally established by the leader in any department/business unit [24]. Leaders can increase the group cohesion and hence employee participation through social dialogue and nurturing healthy interpersonal relationship and empowering the teams [25].

Hence, the purpose of this paper is to investigate how an empathic leadership style might foster team cohesion, which in turn increases employees' sense of psychological security and engagement. Also, the paper reveals the different leadership challenges arising from managing the hybrid workforce during the post-COVID new normal.

Research questions

1. What are the issues that organizations face under turbulent work setting?
2. What are the contextual factors that increase the importance of practicing empathetic leadership in contemporary high pressure work setting?

3. What is the role of leaders in creating a cohesive organizational ecosystem and improving employee engagement under a BANI environment?

4.3 Methods

The research paper is evidence-based. The data was gathered from active industry leaders in India, from diverse sectors, who have 10–30 years of experience and supervised a team of 5–20 individuals. The paper used a qualitative approach to understand the different and varied challenges faced by leaders in the contemporary context. A total of 40 people were approached from the administration level, but only 32 senior leaders agreed to provide their interviews. Table 4.1 represents the data collected from the field of inquiry using convenience sampling. In-depth interviews with leaders from various industrial sectors were conducted. The respondents were all from the senior management level, namely, regional heads, deans, directors, team leaders, operations, and business heads. The respondents were from the private sector, representing various work profiles such as sales and marketing, people management, banking, information technology, and business operations.

A careful assessment of relevant information directed the interview technique, through a semi-structured questionnaire aimed at the issues and difficulties connected with the leadership and management in the post-COVID scenario. The recorded interviews were transcribed into textual form. The qualitative responses were examined using thematic analysis, a technique that identifies and investigates groups or themes within data. We first developed codes by categorizing the answers into shared themes, and then identified the significant themes. Three coders develop the codes covering

Table 4.1 Demographic distribution of respondents

Demography	Description	Number of respondents	Frequency (%)
Age	35–40	15	47
	40–50	13	40
	Above 50	4	13
Gender	Male	21	66
	Female	11	34
Educational qualification	Undergraduate	4	13
	Postgraduate	23	72
	Above PG	5	15
Work experience	10 years–less than 15 years	22	69
	15 years above	10	31

Source Primary data

a broad spectrum of issues like leadership style post-COVID, employee engagement initiatives taken by the organizations, leadership challenges, role of leaders to improve employee engagement post-COVID, etc.

We posed several questions to these leaders, including: 'Traits that modern leaders need more to tackle these contemporary challenges'; 'contextual factors influencing leadership styles, post-COVID'; 'experience sharing on initiative toward employee engagement'; 'understanding and regulating others' behavior'; 'policy change in hybrid work settings'; and 'role of leaders in minimizing the BANI effect of workplace setting and facilitating engagement'. The following section discusses the received responses.

4.4 Results and Discussion

Following the interviews with modern leaders, research revealed, the changing topography of the current environment has shifted the leadership styles. Today's leaders face common challenges such as work from home, employee retention, intense competition for pay scales, adhering to set hours, returning to the office, and talent shortages. In response to these challenges, leaders and organizations are implementing strategies aimed at enhancing employee engagement and reducing the negative effects of the BANI environment. The study is divided into three major areas of discussion to understand the research questions.

- Organizational challenges in the new normal.
- Changing leadership traits and strategies.
- Managing the BANI environment.

Based on the experiential interactions with corporate leaders, which trace the contemporary leadership challenges in the new normal work settings and identify key leadership traits for building effective workplaces for the future, a conceptual model (Fig. 4.1) has been designed.

Since 2000, the changing business environment had already posed a great challenge to the organization's management. Advent of the pandemic worsened the situation. Post-COVID, adaptability became the key characteristic of organizations which was reflected in the implementation of work from home policy or working in a hybrid environment. In one word, the situation was described as a BANI environment. The following section discusses the challenges faced by organizations during and post-pandemic era together with the leadership strategies for creating a sustainable workplace and managing workforce in an integrated way.

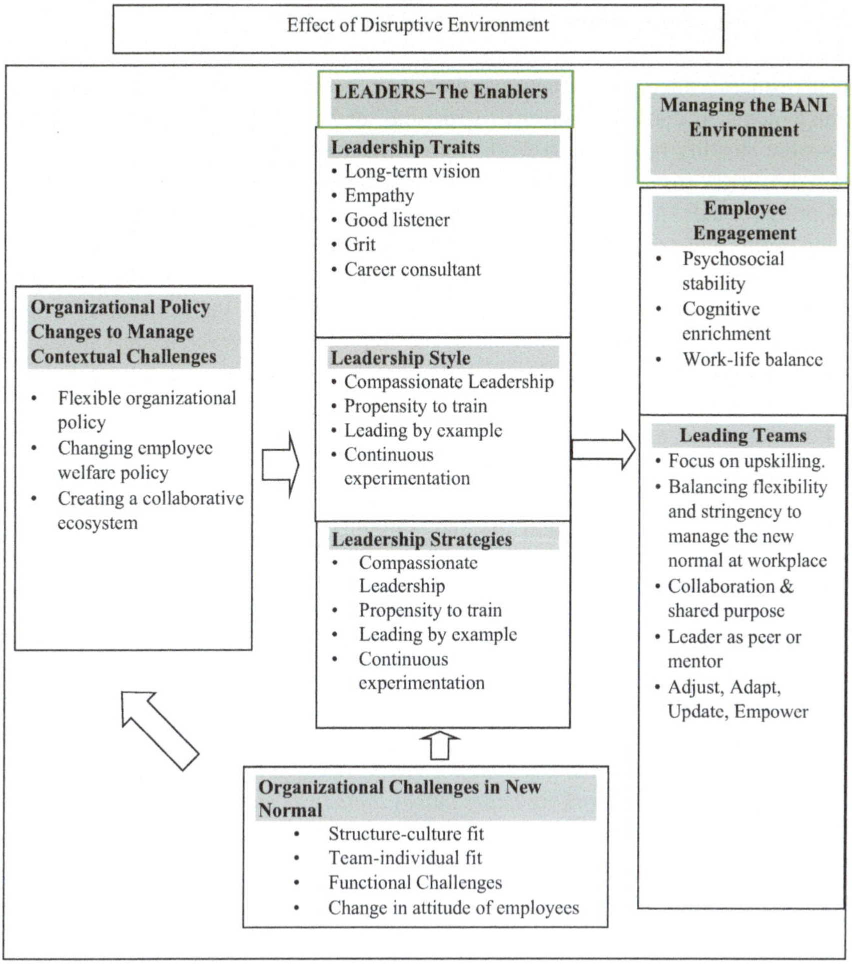

Fig. 4.1 Conceptual model depicting the role of leaders in emerging workplaces. *Authors'* compilation

4.4.1 Organizational Challenges in the New Normal

The first factor identified during the interview process is organizational challenges. The changes experienced by the business leaders were structure-culture fit, team-individual fit, functional challenges, and change in attitude of employees.

4.4.1.1 Structure-Culture Fit

Most business leaders said structure-culture fit concerns were rooted in hybrid work arrangements. It caused functional, structural, and relational difficulties. Companies were steadily returning to work from office mode, although many employees desired to work from home. The hybrid mode was posing serious challenge to the HR professionals. Many structural difficulties were emerging, such as communicating norms and expectations to employees, identifying and developing succession planning competencies, and so on. Though flexible work arrangements help workers manage work and family responsibilities simultaneously, at the same time, longer hours, stress, and overwork overwhelmed the employees. However, leaders struggled with staff engagement, productivity, and integrating organizational culture, values, and rules within the teams.

Post-pandemic, a further challenge, faced by the organizations, was reviving their businesses and returning to profit. Organizations forced workers to chase a high objective to increase revenue and profit, causing mistrust, job overload, and employee tiredness. Leaders were crucial in aligning individual and organizational goals in this critical scenario.

Quotations:

- *People are not entirely ready to shift to a work from office model anymore. Many female employees want to continue in work from home mode only as they can give support to their families simultaneously.*
- *Some employees show dissatisfaction with flexi shift as then they complain to do more work than completing within normal working hours. Sometimes they have to work even beyond working hours makes their lives miserable.*
- *A lot of dependency is on technology and digitization now.*
- *It was hard for me to make my employees understand that we need to keep a steep target to manage the losses.*
- *People were exhausted with long hour meetings and work overload.*
- *We failed to provide a clear communication and direction how to do the work.*
- *We forced the marketing team to generate revenue by chasing the dues in the market; but customers were also not ready to disburse the fund during those days, and it created problems.*

4.4.1.2 Team-Individual Fit

22% of the respondents reported that maintenance of trust, allocation of duties, creation of emotional bonding with the team, and integration of the employees in a team are challenges to the contemporary leaders. According to the respondents, a major challenge faced by the leaders during and immediately after COVID was ***maintaining trust, allocating duties, and integrating employees***. In a hybrid system, employees do not meet at the office regularly and every day, which creates an emotional distance within the team. Finally, productivity can sometimes be hampered. At times, individuals may hesitate to seek solutions from their leaders. A proactive and empathetic leadership style has proved to be useful in these stringent situations.

4.4.1.3 Embracing Diversity

Is the new normal for any business organization. Whether in physical mode or in hybrid mode, managing diversity has always been a sensitive issue and an imperative challenge to the leaders. Sometimes, individuals conceal a disability, gender identity, or sexual orientation to avoid social disapproval. Concealing such personal identities might hurt the mental and physical health, as well as the productivity, of the employees. Sometimes, hybrid mode has found to be a better opportunity to create an environment where people may mask it successfully. But the mental agony remains the same. Managing diversity, sensitizing employees to embrace diversity, and creating an equitable culture have become major challenges for the organizations.

Diverse structural, mental, and emotional issues created problems to tie up the team and challenged management to ***keep team spirit*** up.

Quotations:

- *The employees who joined during COVID period, we do not know them personally. It created a lack of bonding. Still, they feel a little uncomfortable in office gatherings.*
- *My manager helped us to create a better understanding within the team through several informal sessions, phone calls and motivating the entire team to help the teammates in their distress situations.*
- *Making the employees, who are working in hybrid mode, sensitive about the company culture is a challenge for us.*
- *We feel bored if only we are working from home as it lacks human connection.*
- *During COVID, everyone was home locked and working from home. But often as a girl, I had to hear that I am enjoying to live with my family. Sometimes, it was hard to manage odd hours office meetings due to my family responsibilities, then I had to hear that my performances are degrading as I am working from home. As a working woman, it is hard to manage office works and family demands while working from home.*
- *Due to hybrid mode of operation, often new employees who joined during COVID period cannot relate themselves with other fellow colleagues.*

4.4.1.4 Functional Challenges Due to Disruptive Business

The respondents' interviews reveal that organizations continue to encounter challenges in areas such as ***attraction*, *onboarding, retention, mobility, and compensation***. Only 11% of the respondents expressed support for this type of challenge, which is significantly less than those advocating for other factors. During the COVID period, the functional challenges were surmounted, whether it was related to salary/compensation, or related to appraisal. Following the pandemic, the focus has shifted to challenges related to attracting and retaining potential talent within the organization. The organizations are now in dire need of sustainable employees who will be loyal and stay in the organization for the long term. High ambition, fast growth, a fat salary, and the need for a luxurious life are pushing the generation Z workforce to switch jobs faster. Learning new skills, loyalty, and adaptability has taken a back seat. Quick learning of new and diverse skills in capsule format is creating a workforce with 'certificates without skills'. Micromanagement of office assignments is becoming a real threat to quality output. Modern leaders have to manage the issues arising from moonlighting, emotionally less connected employees as well as employees with limited expertise.

Quotations:

- *We faced huge salary cut during the COVID period. There was huge job cut in many organizations. It led to huge tension among the employees.*
- *The new employees are coming with degrees, but they are very poor in their knowledge and skill level.*
- *Employees do not want to learn. They don't want to get into the deep of any issues/ problems/ knowledge. To learn they more rely on YouTube videos, micro learning modules,* etc.
- *Many employees are moonlighting to earn more. It is creating issues like authenticity, reliability, and getting 100% productivity from the employees.*
- *Managing employees' performances is becoming a challenge, particularly those who are working from home.*
- *Onboarding of the new joinee employees is becoming a challenge, as it is lacking personal touch.*
- *Appraising people in hybrid setting becomes a real challenge as people are working in different work settings, other challenges are: alignment, capability development, learning agility.*

4.4.1.5 Change in Attitude of the Employees

Managing the Gen Z workforce, who are also digital natives, is very different from their predecessors. They argue, ask questions, and try to manage assignments using AI. They want empowerment, a big salary, and congenial and luxurious facilities to live in. They are always on social media and sensitive to words. They are less connected to the team. They love flexibility and are mostly happy to work from home, casual about the dressing norms and stringent organizational rules. The leaders face a big challenge to manage diverse generations of employees and their gaps.

Quotations:

- *New employees are always arguing. They do not have respect for people, and process.*
- *The young fresh students are generally having big degrees, but they lack in functional skills.*
- *If the manager is scolding due to any valid reason they take it on their ego.*

- *Employees became very much unpredictable in terms of their aspiration, expectations and behaviour.*

4.4.2 Organizational Policy Changes to Manage Contextual Challenges

The second factor emerged out of interview is organizational policy. Many organizations addressed the challenges with resilience, often modifying their policies and strategies. The major factor has been constructed by the minor factors: flexible organizational policy, changing employee welfare policy, and creating a collaborative ecosystem.

4.4.2.1 Flexible Organizational Policy

Family obligations were being prioritized after the COVID-19 epidemic. The workforce became increasingly appreciative of a friendly atmosphere and work-life balance. Flexible shifts helped them balance work and life. Gen Z workers enjoy flexibility in job, time, and environment. The flexible work schedules were being decided mutually. Clear expectations, rules, and norms were communicated and continuously updated over time physically or virtually. Remote recording of attendance through electronic monitoring and ensuring business meetings to be conducted within stipulated working hour secured transparency, mutual trust as well as prevented burnout situation for employees. So, to embrace the changing contexts of business, organizations are becoming flexible in terms of policies, procedures, and norms.

Quotations:

- *Flexi-shift policy is good as it gives time to manage the family obligations.*
- *Junior employees feel the organisation must provide flexibility in work timing, space and policy as it helps them to continue their passion while working for an organization.*
- *We are electronically monitoring attendance while they work from home.*
- *We have initiate leave bank facility where the employee can take leave in loan and adjust later.*
- *Policies have become flexible in the new normal.*
- *We have moved into hybrid model in our operations, encouraging flexible work hours, work from anywhere and remote reporting systems.*
- *We ensure strict adherence to the working hour.*

4.4.2.2 Changing Employee Welfare Policy

The welfare policies in the new normal have become more flexible and dynamic. Many organizations were designing innovative and flexible welfare policies to retain potential employees. Companies were prioritizing psychological well-being of their employees and organizing many wellness programs. Rewards and recognitions were more valued. Many organizations were advocating inclusion of people with disabilities and the LGBTQ + community in their organizations. Hence, organizations were designing an equitable and just organizational culture. Prompt and comprehensive assistance during times of crisis was primarily focused to enhance the comfort and satisfaction of the employees. Though inclusion of childcare leave for male employees created an atmosphere of debate in some cases. It is evident from the interview that the HR process became open and resilient to cater to the physical, mental, and emotional well-being of the employees.

Quotations:

- *Encouraging family welfare as essential part of employee wellbeing.*
- *I agree that empathy and compassion go a long way in the new normal as employees started to prioritize wellbeing much more than ever after the pandemic.*
- *Yes. It's very much required for the current situation to retain quality manpower.*
- *We are providing one day wellness leave which the employees can take, once in a month or as per the company policy if they are feeling blue or feeling sick.*
- *Wellbeing being the primary focus of employee engagement in our organizations.*
- *We also have analytics on attrition for our understanding and talent shows that impact the employee mood. Like recently for the festivities we had employees voluntarily participate in talent showcasing and it turned out to be a success given the upliftment of mood.*

4.4.2.3 Creating a Collaborative Ecosystem

In a hybrid mode of operation, boundaryless organizations were fostering an open and dynamic work culture to ensure a collaborative ecosystem in the workplace. During the pandemic, numerous organizations encountered a significant challenge in maintaining trust within their groups. Continuous interaction, involving employees in problem solving, management development programs, one-to-one interaction, incorporating peer-to-peer recognition, and virtual coffee with the team were some innovative practices helping to level up the employee experience and create a collaborative ecosystem.

Quotations:

- *The organizations are getting more flexible in terms of work culture. Empathy and compassion are surely going to be game changers in enhancing employee engagements.*
- *Creating a collaborative eco system is a very vital part of any organization.*
- *when organizations go boundaryless and people working in a team comprising of remote stakeholders, organizational policies also need to be aligned and respectful as per regional trends, cultures and habits.*
- *Individual and team autonomy, peer to peer recognitions.*

- *For hybrid team, frequent meeting is conducted online to have the proper engagement. New assignments are given to build the sense of ownership among the team members.*
- *We organize brainstorming sessions among team members and motivate them to think out of box.*
- *Regular communication and check-ins are vital to maintain contact, avoid isolation and to provide mutual support for the well-being of team-members and also their family members.*
- *We ensure open and effective communication is a must for a happy and satisfied team.*

Therefore, it is clear from the above discussion that organizations were utilizing innovative concepts to engage employees, foster a collaborative team, and prevent isolation and distress. To gain a deeper understanding of the role of leaders, the second part of the discussion concentrated on the leadership traits and strategies employed by these leaders to manage teams in the BANI situation. The following part examines the role of leadership in enabling the BANI toward creating a sustainable workplace.

4.4.3 Leaders—The Enablers

To manage a BANI situation, leaders play a very important role. They enable the organizational policies and strategies through their styles and approaches. Research has proven the role of leaders in creating a sustainable workplace environment. In the context of a volatile and uncertain situation, leaders play more crucial role to keep up the employee motivation and ensure productivity. During and post-pandemic, effective leaders have displayed some exemplary traits. Leadership traits are comprehended here through three minor factors: leadership traits, leadership style, and leadership strategies.

4.4.3.1 Leadership Traits

During COVID, the traditional leadership approach was inadequate to deal with the complicated, stressful, and social/organizational difficulties. During the pandemic, it was identified that empathetic leaders who supported their teams and families during times of suffering performed better. Empathy, grit, growth mentality, and active listening were among the leadership qualities that became increasingly important during times of crisis and distress. Leaders who refused to address their team members' challenges in complex settings rarely achieved productivity. The forgotten human values such as honesty, integrity, humility, and transparency in communication were found to be important leadership traits for managing a hybrid workforce.

Quotations:

- *VUCA is nothing but a type of Risk; to mitigate the same we as leaders need to change our mindset & approach of facing the challenges.*
- *Leaders should showcase grit and growth mindset so as to lead the race in these unprecedented times and ensure to balance flexibility and stringency to manage the new normal at workplace.*
- *You have to Adjust your plans, Adapt to the challenges & apply the greatest virtue of Flexibility.*
- *Leaders should be understanding and active listeners: The approach "WE HAVE ALWAYS DONE IT THIS WAY" – need to be avoided, CHANGE is must with times.*
- *The traits which are required are: Trust & respect for individuals working in the team and the organisation, empathy and proper communication.*

4.4.3.2 Leadership Style

Successful leaders spend the majority of their time working closely with team members to understand their needs, showing them a growth path, and motivating them to achieve the same. Evaluating team members' effectiveness and guiding them to better output has been a popular approach among leaders to address employee detachment and improve productivity. Research has shown that an empathetic leadership style is more effective in handling chaotic situations. Under stressful situations, leaders enabled the team to work in a more meaningful way by fostering hope and faith, altruistic love for work, self, team and creating a vision. Though the leaders maintained a compassionate approach, they also implemented strict rules and had a robust control mechanism to manage organizational issues. A sustainable approach to improving cohesion within the team, as advocated by the respondents, was to trust the teammate's abilities and acknowledge their contributions.

Quotations:

- *Leaders must be the spokesperson and promote long term vision among the team. I mostly rely on sharing impactful incidents that denotes the day-to-day decision-making effectiveness.*
- *Leaders must be visionary – As a leader, while we lead a life of charisma externally, we must lead a life of character internally.*
- *Be a career consultant and generate trust and ownership in the team.*
- *Aware about market data analytics, propensity to train, guide and motivate.*
- *Modern leaders need to lead by example and also need to be empathetic.*
- *They need to balance the team in such a way that every team member is heard. The leaders need to also understand and bring out the best in every team member.*

4.4.3.3 Leadership Strategies

Following the individual leadership style, leaders designed innovative leadership strategies for managing people in the face of contemporary challenges. Based on responses obtained from interviews, leaders adopted tailor-made approaches that were flexible and resilient. Creating high-esteem team culture ensured individual and team autonomy. One-to-one communication, a transparent reward and recognition policy, and the celebration of diversity fostered an engaged team. Leaders were responsible for ensuring that employees had equal access to opportunities at work. The role of leadership was to create psychologically safe spaces for employees to be able to speak freely and bring their whole selves to work. They ensured well-being of employees through hybrid work schedules, online training programs, and employee support by decentralizing power.

Quotations:

- *Focus on nonconventional Trainings, MDPs to inculcate new skills.*
- *Employee engagement sessions: weekly recreational activities, team dinners, brainstorming meetings, team building games (ex: skills cafe Flo games), some gamified learning* etc.
- *Creating a high esteem team culture, peer to peer recognitions, corporate reward and recognition.*
- *A culture of communication-care and compassion-collaboration.*
- *Maintaining transparent control mechanisms by communicating clear target/ KRA/Goal sheet and a proper informal review mechanism which will help regulate employees' behavior and they will understand the context of leaders perspectives.*
- *Charting: A friend in- need approach, regular communication, check-ins, provide mutual support for the well-being of team-members and also their family members. Indulging in open and effective communication.*

4.4.4 Managing the BANI Environment

The final section of the study concludes with the impact of changing organizational policies and leadership strategies on management of individual and team to create a sustainable workplace environment.

4.4.4.1 Employee Engagement

Employees are highly educated and they had their own priorities and aspirations. To manage a diverse team in a hybrid mode, leaders had to focus on individual employee engagement and management of team more tactfully and sensitively. In modern organizations, leaders focused on psychosocial stability, cognitive enrichment, and work-life balance to improve employee engagement.

Quotations:

- *Reduce burnout through regular communication.*
- *Leading by example, leaders can create clear set of expectations and norms to follow.*

- *Encouraging break and maintaining office time.*
- *Ensuring open and nonjudgmental discussion related to any issues/ strategies/ plan.*
- *Initiating internal knowledge transfer sessions, involving them in knowledge sharing to outer world, encouraging them mentoring the interns, involving them in research practices.*

4.4.4.2 Leading Teams

To lead teams effectively, leaders concentrated on upskilling the teams, balanced flexibility and stringency to manage the new normal at workplace, fostered collaboration & shared purpose within the teams. Leaders worked as peers or mentors and tried to adjust, adapt, and update oneself regularly and empower team to take decisions.

Quotations:

- *Leaders need to adopt flexible, creative and empathetic approaches.*
- *Organization needs to be agile and resilient.*
- *Fostering innovation culture through design thinking.*
- *Foster individual skills and create synergetic body of knowledge.*

The thematic study encodes four major factors from the minor factors obtained from the interview. Table 4.2 presents the list of major and minor factors addressed in the present study.

Based on the in-depth interviews, a schematic diagram (Fig. 4.1) is prepared, which shows the inter-relationship between factors.

According to Fig. 4.1, in today's competitive economy, leaders worry about managing a hybrid workforce. Engaging employees goes beyond keeping the basics right and incentivizing them regularly. Leaders must consider their emotional and

Table 4.2 List of major and minor factors in a disruptive environment

Major factors	Minor factors
Organizational challenges in new normal	• Structure-culture fit • Team-individual fit • Functional Challenges • Change in attitude of employees
Organizational policy changes to manage contextual challenges	• Flexible organizational policy • Changing employee welfare policy • Creating a collaborative ecosystem
Leadership traits and strategies	• Leadership traits • Leadership style • Leadership strategies
Managing the BANI environment	• Employee engagement • Leading Teams

Source Primary data

psychological needs and monitor results. Today's workplace values people analytics as well as empathetic abilities. Working away from the workplace may increase insecurity; therefore, leaders must offer equal space in hybrid mode and ensure equitable growth chances for all. Leaders highlighted building a free and fair work and communication environment for employee and family well-being in hybrid work schedules.

4.5 Conclusion

Rapid technological upgradation, competitive businesses and quality-conscious customers were already a challenge. These were further strengthened by the outbreak of the pandemic and it left the entire society bewildered. The scenario became intensely ambiguous, complex, and frightening due to severe business losses, rapid digitization, work from home assignments, man management issues, and panic and anxiety about one's own and family members' health. The lack of direct interaction with peers, social distancing, and social anxiety have exacerbated the severity of the situation. As discussed above, organizational challenges, leadership traits and strategies, and managing the BANI environment were the key focus areas in the post-COVID era. The business leaders admitted that increased work pressure to cope with the losses made during the pandemic and overdependence on technology resulted in employee dissatisfaction. The team leaders or managers confronted hardships to address the emotional needs of the employees and create a well-integrated team. Despite challenges and obstacles, leaders put in their consistent effort to bridge the gaps of emotional connect, made sincere efforts through policy formulation, and embraced diversity through different sensitization programs for creating engagement with the system and retain existing employees. Post-pandemic recruiters encountered an additional hardship of new segment of potential candidates who had imperceptible engagement with the organization and inadequate knowledge and skills. Revisiting organizational policies and designing flexibility in the system were imperative. Employee welfare policies became the focal point of all managerial discussions as it was a key tool to generate trust and engagement among the workforce. Leaders were thoughtful of creating a collaborative ecosystem through empathetic conversations, generating employee ownership in tasks by brainstorming assignments, regular engagement sessions, and extending well-being support to employees and their family members. Through their adaptive styles, winning approaches, productive plans, and leadership attributes, the modern leaders embraced change gracefully to address the BANI environment and employee expectations. Informal performance reviews and a friend-in-need approach proved to be effective in the new normal. Reducing burnout, ensuring open and non-judgmental discussion, and fostering innovative culture were reflected in the creative and empathetic approaches of the effective leaders. Thus leaders became mentors and focused on team management, constructive feedback, trust, and faith. Leadership has been observed to be crucial in forging a united front in the face of daunting external risks. The paper investigated how empathic leadership style fosters team cohesion in a hybrid setting, which in turn ensures employees' sense of psychosocial stability, thus enhancing their level of engagement with the organization.

The current study substantiates from the interviews that though the pandemic instilled a lot of stress, anxiety, and insecurity among the workforce, effective leadership strategies resulted in positive regulation of the psychosocial state of minds of the employees, resulting in satisfaction and engagement. However, the study can be further explored in future with quantitative reinforcements and investigate the interrelationships of the major and minor factors in a disruptive environment through advanced analytical evidences.

References

1. Kirchner, K., Ipsen, C., & Hansen, J. P. (2021). COVID-19 leadership challenges in knowledge work. *Knowledge Management Research & Practice, 1*(1), 1–8. https://doi.org/10.1080/14778238.2021.1877579
2. Nataliia, H., & Olena, M. (2023). The key administrative competencies of managers required for company development in the Bani world. *Economics (Bijeljina), 11*(1), 289–305. https://doi.org/10.2478/eoik-2023-0012
3. Kaul, V., Shah, V. H., & El-Serag, H. (2020). Leadership during Crisis: Lessons and applications from the COVID-19 pandemic. *Gastroenterology, 159*(3), 809–812. https://doi.org/10.1053/j.gastro.2020.04.076
4. Shukla, B., Sufi, T., Joshi, M., & Sujatha, R. (2022). Leadership challenges for Indian hospitality industry during COVID-19 pandemic. *Journal of Hospitality and Tourism Insights, ahead-of-print* (ahead-of-print). https://doi.org/10.1108/jhti-08-2021-0217
5. Andrea, Medina-Ricaurte, G. F., & Jimenez-Hernandez, P. R. (2024). Organizational elements to confront turbulent and fragile VUCA to BANI Scenarios. *Advances in Human Resources Management and Organizational Development Book Series*, 20–43. https://doi.org/10.4018/979-8-3693-0720-5.ch002
6. Evseeva, S., Evseeva, O., & Rawat, P. (2022). Employee development and digitalization in BANI world. *Communications in Computer and Information Science, 253–264*,. https://doi.org/10.1007/978-3-031-14985-6_18
7. Hersey, P., Blanchard, K. H., & Johnson, D. E. (1996). *Management of Organizational Behavior*.
8. Franken, E., Bentley, T., Shafaei, A., Farr-Wharton, B., Onnis, L., & Omari, M. (2021). Forced flexibility and remote working: Opportunities and challenges in the new normal. *Journal of Management & Organization, 27*(6), 1–19. https://doi.org/10.1017/jmo.2021.40
9. Kumar, S., & Kumar, A. D. (2024). Human Resource Development as a Tool to Enhance Knowledge to Thrive in VUCA Situations [Review of *Human Resource Development as a Tool to Enhance Knowledge to Thrive in VUCA Situations*]. In *Organizational Management Sustainability in VUCA Contexts* (pp. 44–62). IGI Global. https://doi.org/10.4018/979-8-3693-0720-5.ch003
10. Shuck, B., & Herd, A. M. (2012). Employee Engagement and leadership: Exploring the convergence of two frameworks and implications for leadership development in HRD. *Human Resource Development Review, 11*(2), 156–181. https://doi.org/10.1177/1534484312438211
11. Mester, C., Visser, D., Roodt, G., & Kellerman, R. (2003). Leadership style and its relation to employee attitudes and behaviour. *SA Journal of Industrial Psychology, 29*(2). https://doi.org/10.4102/sajip.v29i2.100

12. Aritzeta, A., Swailes, S., & Senior, B. (2007). Belbin's team role model: Development, validity and applications for team building. *Journal of Management Studies, 44*(1), 96–118. https://doi.org/10.1111/j.1467-6486.2007.00666.x
13. Goleman, D., Boyatzis, R., Davidson, R. J., Druskat, V., & Kohlrieser, G. (2021). *Empathy*.
14. Holt, S. (2021). Nurturing empathy. *Innovative Leadership in Times of Compelling Changes*, 117–131. https://doi.org/10.1007/978-3-030-67258-4_8
15. Van Bommel, T. (2021). *The power of empathy in times of crisis and beyond (Report)*. Catalyst. https://www.catalyst.org/reports/empathy-work-strategy-crisis
16. Moore, K. J., Xiong, S., Bhattacharya, M., Bustamante, G., & Calvert, C. (2020). Beyond diversity: Focusing on and enhancing inclusion in the society for epidemiologic research. *American Journal of Epidemiology, 189*(10), 1042–1046. https://www.ncbi.nlm.nih.gov/pmc/articles/PMC7666410/
17. Marques, J. (2015). The changed leadership landscape: What matters today. *Journal of Management Development, 34*(10), 1310–1322. https://doi.org/10.1108/jmd-02-2015-0010
18. Salovey, P., & Mayer, J. D. (1990). Emotional intelligence. *Imagination, Cognition and Personality, 9*(3), 185–211. Sagepub. https://doi.org/10.2190/DUGG-P24E-52WK-6CDG
19. Kock, N., Mayfield, M., Mayfield, J., Sexton, S., & De La Garza, L. M. (2019). Empathetic leadership: How leader emotional support and understanding influences follower performance. *Journal of Leadership & Organizational Studies, 26*(2), 217–236. https://doi.org/10.1177/1548051818806290
20. Kalshoven, K., Den Hartog, D. N., & De Hoogh, A. H. B. (2012). Ethical leadership and follower helping and courtesy: Moral awareness and empathic concern as moderators. *Applied Psychology, 62*(2), 211–235. https://doi.org/10.1111/j.1464-0597.2011.00483.x
21. Zivkovic, S. (2022, April 21). *Empathy in leadership: How it enhances effectiveness*. ResearchGate. https://www.researchgate.net/publication/361952690_Empathy_in_Leadership_How_it_Enhances_Effectiveness
22. Edmondson, A. C., & Lei, Z. (2014). Psychological safety: The history, renaissance, and future of an interpersonal construct. *Annual Review of Organizational Psychology and Organizational Behavior, 1*(1), 23–43. https://doi.org/10.1146/annurev-orgpsych-031413-091305
23. Kelloway, E. K., Turner, N., Barling, J., & Loughlin, C. (2012). Transformational leadership and employee psychological well-being: The mediating role of employee trust in leadership. *Work & Stress, 26*(1), 39–55. https://doi.org/10.1080/02678373.2012.660774
24. Northouse, P. (2019). *Leadership: Theory and practice* (8th ed.). Sage Publications.
25. Riisla, K., Wendt, H., Babalola, M. T., & Euwema, M. (2021). Building cohesive teams—The role of leaders' bottom-line mentality and behavior. *Sustainability, 13*(14). mdpi. https://doi.org/10.3390/su13148047

Soma Bose Biswas is an enthusiastic instructor who wants to establish a management knowledge foundation with a PhD (Human Resource Management) or MBA and a PhD guide. She has taught management courses of MBA and BBA students at leading business colleges of Kolkata and Bangalore for over two decades. She has also written or presented research papers in prestigious publications and at prestigious conferences. She has published many papers in national and international journals and books. As session chair, she has supported many FDP and conference lectures. Doctor Bose Biswas is affiliated with the NIPM, NHRD, ISTD, and others. She passionately promoted management education through lectures, projects, case studies, seminars, and workshops.

Ipsita C. Patranabis is a passionate academician and researcher in the field of Organizational Behvaiour and Human Resource Management. She completed her Masters in Applied Geology and MBA from Jadavpur University. Following her MPhil in Management, she gained her PhD degree from the Department of Business Management, University of Calcutta. She has significant contributions in research paper presentations and publications across various national and international conferences and journals. Her teaching career spans across two decades and she also serves in the journal review boards of eminent research journals of India and abroad. She has research scholars pursuing their PhD under her guidance from reputed Universities. She is also the life member of distinguished professional bodies namely, NHRDN and ISTD. Dr. Ipsita believes in collaborative learning and research and maintains high ethical standards in each and every endeavour.

Indranil Bose Associate Dean and Professor- School of Management, MIT University of Meghalaya, Shillong, India; A Fellow of the Higher Education Academy (FHEA, UK) and a Certified Management and Business Educator (CMBE) by the Chartered Association of Business Schools, Indrani Bose has a PhD in management. His nearly 23 years in academics and leadership have contributed to teaching, mentoring, and research in HRM and general management in India and the UAE. He has taught certified UK, USA, UAE Ministry of Education, and UGC-approved Indian university programs. Dr. Bose has produced many ABDC-listed and SCOPUS, Web of Science, and other database-indexed papers on AACSB and ACBSP-accredited business programs. He reviews and edits for Emerald and SAGE and has presented at several international and national conferences. The Harvard Business Review International Advisory Council includes him. Dr. Bose has received the Turn-it-in Global Innovation Honorable Mention (2017), Dewang Mehta Best Professor Award (2012), and MTC Global Distinguished Teacher Award (2016) in India, among others. He received the ICFAI Business School Dehradun Best Professor Award and ATHE UK Star Assessor Award 2021. He also reviewed management PhD theses in India and the UAE.

Chapter 5
Navigating Uncertainty: Assessing the Financial Resilience and Competitive Adaptation of DMart in Response to the Covid-19 Pandemic

Hanzala Awais and Darshna Bhatter

Abstract The impact of the Covid-19 pandemic was felt across various domains which included the retail business sector. The retail business sector was the one which was impacted widely but at the same time an opportunity was created for growth and expansion. Those who were able to adapt to new technologies got the competitive advantage. Slowly most of the players in the retail business sector were forced to adopt new technologies for survival even after the Covid-19 pandemic as it had changed the customer's behavioural pattern on a permanent level. DMart, a one stop supermarket chain, was one of the first few to adopt to new technologies to serve the customers and gain the competitive advantage at the same time. Thus, this study has been carried out to analyse the operations carried out, how efficiently were they carried out, if there were any limitations, and what were the steps of improvement undertaken by DMart during the Covid-19 pandemic period. This study is completely based on data collected from secondary sources. The analysis has been carried for the period 2019–2024 with the help of the annual reports published by DMart for the above-mentioned years. The analysis reveals that DMart has been able to deal with the Covid-19 pandemic efficiently by adopting and implementing newer strategies at the correct time like introduction of DMart App, etc. The study concludes that DMart had maintained its profitability throughout the period 2019–2023 and this will provide a stable financial foundation for future growth of the company.

Keywords DMart · Covid-19 pandemic · Financial resilience · Fundamental analysis · Retail innovation · DMart ready · Profitability · Liquidity · Solvency

---

H. Awais (✉)
Department of Commerce, St. Xavier's College (Autonomous), Kolkata, West Bengal, India
e-mail: hanzala.awais@sxccal.edu

D. Bhatter
St. Xavier's College (Autonomous), Kolkata, West Bengal, India

S. Gangopadhyay et al. (eds.), *Pandemic Diaries*,
https://doi.org/10.1007/978-981-96-5415-4_5

5.1 Introduction

The Covid-19 pandemic had significantly impacted various fields across the globe bringing everything to an uncertain position. The retail business had also faced a huge impact of this pandemic. The whole sector went through a dynamic change during this period. Eventually, a global shift towards new technologies was seen in the sector for survival in the industry. The lockdowns had altered the consumer as well as the seller practices. Thus, adapting to the 'new normal' became imperative. Over nearly two years, the industries had witnessed numerous developments to sustain during the pandemic period. The industries were dealing with a risky future as predicting the long-term effects of this Covid-19 pandemic on the consumer behavioural pattern was challenging for them which resulted in them taking a cautious approach towards their investments and spending. The companies were in a position where they were facing lower or no financial growth but were fighting in the industry for survival.

Online retailing was not a new concept but was not adopted by many companies in the retail sector as the customers preferred the offline channel. The Covid-19 pandemic compelled widespread adoption of innovative business practices, transforming the landscape of retail sector from offline to offline and online both with online now fetching them more profit as due to lockdown and Covid-19 protocols, customers avoided moving out of their house for purchasing goods. Despite mixed feedback, businesses across the retail industry were increasingly transitioning to technology-driven online modes of operation. Presently, the Indian Retail Sector is contributing 10% of the country's Gross Domestic Product and is considered a key contributor in the economy (IBEF Report, 2021).

One such retail company is Avenue Supermarts Limited which is doing business under the trade name 'DMart'. It is a supermarket and hypermarket chain which was incorporated in the year 2000 by Mr. Radhakishan Damani and his family to cater to the growing needs of Indian households. The first store was opened in 2002 in Mumbai, Maharashtra, and slowly it went on to expand to other states like Gujarat, Karnataka, Telangana, etc. By the year 2013, it ranked as the third largest retail company in terms of revenue after Reliance Retail and Future Group. It continued its business wholly through the offline channel till the year 2016 when it entered the e-commerce (online) market with its own App 'DMart Ready'. Initially, the app was not very successful but it gave an edge to the company during the pandemic period. It is currently present in 200 + locations with 365 stores across 11 states and 1 union territory in the country. DMart has been growing sustainably over the years with a mission of being the lowest priced retailer wherever it operates. DMart's prosperity revolves around three main pillars—Customers, Vendors, and Employees. The DMart store offered home utility products which included variety of garments, electronics, kitchenware, cosmetics, packaged foods, etc., all of which were available at a very competitive price. It was not a grand business strategy, but rather only a hunch backed by its promoter. The Strategies adopted by DMart included selling large quantities in low margin, purchasing their store space instead of renting it, eliminating intermediaries, offering private brands at huge discount, and giving the customers a luxury feel while shopping.

The online groceries market was very slow in gaining importance till after the Covid-19 lockdown which had acted as a boost for the growth of this channel of business. Online groceries are at present an under-penetrated segment within the e-retail line of business but had gained a significant importance and dependency in the last few years as more and more consumers have been shifting to online marketing of household grocery and other items.

Importantly, DMart has recently begun to have some of its locations open 24 h a day to assist customers in purchasing imported goods during the emergency lockout. Despite this initiative, the company has had to shut down over half of its outlets and cut down its employees in a couple of cities by an estimated 80–90%. Thus, this study has been carried out to examine the financial and operational impact of the Covid-19 pandemic on DMart for the period 2019–2024, including changes in revenue, profitability, and operational efficiency. This study also analyses DMart's strategies for addressing challenges posed by the pandemic, with a particular focus on its transition to online platforms, and to identify key factors influencing the success of these strategies and investigates the factors which are contributing to DMart's growth and increase in profitability with a specific emphasis on understanding its competitive positioning within the retail market and identifying its main competitors.

5.2 Review of Literature

There has been a dynamic growth of the Indian retail industry over the years and currently India is considered the fifth-largest shopping destination (IBEF Report, 2021). The Indian retail sector has witnessed a revolution during the Covid-19 pandemic [1]. The customers had started undergoing a shift in their behavioural pattern for grocery and other household purchasing from the year 2015 which was comparatively very slow but the pandemic acted as a catalyst to this shift and by the end of the pandemic, this shift to online shopping had become normal for the customers [2]. A similar study was carried by Louise [3] where he mentioned that the companies who could adapt to this change and take advantage of this opportunity created were the ones who survived and thrived during the pandemic and are in a competitively advantageous position after the pandemic.

Spencers had started the hypermarket concept in the year 2001 while DMart has started the same segment in the year 2002. DMart's has been successful in providing customers with a wide product range at competitive prices and this has shown the path to many other retail corporations in adopting the low margin-large scale sales approach as their business strategy [4]. The pillars of DMart's financial and market success are identified as customers, vendors, and employees [5]. DMart has been able to sustain its operations during the economic crisis [6]. As compared to Spencer's, DMart has been more profitable throughout [7]. The study conducted by Kowsalya and Gandhimathi [8] delved into the influence of advertisement and sales promotion strategies which were employed by DMart during the pandemic and showed how it can impact in increasing profitability of the company. Their research also highlighted DMart's adeptness in leveraging these strategies to bolster its presence and competitiveness in the retail market. During the pandemic, DMart faced a decline in profits attributed to store closures and limited sales of essentials. The second lockdown in India further impacted the profitability, leading to increased year-on-year expenses [9]. Roggeveen and Sethuraman [10] in their study discussed the transformative impact of the pandemic on the retailing world and emphasised the need for retailers to address both immediate, short-term concerns and long-term strategies for a post-Covid-19 market while Lee and Jong [11] underscored the post-pandemic retail sector's need for creativity and uniqueness, offering practical strategies for retail companies to mitigate the pandemic's adverse effects. The studies conducted by Pradhan and Bhatkal [12] and Chowdhary and Sekhar [13] on DMart examined its financial health using ratios, focusing on the unique features of DMart's business model. It highlights the utilisation of the DMart Ready online app and DMart on Wheels during the lockdown to cater to customer needs, showcasing adaptability in a changed business environment.

The review of literature showed that the Indian retail sector has been growing and has transformed during the Covid-19 period. DMart being one of those companies who were the pioneers in adopting new strategies and technologies was successful in remaining profitable throughout the period and thus there was a need to study the factors for which led to this success.

5.3 Methods

The research methodology employed in this study relies completely on secondary sources of data for the analysis of the financial resilience and competitive adaptation of DMart in response to the Covid-19 pandemic. Additionally, the methodology encompasses an examination of the broader impact of the pandemic on the retail industry, including trends, challenges, and shifts in consumer behaviour. To gather relevant data, a comprehensive review of past and present literature has been conducted, encompassing academic journals, industry reports, news articles, and research papers. The financial analysis has been carried out with the help of the annual reports published by Avenue Supermarts Limited for the period 2019–2023. This approach enabled the synthesis of diverse perspectives and insights into DMart's response to the pandemic and its evolving operational strategies.

5.4 Results and Discussion

5.4.1 Landscape of Retail Business Post-Covid-19

At the point when the virus started to spread, a few organisations talked about Covid-19 yet depicted it as an impermanent burden or a production network challenge. However, within a month of its onset, everyone realised the brutality of Covid-19 and soon everything changed.

Before the pandemic, a major piece of shopping was the active experience: Eating tests at the supermarket, going through an evening in the fitting room, attempting a lipstick at the magnificence counter, and swiping a credit card on a long-wanted buy. Numerous retailers are currently attempting to stand apart for the contrary explanation: They are making it simpler for shoppers to keep away from imparted surfaces and limit associations to workers or different clients that has directed to invest and develop contactless alternatives especially by grocery stores. Clients shopped online before the pandemic too, yet a compulsion to restrict outings to stores released new interest and urged retailers to bring out new options through technology. Some creative retailers are answering the truth of Covid-19 by straightforwardly putting their products where clients are—with spring up stores, even working out of trucks.

A pattern was followed once the pandemic retreated, and a select gathering of retailers were consummating it then too. The main aim was that the right products are available at the right time, since during that time, the buyer was the king, the employees were the backbone, and the goodwill was the fuel that kept the business running.

5.4.2 DMart Ready App

During the lockdown, DMart had made a stride further. It delivered fundamental and essential products to the doorstep of purchasers by means of its own application 'DMart Ready App'. Though this App was launched in the year 2016, it gained importance only during the pandemic. The App had two options for delivery which made it stand out from the other companies:

5.4.2.1 Self-Pick-Up

DMart had tied up with offline stores which became the pick-up points for the customers. The customers could place an order online from the application and get their requested items delivered to a DMart Ready Pick-Up point which was closest to their home. The customers could collect the order from the pick-up point as per their convenience. For this, the company was not charging any conveyance fee for delivering the items.

5.4.2.2 Home Delivery

The second option provided to the customers was direct to home delivery. The customers could place their request on the website/application and get it delivered to their home address at the preferred date and time. This option had an ostensible conveyance expense which the customer had to pay to the company.

Currently, this application is active in Gujarat, Maharashtra, Punjab, Rajasthan, Tamil Nadu, Telangana, Uttar Pradesh, Chhattisgarh, Andhra Pradesh, and Karnataka. The company is in the process of launching it in the other states soon.

5.4.3 DMart on Wheels

DMart has presented this thought where-in a customer can allow the company to leave their truck in the public complexes and individuals can purchase fundamentals from them. To reach a larger customer base, the organisation had begun online business home conveyance, but it could not make a mark in this space where it had been wary and less forceful as contrasted by its competitors Big Basket, Grofers, Swiggy Instamart, and Amazon Fresh. It has likewise chosen a different concept to reach the customer base by offering the DMart on Wheels and making the goods available to the customers in their complex buildings where the customers can buy the goods after checking it personally without having to go to the Supermart and save time and money both.

5.4.4 Revenue Drivers (DMart's Revenue Model)

DMart generates revenue mainly from three sections: food, non-food, and general merchandise and apparel. It also increases its revenue by reducing its costs from different avenues such as low labour expenses, economies of scale, own brand products, no intermediaries, etc. The different avenues have been discussed below:

5.4.4.1 Slotting Fees

DMart, being a grocery store chain, charges a 'Slotting Fee' from the manufacturers. It is an instalment made by the manufacturer of the merchandise to the superstore to reserve a space on the rack where it can display its merchandise or goods to the customers. The Slotting Fees for the middle racks are higher than the ones at the bottom and at the top. Additionally, DMart charges a section charge for the items, which are held in the grocery store.

5.4.4.2 Minimal Expenses

DMart follows an approach of minimising cost in all aspects. It uses its space efficiently by placing more items in less space; in this way, it makes space for additional items; it has minimum number of charging counters, thus diminishing the necessity of more labour force, which subsequently reduces workers cost; an exceptionally essential and low-support inside of the store is another way through which DMart has controlled its costs.

5.4.4.3 Volume Sales

DMart offers products at low prices or at deep discounted prices to the customers by purchasing directly from the suppliers, thus eliminating intermediary cost. It also offers its own brand products at low cost and keeps a huge inventory to cater to more number of customers. Low valued products lead to high footfall in the stores prompting high deals volume. As the Sales increase, it draws an ever-increasing number of sellers to keep their items in DMart.

5.4.4.4 Owned Outlets

DMart purchases its stores and it mainly purchases space in residential or market areas which are accessible by the customers easily. As a result of operating in its own stores, it is able to operate as an organisation with no fixed obligations and is therefore more financially stable. Further, no rental expense helps in high sure incomes, which are re-invested in the business for opening more stores. Among the relative multitude of existing stores to date, around 80% are self-claimed.

5.4.4.5 Low Debt

DMart follows a conservative approach when it comes to debt. It currently has a debt-equity ratio of 0.7 which helps DMart generate a higher Return on Equity as there are low financial fixed charges for debt. This approach has helped DMart to grow at a faster pace than its competitors.

5.4.5 SWOT Analysis

5.4.5.1 Strengths

Power is defined as the ability of any business to be efficient in its operations so that it can provide more power than its competitors.

The following are the strengths of DMart:

Everyday Low Pricing

DMart follows the strategy of providing goods at Everyday Low Pricing which helps it to bring in the middle-class customers in masses.

Deep Discount Policy

DMart purchases directly from the suppliers and gets a high margin on MRP. It keeps a low margin and uses this margin to offer products well below the MRP at deep discounted price. This low price attracts a lot of customers which increases its revenue.

Human Resource Management

DMart maintains long-term relations with its suppliers and vendors which indicated that it has a strong relationship with them. It also has a large number of employees with low turnover rate which indicated employees are happy in the working conditions in the company. DMart trains its staff to give a friendly shopping experience to customers which helps them coming back to the store for their shopping experience.

Efficient Supply Management

DMart has a smooth Supply Chain Management because of which it doesn't require huge inventory storage space while making sure that the goods are available on time for the customers. Thus, customers never face a shortage in the store for their demand which creates a goodwill for the company.

Huge Variety of Products

DMart stores offer a wide range of products in all three segments: food, non-food, and merchandise. It caters to the needs of everyone in the family. Thus, it makes shopping a family experience for the customers.

Digital Platform

Customers can now order products through their online ordering app, 'DMart Ready' App. Thus, customers can order for their need on the go and get it delivered at their doorstep. DMart is now on the fingertips of the customers.

5.4.5.2 Weakness

Weaknesses are used to express areas of a business or product that need improvement.

Focus on a Specific Area

Most of the DMart stores are located in the western part of the country. It has not been able to expand to other regions. There are just a few stores in the southern and eastern market. Thus, it is focused on just a small region with no vision of expanding.

Not a Global Player

DMart is based in India and it doesn't have any global presence but within the country, it is exposed to risks from foreign players such as Amazon Fresh, Flipkart Grocery, etc.

Slow Growth

DMart was founded 20 years ago but given the time period of two decades, it has not yet achieved the success it should have. This is mostly because of its over cautious approach and its focus on long-term projects.

Focus on Low Prices in the Long Term

In order to provide low prices to customers, DMart negotiates with suppliers and enters into long-term contracts with them. Any problem with the suppliers can affect their supply chain management and have an impact on Sales too.

No-Frills Approach

In order to reduce the costs, DMart follows a no-frills approach, i.e., it follows basic layout and utilised the space efficiently so as to offer more products in the given space. Some customers find it compact and don't get a luxury experience while shopping.

5.4.5.3 Opportunities

DMart is a supermarket and thus there are many opportunities to increase their profits.

Growth Opportunities

DMart stores want to be the most valuable store in India for its customers. Thus, they should be expanding their business in the other untapped regions of North and East India either by investing in stores themselves or by tying up with other small businesses.

Improving Technology

Improving technology will improve customer experience in the market. Self-checkout Kiosk can be introduced in the stores which can reduce the wait time in the checkout line and improve the customer experience too. Also, the warehouses can be automated which can help in reducing costs.

Personal Service

Sellers like DMart should take advantage of spending more and continue improving the quality of service. Customers always look for personal service and are ready to pay extra for that personal touch. So, a personal staff can be assigned on request by the customers when they enter the store by paying some extra amount.

Advertisement

Online markets are open and DMart has more and more opportunities to capture these growing markets. It should spend more on advertising and increase its online presence which can help in increasing their revenue. It should also invest more in online delivery system to reduce the delivery time.

5.4.5.4 Threats

DMart also has some threats to its business from the outside.

Online Competition

Customers are becoming more and more dependent on online shopping nowadays. Online rivals like Amazon Fresh, JioMart, Blinkit, etc. and local platforms are becoming tough competitors for DMart.

Open Market

DMart operates in a market where there is no barrier to entry of new companies and thus DMart faces the threat of new entrants. It also faces competition from local stores and supermarkets.

Government Regulations

Changes in Government regulations or increase in corporate tax or increase in inflation can have a negative impact on DMart's revenue.

5.4.6 Fundamental Analysis of DMART

The fundamental analysis of DMart has been done for the period 2019–2024. For the analysis, profitability ratios, liquidity ratios, and solvency ratios have been calculated and interpreted.

5.4.6.1 Profitability Ratios

Operating Profit Margin

See Fig. 5.1.

Interpretation: Operating Profit Margin is calculated by dividing Operating Profit by Total Revenue. This ratio measures the company's ability to generate profit through sales. It is also known as Return on Sales. It can be seen in the above chart that the Operating Profit Margin has remained consistent between 5 and 7%. It indicates that the company is generating enough profit to cover its operating costs. Also, it can be seen that its profitability was impacted by the Covid-19 pandemic and it went down to 5.52% in 2020–21, but it gradually increased to 6.46% in the year 2022–23 and 7% in 2022–23. Again, it has seen a decrease of 6.74% in 2023–24 in its operating Profit Margin as compared to 2022–23. It may be because of its increase in cost or raw materials in the country.

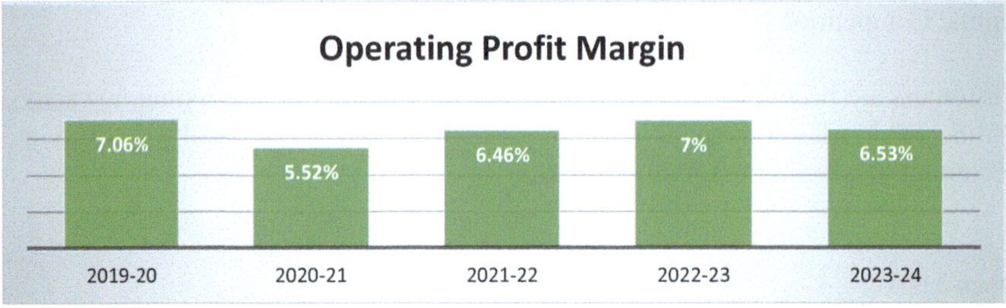

Fig. 5.1 Operating profit margin. *Source* Generated from the annual reports of DMart for the period 2019–2024

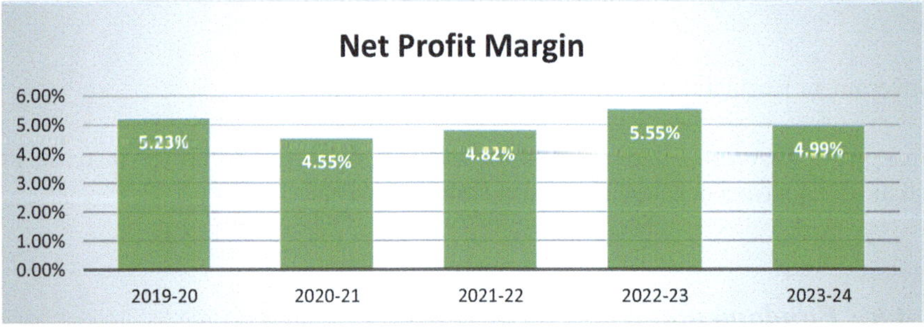

Fig. 5.2 Net profit margin. *Source* Generated from the annual reports of DMart for the period 2019–2024

Net Profit Margin

See Fig. 5.2.

Interpretation: Net Profit Margin is calculated by dividing Net Profit by Total Revenue. This ratio measures the profitability of the company and its ability to generate profit through the management of its expenses. It can be seen in the above chart that the Net Profit Margin has remained consistent from 4 to 6%. But this margin should be higher than 10%. A Net Profit Margin less than 10% indicates that a company is not able to control its cost and generate enough profit. Also, it can be seen that its profitability was impacted by the Covid-19 pandemic and it went down to 4.55% in 2020–21, but it gradually increased to 5.55% in the year 2022–23. Again, it has seen a decrease of 10.05% in 2023–24 in its margin which may be because of its increase in cost or more competition from other players in the retail industry.

Return on Capital Employed

See Fig. 5.3.

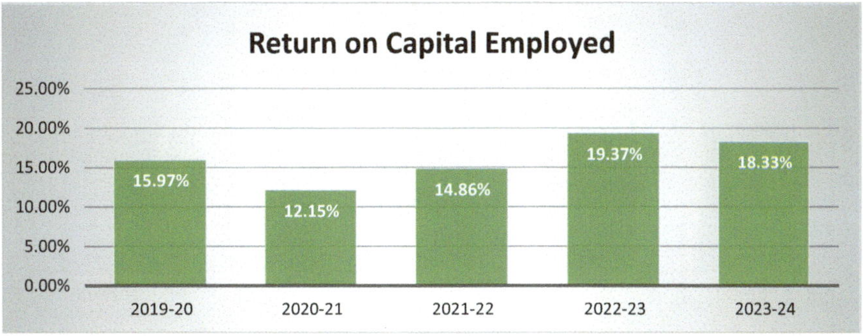

Fig. 5.3 Return on capital employed. *Source* generated from the annual reports of DMart for the period 2019–2024

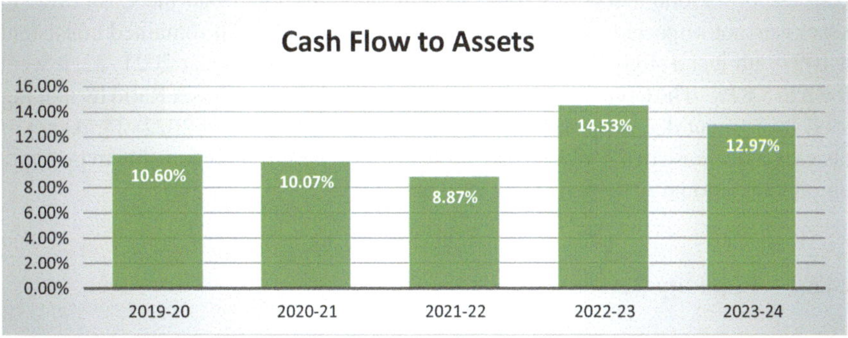

Fig. 5.4 Cash flow to assets. *Source* Generated from the annual reports of DMart for the period 2019–2024

Interpretation: Return on capital employed (ROCE) is calculated by dividing Earnings Before Interest and Tax (EBIT) by Capital Employed. This ratio measures the long-term profitability of the company and finds out if the company is utilising its capital efficiently for generating profits. It can be seen in the above chart that the ROCE was impacted by the Covid-19 pandemic and it went down to 12.15% in 2020–21, but it gradually increased to 19.37% in the year 2022–23. Again, it has seen a slight decrease of 5.36% in 2023–24 from the ROCE of previous year. Overall, DMart has been able to generate good return on its capital employed for the given period.

Cash Flow to Assets

See Fig. 5.4.

Interpretation: Cash Flow to Asset ratio is calculated by dividing Cash Flow from Operations by Total Assets. This ratio measures the company ability to generate cash

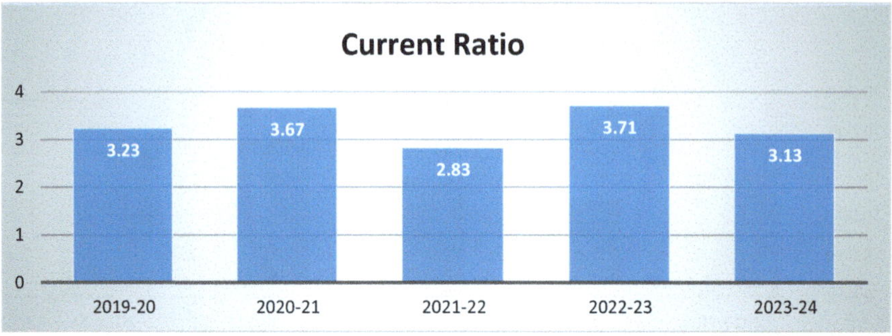

Fig. 5.5 Current ratio. *Source* Generated from the annual reports of DMart for the period 2019–2024

from use of its total assets. It can be seen in the above chart that the Cash Flow to Asset was not impacted much by the Covid-19 pandemic as it remained consistent at 10% with just a slight decrease in 2020–21, but in the next year 2021–22, it went down to 8.87%. The Company then recovered its Cash Flow to Asset Ratio by making good Investment decisions and the ratio increased to 14.53% in 2022–23. Again, it has seen a decrease of 12.97% in 2023–24. Overall, DMart has been able to generate a moderate amount of cash from its total assets.

5.4.6.2 Liquidity Ratios

Current Ratio

See Fig. 5.5.

Interpretation: Current ratio is calculated by dividing Current Assets by Current Liabilities. This ratio measures the company's ability to repay its short-term obligations. It can be seen in the above chart that the Current Ratio was not impacted much by the Covid-19 pandemic as it remained consistent at around 3% with just a slight decrease in 2021–22. Overall, DMart has been able to keep the current ratio well above the standard requirement of 2 which indicates that the company is in a very good position when it comes to short-term obligations.

Quick Ratio

See Fig. 5.6.

Interpretation: Quick ratio is calculated by dividing Liquid Assets by Current Liabilities. This ratio measures the company's ability to repay its very short-term obligations. It can be seen in the above chart that the Quick Ratio has been fluctuating in the given period. It has recovered its Quick Ratio from 2022–23 onwards after the sharp decrease in 2021–22. Though the ratio has gone down to 1.15 in 2023–24,

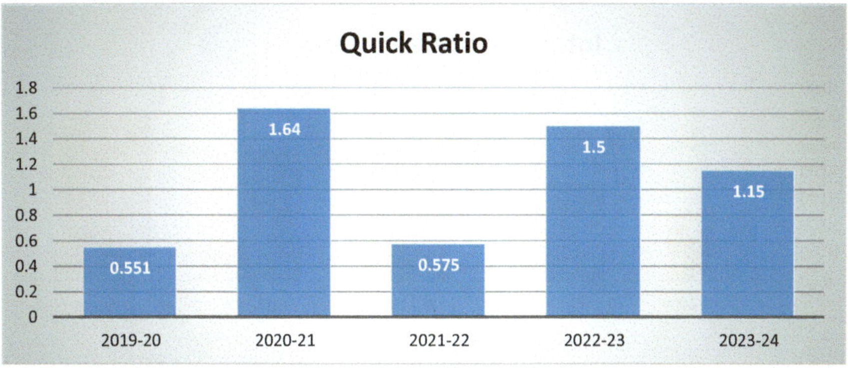

Fig. 5.6 Quick ratio. *Source* Generated from the annual reports of DMart for the period 2019–2024

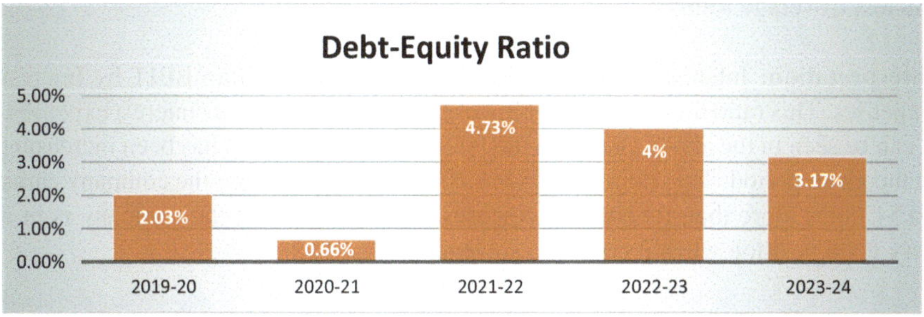

Fig. 5.7 Debt-equity ratio. *Source* Generated from the annual reports of DMart for the period 2019–2024

it is above the standard requirement for this ratio, i.e., 1, which indicates that the company is in a very good position when it comes to very short-term obligations.

5.4.6.3 Solvency Ratios

Debt-Equity Ratio

See Fig. 5.7.

Interpretation: Debt-Equity ratio is calculated by dividing Total Debt by Total Equity Share Capital. This ratio measures the company's ability to repay its long-term obligations. It can be seen in the above chart that the Debt-Equity Ratio has increased after the Covid-19 period. The Debt-Equity Ratio is 3.17% in 2023–24 which indicates that the company is in a very good position when it comes to long-term obligations as a capital comprising such less debt gives the company an edge over its competitors and also makes the company financially strong.

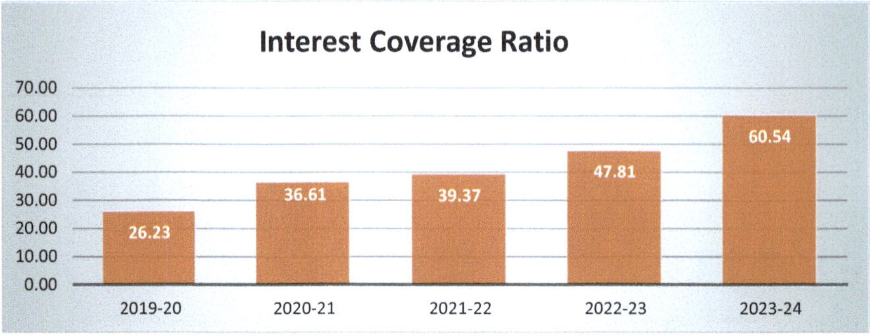

Fig. 5.8 Interest coverage ratio. *Source* Generated from the annual reports of DMart for the period 2019–2024

Interest Coverage Ratio

Interpretation: Interest coverage ratio is calculated by dividing EBIT by Interest Expenses. This ratio measures the ability of the company to pay its interest expenses. It can be seen in the above chart that the Interest Coverage Ratio has been increasing in the given period. This ratio of 60.54 in 2023–24 indicates that the company earns 60.54 times more than its interest expenses which shows that the company is in a very good position when it comes to solvency.

5.4.6.4 Revenue Mix of DMart

See Fig. 5.9.

Interpretation: The food industry accounted for around 56% of the total sales share of DMart, an Indian low-cost value retailer in both 2022–23 and 2023–24. FMCG contributed to almost 21% of the total revenue for both the years. Merchandise contributes around 23% of the total revenue for both the years. This shows that the

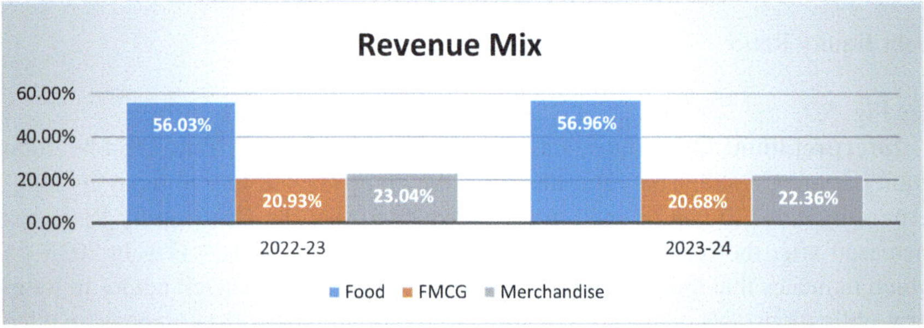

Fig. 5.9 Revenue mix. *Source* Generated from the annual reports of DMart for the period 2022–2024

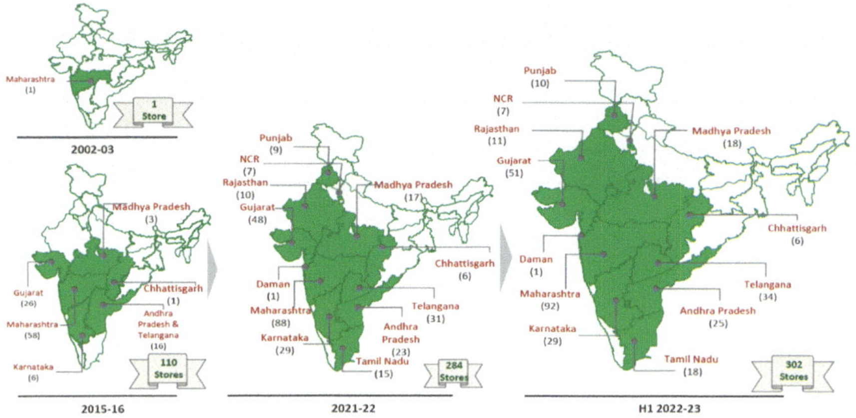

Fig. 5.10 Geographical distribution of DMart stores. *Source* Map generated by the author using company's data

company has been able to maintain its revenue from all the segments for the given period. The revenue is mostly from the food segment. There is a huge scope of improvement in the FMCG and the Merchandise segment from where it can increase its revenue.

5.4.6.5 Expansion

See Fig. 5.10.

Interpretation: DMart being a big retail business in India has till date only expanded its stores in the western and the southern side of the country as can be seen in Fig. 6.5. They have been opening stores at a very high speed. They reached from 1 store to 110 in 12 years approximately and despite the Covid-19 pandemic, they kept on expanding and finally have reached 302 stores by the year 2023. They are yet to capture the untapped market of north and east India which can help in generating more revenue for the company (Table 5.1).

Interpretation: After the pandemic, a lot of companies entered into this market with varied facilities like JioMart, Big Basket, Blinkit, and Swiggy Instamart. Even foreign players like Amazon Fresh and Flipkart grocery have been doing in the online retail industry. One of the biggest competitions for DMart will be the Zomato acquired Blinkit in the upcoming years because it has covered a larger area in terms of expansion and takes lesser time than DMart to deliver. Hence, DMart needs to focus more on its delivery management system so that it can cater to more customers and is able to deliver at a faster pace. With this, DMart also needs to reduce the minimum order amount from ₹500 to compete with other players as customers are having options of no minimum order amount in the sector.

Table 5.1 DMart's competitive advantage

Analysis	DMart	JioMart	Big Basket	Blinkit/Swiggy Instamart	Amazon Fresh
Delivery time	1 Day	Few hours to same day	1 h–2 h	Few minutes	2 h–24 h
Minimum order	₹ 500	No minimum order	No minimum order	No minimum order	₹ 249
Delivery fees	₹ 49	Free	Free	₹ 15	₹0–₹ 59
Bank offers	Yes	Many	Yes	No	Yes
No of stores	361	260	200	Online	Online

Source Table created by author using the respective company's data

5.5 Conclusion

DMart achieved significant growth in profitability and revenue diversification in the face of the Covid-19 pandemic which will have a significant and lasting impact on the overall development of the company.

First of all, the increase in profit not only demonstrates the good performance of DMart but also provides a stable financial foundation for future growth and investment. Successful financial development allows the company to allocate resources to strategic initiatives such as technological innovation, market expansion, and business diversification, thereby strengthening its competitive position in the market.

Additionally, the strategy focuses on food and most of the revenue comes from this segment, proving that DMart is a strong player in the industry. By strengthening its position in the food, DMart not only encourages loyal customers but also drives revenue and business growth by opening sales channels and engagement opportunities hand-in-hand with vendors.

Furthermore, DMart's expansion into new regions, especially the East, demonstrates the company's commitment to catching up with products, not copying any talent and expanding its footprint. This geographic diversity not only reduces the risks associated with regional dependency, but also allows DMart to reach different customers and preferences, thus expanding its customer base and supporting long-term growth.

As a result, increased profitability, revenue diversification, and geographical expansion will increase DMart's success and market leadership in the post-pandemic period. By leveraging its strengths, adapting to changing consumer markets, and driving innovation, DMart is well positioned to become a major force in the retail industry, creating value for stakeholders and contributing to the broader economy.

DMart has been able to serve the customers with low priced products for the last two decades which shows how efficient it has been in facing competition and maintaining low operation and supply and distribution cost. The success of DMart will be a source of inspiration for many new entrants in the Indian retail sector.

References

1. Ahmad, S., Akhtar, M., & Farooque, A. (2020). Study of the rise and growth of the Indian retail industry. *Catalyst – Journal of Business Management (CJBM), 2(1),* 1–12.
2. Sayyida, H., Sri, G., & Husin, S. (2021). The impact of the Covid-19 pandemic on retail consumer behavior. *Aptisi Transactions on Management (ATM), 5*(1), 79–88. https://doi.org/10.33050/atm.v5i1.1497.
3. Louise, G. (2022). Lessons from the COVID19 pandemic: The case of retail and consumer service firms. *Journal of Retailing and Consumer Services, 68,* 1–10. https://doi.org/10.1016/j.jretconser.2022.103012
4. Begum, R., & Varalakshmi, A. (2024). A comprehensive study on DMart services: Understanding consumer perceptions and market dynamics. *International Journal of Engineering Technology and Management Sciences, 2*(8), 307–317. https://doi.org/10.46647/ijetms.2024.v08i02.039
5. Sangvikar, B., & Pawar, A. (2019). Comprehending the pre and post economic turbulence calamity of India: The realization message from the Nineties. *Journal of Critical Reviews, 6*(6), 345–349.
6. Sharma, V., & Sonwalkar, J. (2013). Does consumer buying behavior change during economic crisis? *International Journal of Economics & Business Administration, 1,* 33–48. https://doi.org/10.35808/ijeba/9
7. Burt, S., Mellahi, K., Jackson, T. P., & Sparks, L. (2002). Retail internationalization and retail failure: Issues from the case of marks and Spencer. *International Review of Retail, Distribution and Consumer Research, 12,* 191–219.
8. Kowsalya, S. & Gandhimathi, S. (2023). Impact of advertisement and sales promotion strategies adopted by DMart. *EPRA International Journal of Research & Development, 8(6),* 239–244. https://doi.org/10.36713/epra2016
9. Nithya, S., & Chirputkar, A. (2020). The analysis of impact of Covid-19 on retail consumer behavior and business models. *Palarch's Journal of Archaeology of Egypt/Egyptology, 17*(6), 4384–4401.
10. Roggeveen, A. L., & Sethuraman, R. (2020). How the COVID-19 pandemic may change the world of retailing. *Journal of Retailing, 96,* 169–171. https://doi.org/10.1016/j.jretai.2020.04.002
11. Lee, Y., & Jong, K. (2009). Customer satisfaction in food retailing: Comparing speciality and conventional grocery stores. *International Journal of Retail & Distribution Management, 37*(1), 63–80.
12. Pradhan, S. & Bhatkal, S. (2021). *DMart: Driving Growth in a Changed Business Environment, 11(3),* 371–389.
13. Chowdhary, G. & Sekhar, S. (2021). A study on role of retail Chain in India: The business model of DMart. *Journal of Interdisciplinary Cycle Research, XIII(III),* 1559–1570.

Hanzala Awais Ph.D. in Commerce, is currently working as an Assistant Professor in the Department of Commerce, St. Xavier's College (Autonomous) Kolkata, India for the past 9 years. Prior to this, she has worked as a Lecturer in the B.Com. Department of Shivnath Sastri College, Kolkata, India. Her area of teaching includes Corporate Accounting, Cost and Management Accounting, Financial Management, Financial Reporting and Taxation. Her research interest lies in Conventional and Islamic Banking and Investigation of Financial Statements. She has presented papers in various national and international conferences and has published articles in reputed journals and chapters in edited books.

Darshna Bhatter a graduate student of B.Com., St. Xavier's College (Autonomous), Kolkata, India is currently working as a Management Consultant at PWC, Kolkata, India. She specialises in Financial Modelling and Research. As a student she had an inclination and passion towards research and has a deep interest in evaluating the performance of companies, understanding their marketing strategies and exploring innovation across industries with a particular focus on the dynamic FMCG sector. Even after working in the corporate sector, she is continuing with her passion for research and is developing cases and articles for publication under the guidance of her professors.

Chapter 6
Covid-19: A Catalyst for Digital Transformation in Medical Representation

Tamal Maity and Shweta Kishore

Abstract Various industries were impacted by the Covid-19 pandemic, and the pharmaceutical industry was no exception. Medical Representatives (MRs), who had previously relied on face-to-face interactions with patients and healthcare professionals (HCPs), had to adjust to the new social distance realities. This unpredicted obstacle augmented up the MR role's digital evolution and required a change in their engagement strategies.

In order to stay in touch with patients and HCPs, MRs resorted to digital means when in-person encounters were not feasible. Online conferences, webinars, and virtual meetings have become popular alternatives for exchanging product details, answering queries, and building connections. Additionally, MRs were able to increase their outreach by using digital platforms to connect with patients and remote healthcare professionals.

Besides supporting the pharmaceutical business in overcoming the obstacles posed by the pandemic, the adoption of digital technologies created newer opportunities. MRs might learn more about patient demands, market trends, and HCP preferences by applying data analytics and AI-powered solutions. They were able to improve targeting, customize messages, and increase overall efficacy. Despite the restrictions of physical distancing, the pharmaceutical industry was able to preserve and even improve its efficacy by implementing digital tools and technologies. Travel was no longer necessary for virtual meetings, resulting in saving of money and time. Better client interactions, more efficiency, and eventually better patient outcomes were the results of this digital transition. Caplet India Pvt. Limited, a pharmaceutical company in Kolkata, has been used as a case study to examine the impact of its digital transformation.

T. Maity
CAPLET INDIA PVT. LTD., Kolkata, West Bengal, India

S. Kishore (✉)
Department of Business Administration, Institute of Engineering & Management, University of Engineering and Management, Kolkata, West Bengal, India
e-mail: shweta.kishore@gmail.com

S. Gangopadhyay et al. (eds.), *Pandemic Diaries*,
https://doi.org/10.1007/978-981-96-5415-4_6

Keywords Covid-19 · Pandemic · Physical distancing · Pharmaceutical industry · Medical representatives · Healthcare · Healthcare professionals · Digital transformation

6.1 Introduction

The Pandemic era of Covid-19 had imposed a deleterious effect on every industry, both financially and psychologically, thereby resulting in a detrimental effect on the working methodology of the same.

The Pharmaceutical Industry is not an exception. However, in this industry, the Covid pandemic has culminated in a strange implication, affecting both internal and external customers at the same time.

The Medical Representatives (MRs) are the only living interface between a pharmaceutical company and the prescribing Doctors. Their job is to generate revenue through prescriptions by sharing information about their respective medication. The fulcrum of their job is their day-to-day visit to the physicians [1].

But despite their effectiveness, the MRs always pose to be the most expensive promotional channel in the pharmaceutical industry. For such reason, the field force controllers have always endeavored to find the most suitable process to extract the best call frequency and coverage from the MRs. The marketers always honed the messaging nitty gritty to shape up the ideal representation model to be conveyed to the medical fraternity by the MRs [2].

This research aims to find out the impact of Covid-19 in the pharmaceutical industry and its outcome on pharmaceutical marketing. Many insights of digital transformation during Covid-19 have researched. This research also includes the phase-wise development of digital marketing promotions. Major focus has been taken on the key marketing people, who are the main link between the pharmaceutical house and doctors. Their conventional process of physical meeting with the doctors during Covid-19 had been restricted and lack of interaction had impacted the growth of the pharmaceutical industry in acute segment. Pharmaceutical industry also suffered from "Work from Home" and the disrupted working pattern created social distancing with the doctors. This situation created a mental toll in the minds of all medical representatives.

This changed social attachment had to be taken care for any pharmaceutical company in the universe. Biggest threat was the acceptance of the doctors and digital transformation had to create a win-win situation for both, i.e. medical representative and doctors. All aspects of digital transformation and gradual changes in digital communications with the doctors have been covered. The given new era of brand management and shifting toward digital marketing of brands have been cited. Digital content of brands and its reach to the doctors have also been included.

Impact of reduced doctors' engagement with the medical representatives and the way pharmaceutical industry improved and used digital platform for digital engagement with the doctors are the major points of discussion in this research. The strategic changes in brand promotion and its result in enhancement in brand growth have been incorporated in this research. This marketing innovation with necessary data has been studied and major impacts have been cited. Confinement due to Covid pandemic and many ways to overcome this restriction by medical representative are the major focus.

The research includes the changes in brand portfolio management and use of digital content for doctors through online media has been studied extensively and its impact in sales performance has also been studied. A case study of a pharmaceutical company and their approach toward digital pathway have been cited. Their growth strategies adopted during this pandemic situation was aimed towards survival and improvement in sales. The research included the digital channel opportunity through pull market strategy and its process and benefits, the utilization of "New Digital Solution" and its benefits, and "Omni Channel Engagement strategy" used by different pharmaceutical companies and its benefits.

The refurbishment of pharma sales through cutting-edge selling mechanics and its impact in terms of sales have been evaluated. Restructuring of meeting process through remote engagement and result of virtual in terms of enhancement meeting more doctors have been studied.

The research includes the process and result of digital medical content and its utilization by the medical fraternity. The impact of more prominent digital drug banks and its properties were more reliable and acceptable for the doctors to study and use.

The effect of digital transformation of a specific pharma company, Caplet India Pvt. Limited, Kolkata, has been taken as case study. The organization is operating in Indian pharmaceutical industry for last 35 years and its product portfolio includes acute and chronic segments. Impact of Covid-19 in both acute and chronic segment was distinct.

Digital shifting created a measurable outcome in terms of sales and the way this transformation reduces overhead expenses of the organization and increases sales growth.

6.2 Organization Details

Caplet India Pvt. Ltd. is a Kolkata-based Pharmaceutical Manufacturing company which had initiated its operation in the year 1989. It has its presence mainly in Eastern India along with Northern India and part of Western India. It has major four types of Product Portfolios mainly in Cardiology, Gastroenterology, Respiratory, and Paediatrics. It also has its presence in the Vitamins and antibiotic market. It operates with more than 200 field forces [3].

6.3 Covid-19 and the Crisis

With the advent of the Covid-19 pandemic, the conventional working methodology in the pharmaceutical industry desisted abruptly. It became impossible for the MRs to meet the doctor face to face and they were condemned to a strange working methodology of "work from home". Such a sudden advent of home-office job ambiance amid an unknown job description nurtured the threat of job insecurity, which took a big mental toll on the MRs [4].

The employers were also in a fix to choose the ideal representation method for the Doctors in this critical situation. In this period, due to the absence of face-to-face calls, the pharmaceutical industry had chosen some digital avenues to add leverage to their meeting of their targeted Doctors [5].

But somehow in this venture, one major factor had been overlooked: the mental status quo of the Doctors in this pandemic period. The entire medical fraternity was also running through a tremendous stressful condition in this period either due to the onset of such an emergency medical situation or due to a sudden fall in patient outpour. So, in such conditions, digital means of pharmaceutical representation like WhatsApp posts, Webinars, Skype calls, etc. did not pose to be quite alluring for them. During this pandemonium, such digital ventures, due to the absence of any relevant value addition, failed to arouse any interest from the doctor population. 41% of Doctors felt that the Webinars done by the pharmaceutical companies have no significance. 65% of Doctors were so irritated by getting messages at odd times that they felt of blocking the numbers [6].

The Pharma houses now realized that for a doctor, it was quite hard to say "No" when an MR stood in front of the physician, whereas it is quite easy enough to refuse a Digital call or to ignore a webinar request. Thereby, the essence of the MRs as the most effective living interface came in float.

However, as the pandemic situation normalized, the doctor's population started facing a new challenge. Many Doctors who were having roaring practice before the Covid era faced a huge cut in inpatient outpour afterward [7]. Many Doctors, who had opted for a digital platform during the pandemic period to maintain access to their patients, had their patient crowd following after the cessation of Covid period. Thereby, the necessity of Digital Marketing for Doctors now emerged in vogue. The Doctors have now started showing more inclination toward the process of updating, designing, and expanding the reach of their own social media and search engine platforms to invite more patient outpour.

Therefore, to meet the new needs of the medical fraternity amid such a changed scenario, the role of MR is now bound to undergo a rapid metamorphosis. Their job now, apart from drug promotion, also demands a more value-added service to such newly emerging professional needs of the Doctors.

Pharmaceutical companies are now compelled to give serious thought to how to blend the effectiveness of the living interface value of the MRs with the digital service to meet the present marketing needs of the medical fraternity. A new era of "Consultative Marketing" is now about to begin.

Meanwhile, reshaping was needed in branding innovation and portfolio management by leveraging data to identify the most productive and lasting shifts. Claims innovation, product innovation, CX innovation, etc. were accelerated. Digital shifting of the brands toward digital channels and behaviors was the need of the hour. Customer Engagement and Content plans were reviewed. Digital content/Digital services/Utility/Commerce were addressed properly. Over the 2 years, substantial challenges with medical visits were faced by pharma companies.

MRs were unable to visit the Doctors at hospitals due to strong restrictions on interpersonal contact. Physical Doctors' engagement has decreased significantly as there was no access and Doctors were busier with the treatment of the Covid-19-positive patients. Due to concerns about contamination, the number of non-Covid patients has drastically decreased even in recent months.

The McKinsey report shows that patient falls in dermatology, oncology, and cardiology decreased by 43%, 14%, and 31% respectively in the number of weekly visits [8].

As per the data from INCROWD, very few Doctors were keen to have in-person meetings due to the continuing pandemic. Supporting the data further, McKinsey reports that 28% of Doctors said that the "Covid-19 pandemic will have a lasting impact on their willingness to have in-person meetings with sales representatives". Intending to maintain communications with healthcare professionals, pharma companies have had to review their approach and communication strategies [8].

In reality, pharmaceutical manufacturers have had to deal with the problem of overstretched schedules, and as a result, the digital channel has rapidly become an important tool in promoting communication between the two parties. The Covid-19 pandemic has accelerated the digitization of the health sector for pharmaceutical companies (via remote medical visits) and patients (through teleconsultation). How can pharmaceutical corporations effectively engage healthcare professionals through a digital approach, given that the traditional foundation of this dual connection is the in-person medical visit?

6.4 Innovative Attempt to Manage the Crisis

6.4.1 Crisis-Driven Digital Shifting

There is now a disparity in the healthcare industry between in-person and remote pharmaceutical medical appointments. Indeed, if over 70% of medical appointments were in-person in the pre-Covid period, this has completely altered. When confronted with a pandemic epidemic, the pharmaceutical industry had to change their strategy and figure out how to reach out to the doctors without engaging in conventional "field" work.

However, will these new norms and expectations that were established during confinement continue in the post-Covid age to the extent that they permanently change how medical visitors communicate? The creation of digital ways for MRs will be a sign of marketing innovation because, in the context of a pandemic, a digital strategy should not merely be a temporary solution.

These days, the main way to communicate with pharmaceutical businesses is through remote medical visiting channels, which are supported by healthcare specialists. Doctors are, in fact, encouraged to use digital communication methods. For so-called "follow-up" medical visits, at least, a "hybrid visits" model is emerging [9].

The digitization of the medical visit is based on three crucial pillars:

- The medical visitors who conduct the meetings remotely,
- The technology that enables the support of digital formats,
- The information that serves as the foundation for conversations between MRs and medical professionals.

Nowadays, when it is just a "copy and paste" of paper papers, the primary problem is the lack of content diversity. According to Euris Group research, medical professionals value theoretical knowledge (clinical data, pathology training, etc.), whereas MRs suggest useful practical information. According to the doctors surveyed, this is by far the least engaging kind of material. As a result, the gap between the content that MRs supply and the content that doctors like tends to grow significantly if no update is done.

The resolution is attainable based on a number of investigations. Presenting content that matches the doctor's expectations and offers genuine benefits is important. This would successfully persuade them to embrace the digital content and make a sustained commitment to it. The content needs to be changed to be more convincing because video conferences demand greater focus than in-person meetings.

6.4.2 Digital Channel Opportunity

6.4.2.1 Pull Marketing Strategy

A more recent tendency is emerging as a result of doctors' increasing need for digital channels: the creation of a pull marketing strategy. The goal of this technique is to position oneself as a receiver of information rather than a transmitter, in light of doctors who are overbooked and have less time to spend with medical visitors. The MRs have to make themselves available to the doctor when he/she needs information and no longer just provide him/her with information when he/she doesn't need or ask for it [10, 11].

Segmentation

Pull marketing strategy can be implemented by doing proper segmentation. Pharmaceutical businesses must take a doctor-centered approach from a business perspective in order to capture their interest and keep it over time. Therefore, it is a question of segmenting the doctor's "target market" using a qualitative database to establish specific action plans according to the profile of each doctor to be able to develop clusters. [12]

The segmentation can be done on the following basis:

1. Digital engage
2. Digital learner
3. Traditionalist

This technique allows for the provision of nearly customized messages for every sub-cluster, improving the experience of doctors in engaging with them until they decide to keep them on board. For instance, relying on the digital channel by preparing virtual events entails thoroughly comprehending everyone's preferences, keeping an eye on the material supplied, examining attendance patterns, and getting insightful feedback to guide future strategic planning.

Benefits

The use of this channel brings advantages such as

1. Increased flexibility
2. Options development
3. The possibility of obtaining up-to-date scientific data.

6.4.2.2 New Digital Solutions

Several businesses have taken advantage of this chance by offering innovative digital solutions like webinars, remote workshops, or even chatbots through the development of a social network designed only for medical professionals.

Information and training distribution to doctors varies, and some healthcare-focused organizations are creating procedures to support laboratories in their promotional approach to physicians.

Benefits

When utilized in closed-loop marketing, the content provided by these modernizations adds significant value.

6.4.2.3 Omnichannel Engagement Strategy

The pharmaceutical industry's omnichannel engagement strategy is favored by the utilization of several digital platforms. Salespeople can thus reach healthcare professionals outside of their office and establish a stronger connection with their target audience.

Benefits

Specialists in the healthcare sector who recognize the value of digital initiatives will be able to deliver the appropriate message to the correct audience at the right time.

The process of digitizing health professionals is underway. However, not everyone is yet comfortable using digital tools, which indicates that there is a genuine chance now. Offering partially modified content that is tailored to digital tools in the form of aesthetically beautiful and useful content is crucial to engaging doctors in this profession in an efficient and long-lasting manner.

The Covid-19 pandemic forced a model to convert physical to digital meetings with doctors as in-person visits were not possible. The situation forced the pharma companies to understand the importance of digitization. A different avenue was the need to communicate about their products to the Doctors. Pharma companies started going digital communication.

6.4.3 Refurbishment of Pharma Sales

Covid-19 shifted the pharma sales structure. In-personal interaction and selling during the pre-Covid period became irrelevant. Today, the scenario demands significant change. Companies tried to adopt a seamless strategy to fulfill the demand for doctors. Here, a specific approach was the need for all pharma companies to withstand the crisis. Reducing engagement with the doctors and refurbishing the old sales model gave cutting-edge selling mechanics to the pharma industry.

6.4.4 Restructuring of Meetings

In-personal meetings were impossible. The industry shifted toward remote engagement with doctors. Work from home by the Pharma representative was the only way to maintain social distancing, survive lockdowns, and retrain from the fear of virus contamination. In-personal face-to-face meetings declined by more than 75% and it was a great challenge for new product launches and increasing brand awareness and brand value. 99% of face-to-face meetings had been shifted to 90% virtual meetings. Making appointments with the doctors for virtual meetings was the main criterion during that time and saved more time. It allowed them to see more doctors virtually. This way, in-person physical meetings had been shifted to virtual meetings. The following were the major platforms for virtual meetings.

1. Google Meet
2. Zoom Meeting
3. Microsoft Team

More than 41% of Doctors become user-friendly and feel comfortable in virtual meetings and now it has given more value for its promotional and informational purposes. Following Virtual meeting tools increased the quality of interactions and information transfer between Doctors, Patients, and MRs [7].

6.4.5 Restructuring of Content

Pre-Covid, MRs were just promoters and sellers of drugs. It was more about promoting the products than adding value to it. During Covid-19, time was a constraint for the doctors for their tireless engagement in their service as front-line fighters against Covid-19 infection. Conveying promotional messages won't be effective. Thus, revamping content is necessary. Characteristics, properties, and aspects of the drugs were more important information to convey. More information on specific molecules and handy digital channels to extract those data were the challenges for MR during Covid-19. The need was easy accessibility of digital content for APIs and generic drugs. The digital adaptation by the doctors and smooth digital transformation were major issues to be resolved. The content must be easy to comprehend as making digital adoption smoother for doctors is the purpose and path for digital transformation. A transformation from a drug seller to a trusted scientific detailer for doctors was a major paradigm shift. This step will make the representative a trusted partner rather than a seller. As per Accenture report "58% of doctors agreed that at least one of the pharmaceutical companies spammed them with digital content during Covid-19". Post-Covid, adding value proposition in the content of drug information was a major shift in understanding the importance of MR and at the same time, pharma companies needed to invest more in analytics to create more engaging content for digital tools. This remote engagement helped the doctors to gather more information than in personal meetings, as there were restrictions in physical meetings. Success in digital excellence is achieved by optimizing new paths of information transfer from introduction to making them experts.

Major tools and USPs to be followed by MR followed the below process to create more engaging content:

1. CRM (Customer Relationship Management) tool for more data
2. DOCTOR segmentation based on their journey status
3. Preference for the digital channel by specific DOCTORs
4. A content management tool for better digital information transfer

5. Tailor-made need-based content framework as per different segments of the doctors
6. Create more insight
7. Design the content as per the channel requirement and characteristic
8. Development of a feedback loop to better future strategy

6.4.6 Restructuring Customer Engagement (E-Detailing)

During pre-Covid, providing information, USP and physical detailing were the major tasks. Physical detailing was impossible in the state of social distancing. E-detailing became the major tool for pharmaceutical companies to engage doctors better. The detailed feature of each brand or molecule, when uploaded in mobile digital media, was known as e-detailing. Initially, uploaded digitized scientific sales content was run as e-detailing. E-detailing became an effective remote digital tool due to its high-end features. This tool made medical information available at any point in time and did not require any pre-downloaded software. Moreover, exclusive storage of scientific information and ease of use made the process hassle-free for Doctors as well as MRs. E-detailing attached doctors to engage and very much evolved with time. The MR began to engage the doctors through digitized physical sales content. Though the process was engaging the doctors, it requires the skill of MR to create valuable scientific content. The interaction became comparatively appealing. Though, the process solely depends upon the capability of sales representatives to produce engaging content.

6.4.7 Role of Information Technology (IT) During and Post-Covid in Pharma

Information technology (IT) made e-detailing to make it user-friendly with the software solutions. IT helped the pharma industry sales by providing an e-detailing application (app) with the help of 3D visualization, computer graphics, and other multimedia software, which helped to provide engaging informative presentations for doctors. These back-end solutions helped the MRs to customize the presentation segment-wise and need-based. It also helped to analyze engaging time, which is otherwise called content's engagement rate and doctor information. Data was collected from each virtual interaction with the CRM of the company and helped to create more engaging solutions for MRs. It also helped to design tailor-made solutions based on different channels used by the doctors.

6.4.8 Doctor's View on E-Detailing Tool

As per the different reports, many doctors were in a belief that MRs helped them to treat patients remotely. As per a media agency, 31% of the clinicians stated that engagement with digital content like ads increased greatly. This was an opportunity for pharma companies to focus on more engaging content useful for the doctors. Effective and more engaging solutions like the Telehealth app helped doctors to practice through virtual treatment, which was contactless and more convenient for the situation [7].

6.4.9 Significance of Virtual Treatment

This situation created an opportunity for MR to fulfill the unmet needs of the doctors, which were not there earlier and they were unable to achieve their (MR) budget. This crisis helped MRs to feel the needs and requirements of the doctors. "Lack of relevance" was taken care of. This helped MR to achieve their sales target.

6.4.10 Digital Solution for the Doctors

The coronavirus forced elder patients to be retrained for hospital visits because the chances of viral contamination were more. The situation demanded a digital treatment solution for the doctors to save lives during social distancing. Doctors were looking for a better way out to combat the consequences created by coronavirus. Doctors needed a path for virtual treatment. At the same time, MRs needed to focus on the needs of the doctors rather than product information. MRs helped rather than assisted the doctors using portals like Telehealth for digital consultation for patients. Digital consultation software helped MR to promote their prescription drugs successfully. Digital pharma tools such as e-banner, e-posters, BOTs, and short videos helped doctors to be more informative about drugs. As per McKinsey's report, doctors have scaled their offerings for patients through digital consultation and started getting 50 to 175 more digital consultations. Using of digital solutions had increased 11% in 2019 to 65% in 2020. Security and workflow integration were the major focus of the progress of change [7].

6.4.11 Restructuring Selling Mechanics of Pharma Companies

Conversion-appropriate lead is important for achieving the sales target. Fulfilling the demand and need of doctors was more important than usual sales talk. Activity and action toward the need were relevant to achieve the sales goal. Solo clinics and speciality research hospitals were two major segments to focus on. Solo practitioners needed more product information than research institutes. Identification of the target audience and tailor-made solution for those audiences was important. Segment-wise tailored solutions helped to push and promote the relevant product with ease.

6.4.12 CRM Software for Doctor Feedback System

The opinions and feedback of the Doctors were the next requirement. Pharma CRM software was the tool. It helped MRs to maintain and nurture doctor relationships. A structured framework helped MRs to keep the sales pipeline from drying out. CRM software helped to enhance operational efficacy and effectiveness, enrich content management, and create more engagement with doctors. The automated CRM system offered a 360-degree view of customers and helped to integrate the data under the cloud. This data under the cloud helped later with more engaging options.

6.4.13 Technology Integration for Medical Representatives

Technology integration in the Pre-Covid created flexibility for the forthcoming challenges. As physical interaction was not likely, the adoption of science was expected in the race. Shifting toward telemedicine portals by engaging the doctors, optimizing content, and providing tailor-made emails was the new normal for MR to the doctors. These had enhanced the new rationale for the MR. But those were new to MR. They were more used to physical meetings than virtual ones. They were habituated to giving sales pitches in person rather than crafting scientific documents. Digitization means rearranging the sales force orientation. Training was required to induct digital mode into the MR. Daily webinars were the need of the hour to train the MR to adopt digital methods. Pharmaceutical companies went to liaison with technology providers to create a new marketing tool for MR.

The pandemic has forced many pharmaceutical companies to digitally transform. Doctors have been offered much-digitized content on patients, treatments, drugs, etc. More relevant and informative data was provided to the doctors than before. The data was relevant and added more value than before.

6.4.14 *Artificial Intelligence (AI) in Reducing the Cost of Promotion*

AI helped in removing the multi-channel cost for product promotion and has assisted in understanding the credentials of the Doctors. AI further helped the MR to understand the channels used by that particular doctor and the MR set their goal to understand the platform for the right customer. Further, AI helped to assist the CRM tool, which helped to automate the process of pre-call analysis and e-detailing. AI helped to generate more business outcomes. The combination of virtual technology and conventional communication tools transformed MR meetings into a new look. This process helped to bring the bond between touch and feel. Digital communications became the only tool for doctors to communicate. Augmented phenomenon occurrences at home further fill the breach of touch and feel. Google ads, search ads, paid ads, blogs, and online forums were the few digital solutions for engaging doctors much better.

Today, the digital processes that are selected by pharma parties lie in the pre-Covid period again. But, the spurring of the process of digitization has started on account of Covid-19. Many pharmaceuticals used this moment and revamped their trade. As we experienced, the digitalization of the process was hard to do. It demands the opportunity to pay. There were challenges and therefore the opportunity to develop was more. Today, Doctors are going to interact more with the MR in addition to what they used to do. Pharmaceuticals are accumulating advantages through this way of digital interaction. That helped MRs to know more customers inside and was a new strategy to stay appropriate.

Making MR future-ready was a hard task and support from the company is in urgent need.

Here are four takeaways that made the MRs future-ready for post-Covid scene.

6.4.14.1 Pharma Field Force Management Software

E-reporting software and its assessment helped the companies to know MR closely and also helped in monitoring sales activities and growth. Geofencing and geotracking helped pharma companies to be more precise and informative regarding field activities. Digital training also helped companies to educate MR effectively and made them information-ready.

6.4.14.2 Smart Digital KPIs

Post-Covid, digitalization played a leading role in pharma industries. The brand strategy became more significant with digital KPIs and became a new success story. Digital modules helped Smart KPIs and e-brochures, e-journals, and e-detailing enhanced customer connectivity and increased sales productivity.

6.4.14.3 Expanding Additional Services for the Doctors

Sympathy toward Covid-19-affected patients and Doctors treating them brought more impactful messages. They helped the patients with their in-home care by providing laboratory reports at their homes and created an impact in the minds of the Doctors during this pandemic situation. This created more engagement with the doctors during that time and made the MR future-ready.

6.4.14.4 Liaising with Digital Solution Providers

Digital solution providers will help pharmaceutical companies and their MRs by providing digital process analytics, risk-based monitoring solutions, supply chain management, and sales and marketing support.

6.5 Use Case: Caplet India Pvt. Ltd.

Caplet India Pvt. Ltd., a leading pharmaceutical company in the Eastern region of India, had adopted most of the above-mentioned process of digital means of marketing procedures and survived the extreme situation of Covid-19. The product portfolio of the company has a mix of acute and chronic therapies and the most needed chronic segment, mainly cardiac brands, given the strong foothold even during the extreme situation of Covid-19.

The only means of communication was digital media, and still company has registered sales growth month after month till September 2020, as shown in Fig. 6.1.

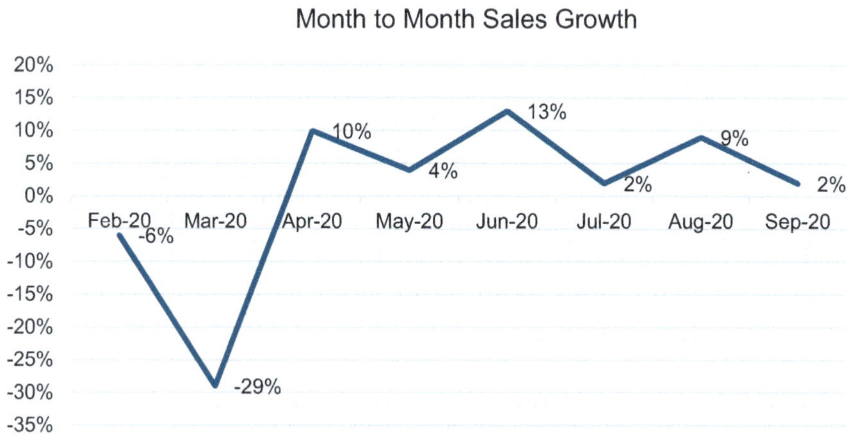

Fig. 6.1 Month to month sales growth. *Source* Caplet India Pvt. Ltd.

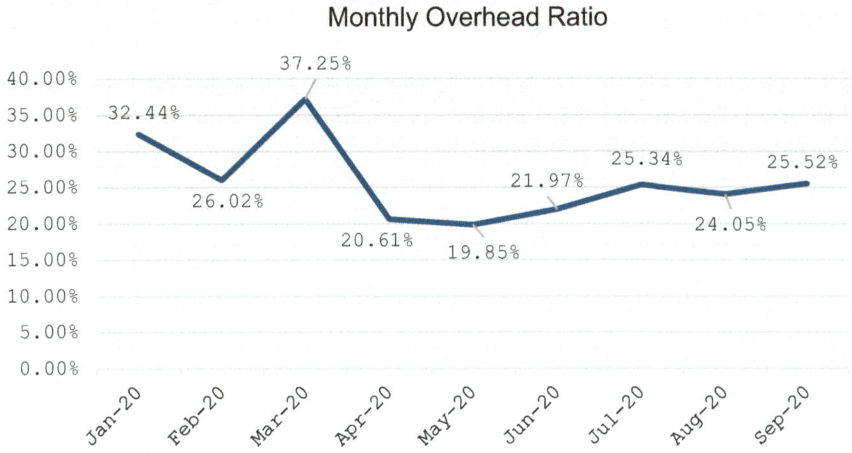

Fig. 6.2 Monthly overhead ratio. *Source* Caplet India Pvt. Ltd.

Remuneration of Pharmaceutical is the sum of Salary plus Travelling and Daily allowances. During Covid-19, physical meeting with the Doctors was not allowed, and as a reason, the company did not pay any Travelling and Daily allowances. So, the overhead ratio also goes down (depicted in Fig. 6.2), giving a profit to the organization, and the company has survived well.

6.5.1 Advantages Obtained from Covid-19 by Caplet India Pvt. Ltd.

6.5.1.1 Digital Transformation of Medical Representatives

With the help of digital media, marketing promotion of the medical representatives of Caplet India turned toward executable and tangible benefits.

6.5.1.2 Enhanced Customer Reach

Digital transformation helped this organization to enhance customer reach and they were able to increase the number of customers through digital ways of connecting the customers.

6.5.1.3 Decreased Marketing Promotional Cost

This transition from physical to digital reduced the promotional cost of organization and less use of printing media reduced the cost further. As a result, less promotional costs indirectly increased profitability.

6.5.1.4 Reduced Overhead Expenses Ratio

Overhead expenses of the organization per medical representative were reduced due to per capita promotional expenditure being reduced with the enhancement in sales.

6.5.1.5 Growth and Profitability Increased

With the help of all the above parameters, the Organization has grown in sales due to increased customer reach, digital way of promotion, excellent CRM software for customer feedback, and increased virtual promotion. Organizations also increased profitability due to less cost of promotion and digital media promotion.

6.6 Conclusion

In addition to overcoming the obstacles presented by Covid-19, Caplet India Pvt. Ltd. has effectively leveraged the power of digital transformation to advance the pharmaceutical sector. The company has lowered marketing and promotional expenses, streamlined operations, and greatly increased its customer base by updating its medical representative team and implementing cutting-edge digital tools. In addition to helping Caplet India, this digital revolution has the potential to completely overhaul the pharmaceutical sector, increasing productivity, improving patient care, and eventually spurring innovation and growth.

References

1. Fahimi, F. (2012). How medical representatives influence prescribers behavior? *Iranian Journal of Pharmaceutical Research: IJPR, 11*(2), 383.
2. Nandy, M., & Pal, B. (2016). Pharmaceutical marketing & product promotion: A paradigm shift in Indian Pharmaceutical Industry (IPI). *Adhyayan: A Journal of Management Sciences, 5*. https://doi.org/10.21567/adhyayan.v5i2.8817
3. *Caplet*. (n.d.). Retrieved November 19, 2024, from https://capletindia.com/
4. Biber, J., Ranes, B., Lawrence, S., Malpani, V., Trinh, T. T., Cyders, A., English, S., Staub, C. L., McCausland, K. L., Kosinski, M., Baranwal, N., Berg, D., & Pop, R. (2022). Mental health impact on healthcare workers due to the COVID-19 pandemic: A U.S. cross-sectional survey

study. *Journal of Patient-Reported Outcomes, 6*(1), 63. https://doi.org/10.1186/s41687-022-00467-6
5. Vora, L. K., Gholap, A. D., Jetha, K., Thakur, R. R. S., Solanki, H. K., & Chavda, V. P. (2023). Artificial intelligence in pharmaceutical technology and drug delivery design. *Pharmaceutics, 15*(7), Article 7. https://doi.org/10.3390/pharmaceutics15071916
6. Gupta, S. K., Nayak, R. P., & Sivaranjani, R. (2016). A study on the interactions of doctors with medical representatives of pharmaceutical companies in a Tertiary Care Teaching Hospital of South India. *Journal of Pharmacy & Bioallied Sciences, 8*(1), 47–51. https://doi.org/10.4103/0975-7406.171695
7. Pandey, S. K., & Sharma, V. (2020). A tribute to frontline corona warriors—Doctors who sacrificed their life while saving patients during the ongoing COVID-19 pandemic. *Indian Journal of Ophthalmology, 68*(5), 939. https://doi.org/10.4103/ijo.IJO_754_20
8. *Telehealth: A post-COVID-19 reality? | McKinsey*. (2021, September 7). https://www.mckinsey.com/industries/healthcare/our-insights/telehealth-a-quarter-trillion-dollar-post-covid-19-reality
9. *Driving digital change during a crisis: The chief digital officer and COVID-19 | McKinsey*. (2020, April 20). https://www.mckinsey.com/capabilities/mckinsey-digital/our-insights/driving-digital-change-during-a-crisis-the-chief-digital-officer-and-covid-19
10. Martin-Barbero, S. (2020, July 21). *COVID-19 has accelerated the digital transformation of higher education*. World Economic Forum. https://www.weforum.org/stories/2020/07/covid-19-digital-transformation-higher-education/
11. Moorman, C., Soli, J., & Seals, M. (2023, January 8). *How the pandemic changed marketing channels*. https://hbr.org/2023/08/how-the-pandemic-changed-marketing-channels
12. Li, C., & Lalani, F. (2020, April 29). *The COVID-19 pandemic has changed education forever. This is how*. World Economic Forum. https://www.weforum.org/stories/2020/04/coronavirus-education-global-covid19-online-digital-learning/

Tamal Maity is a Pharmacy Graduate and he had an experience of 27 years in different spheres of Pharmaceutical Industry. His initial career started with pharmaceutical manufacturing followed by pharmaceutical sales with the companies like Menarini Raunaq & Ranbaxy. He has knowledge of computer programming and completed Medical Transcription from Jadavpur University. He has a 21 years' experience in pharmaceutical marketing and launched many new formulations successfully. He is working with M/s Caplet India Pvt Ltd from 2017 as Marketing Head. He looks after Operation, Sales and Marketing. He completed 'EPGSBL' from IIM, Kolkata and presently the Alumni member.

Shweta Kishore, a seasoned academic with 15+ years of experience, brings a blend of theory and practice to Marketing, Operations, and Supply Chain Management. She holds degrees in Engineering, Management, and a Ph.D. She is also NET qualified. Currently, she is working as an Associate Professor at the Institute of Engineering & Management, University of Engineering and Management, Kolkata. Committed to lifelong learning, she's certified in Lean Six Sigma and has completed courses from Harvard Business School Online and Coursera. Her research interests include automobiles, electric mobility, and general management. She has several publications in reputed journals.

Part II
Breaking and Making: Functional Management and COVID-19

Chapter 7
Assessment of Consumers' Purchase Intention of Men's Clothing in the New Normal in Asian Market with Thematic Analysis

Pritha Ghosh, Rabin Mazumder, and Swati Mukherjee

Abstract Retail is one of the sectors most severely impacted by the recent pandemic caused by the highly infectious COVID-19. The clothing industry has undeniably suffered unprecedented challenges with the pandemic outbreak. The clothing industry was already facing sluggish demand when the pandemic, followed by restrictions of the government regulations during the lockdown, further deteriorated the situation—consumers' reluctance to visit stores adds to global challenges for retail store managers in the new normal. Store managers had to rethink their business plans to innovate for sustenance. The qualitative study attempts to identify, understand, and describe the essential characteristics of store brand consumption behavior in an emerging economy. This paper demonstrates a novel approach to recognizing, understanding, and defining the crucial factors of consumers' decision-making for men's clothing store brands in an Asian market. The study follows a series of steps to conduct a thematic analysis. Forty-one participants responded to the semi-structured interviews. Three themes identified are purchase, usage, and discard. A study of the themes determined the nine categories: price sensitivity, women's decisive role, online shopping, ready-to-experiment, social status, brand sensitivity, restocking, recycling, and donating preference. A conceptual model demonstrates the findings. The findings indicate that middle- to high-income consumers are brand-conscious and possess adequate knowledge to make informed decisions about the properties and quality of store-brand clothes. Retailers can use these insights to develop a private label strategy that aligns with consumers' decision-making behavior. A conceptual

P. Ghosh
University of Engineering & Management, Kolkata, India

R. Mazumder (✉)
Department of Business Administration, Institute of Engineering & Management, University of Engineering & Management, Kolkata, India
e-mail: rabin.mazumder@iem.edu.in

S. Mukherjee
College of Management, Institute of Engineering & Management, Kolkata, India

S. Gangopadhyay et al. (eds.), *Pandemic Diaries*,
https://doi.org/10.1007/978-981-96-5415-4_7

model summarized the findings about the essential characteristics of men's clothing store brand consumption.

Keywords Clothing industry · Men's clothing · Store brands · Thematic analysis · Decision-making behavior · Current normal · Consumer behavior · Asian market

7.1 Introduction

In recent years, store brands of men's clothing have become increasingly popular among consumers due to their affordability and accessibility. While previous research has explored factors such as price, quality, and style preferences that influence consumer behavior in purchasing store brands, a deeper understanding of these factors is necessary to develop effective marketing strategies that target this consumer segment. Thematic analysis is a qualitative research method to uncover patterns and themes within data sets, such as consumer reviews and feedback, to gain insights into consumer behavior. This approach allows researchers to explore consumers' subjective experiences and perspectives, which can be particularly useful when studying complex and multifaceted phenomena such as consumer behavior.

Consumer behavior influences motivation to purchase and how they define value, remarkably of price [1, 2]. The purchase behavior of individuals can vary depending on the situation [3]. Peer pressure can also affect purchasing decisions related to clothing, with some consumers prioritizing quantity over quality to keep up with the latest fashion trends on social media. In urban areas, dressing well is often seen as a way to develop a particular lifestyle, showcase personality, and increase mental satisfaction and confidence. Furthermore, certain situations may require specific dress codes, such as wearing formal attire to a job interview or ethnic clothing to a wedding. In such cases, consumers may be willing to invest considerable resources to comply with these expectations. Store brands can significantly impact purchasing decisions in emerging economies with diverse cultural backgrounds. Prior research has identified factors influencing consumer behavior in these contexts, including price, quality, style, fit, and brand loyalty [4].

7.2 Survey of Literature

Consumer decision-making involves eight key features. These include being conscious of quality, price, and brand, shopping for recreation, needing clarification over the many choices available, being impulsive and desiring a novel fashion approach, and being brand loyal. Consumers consider prices concerning quality and often prefer renowned brands over generic ones. However, they are unwilling to compromise on product quality, and mediocre-quality products may not find takers. Consumers also choose a variety of options to be able to make informed choices.

While some consumers may be impulsive, others may delay purchases and seek discounts or promotional items. In some cases, these results may contradict the findings [5].

The theory underpinning the present study is the Theory of Reasoned Action (TRA) [6, 7]. TRA suggests that individuals intentionally behave according to the situation they find themselves in [6, 7]. According to TRA, attitudes and subjective norms are the antecedents that influence consumer behavior [8]. TRA also posits that positive behavior outcomes lead to its future continuation [7]. Personal and social factors influence consumers' purchase intention [6]. Attitude toward a behavior is an individual characteristic, while subjective norm, which refers to how a consumer perceives social pressure, is the social aspect [6].

Therefore, the study aims to examine the consumption behavior of men's clothing store brands and identify their key features. It seeks to categorize the consumption features of these brands and analyze their implications for the Theory of Reasoned Action (TRA). By addressing these objectives, the study aims to provide a comprehensive understanding of the factors influencing consumers' purchase of store-brand men's clothing and how these decisions impact the TRA. To the best of the author's knowledge, there is no qualitative analysis of the objectives of the current topic. In a study in Bangladesh, thematic analysis developed a conceptual model that identifies the key factors influencing customers' attitudes. They used focus group methods to determine the cues that impact buying behavior in local apparel stores [9]. The present research contributes to the literature on consumer behavior by recognizing, understanding, and defining the crucial characteristics of store brand consumption behavior for men's clothing in an emerging economy.

The following section outlines a process for evaluating the retrospective data provided by participants. The article's uniqueness lies in its genuine account of the participants' viewpoints. The paper ends by discussing the implications and research limitations and proposes a potential avenue for future research.

7.3 Methods

7.3.1 Study Participants

Obtaining a random sample for qualitative research is generally not feasible as it may yield unreliable results [10]. Therefore, for the present study, a purposive sampling method was used [11]. The sample consisted of 41 male consumers who were mid-level to senior managers in various companies located in urban areas of Kolkata, the capital city of West Bengal, India. The participants were selected based on age, qualifications, and monthly income. Specifically, 21 participants were between 25 and 35 years old, 31 had postgraduate degrees, and 17 had a monthly household income ranging from US$1000 to US$2000, as detailed in Table 7.1. The selection criteria included a passion for clothing and styling, knowledge and sensitivity to brands,

Table 7.1 Demographic profile demonstrates the number of male participants in the different categories for age, educational background, and monthly household income

Features	Frequency
Gender	
Male	41
Age (years)	
25–35	21
36–45	11
46–55	5
56–65	4
Educational status	
Graduate	3
Postgraduate	31
PhD	7
Monthly Household Income	
US$501–US$1000	18
US$1001–US$2000	17
>US$ 2000	6

Source The authors analysis

and voluntary agreement to share their comprehensive experiences and thoughts to contribute to the study.

7.3.2 Data Collection

During the study, participants were interviewed in-depth for approximately two hours, each with a semi-structured format. These participants had prior experience purchasing men's clothing from various categories of store brands, including casual, formal, and sports. The participants were aware of the purpose of the research and how their opinions could add value to the study. The interviews were recorded and transcribed [12]. The survey asked, "Why do you buy clothing from store brands?" "Can you describe your experience purchasing clothing from store brands?" "Who decides to purchase clothing products in your family?" "How do you obtain the store-brand clothing products that you want?" and "When and how do you dispose of your used clothing?".

7.3.3 Methods

Thematic analysis is a well-known and widely used qualitative method for analyzing data [13, 14]. It involves systematically identifying and classifying themes in textual, visual, or audio data to understand better the data and the phenomenon being studied

[15, 16]. Thematic analysis allows for flexibility in data analysis, enabling researchers to identify and report patterns in the data while also allowing for theoretical interpretation [12]. Researchers typically follow a series of steps to conduct thematic analysis, which includes familiarization with the data, generating initial codes, searching for themes, reviewing themes, defining and naming themes, and producing the final report.

The first step in conducting a thematic analysis is to get acquainted with the data. It includes reading and re-reading the data, taking notes, and highlighting important passages. Additionally, one can create a coding scheme or a list of potential codes based on their initial impressions of the data. The second step involves generating initial codes and assigning codes or labels to each meaningful data segment. These codes typically describe and capture the essence of the content. Inductive coding codes are based on the data rather than a pre-existing coding scheme. In the third step, we group similar codes and patterns or themes within the data. Themes are identified based on the frequency of codes, their co-occurrence, or their relevance to the research question. In the fourth step, we review the themes to reflect the data accurately. The themes may also be refined or combined based on relevance and coherence. In the fifth step, themes are defined and named. Each theme has a clear and concise description that captures its essence.

The final report provides a detailed account of the thematic analysis process, including the themes identified. Additionally, the report may contain quotes or examples from the data that help illustrate each theme.

7.4 Results and Discussion

The study identified three main themes: purchase, usage, and discard, along with nine categories that fall under these themes. These categories include price sensitivity, the role of women in decision-making, online shopping habits, willingness to experiment, social status, brand loyalty, restocking tendencies, preferences for recycling, and donation habits.

7.4.1 Themes and Categories

Theme 1: Purchase

Category 1: Price Sensitivity

In an emerging economy, consumers are careful in their utilization and management of resources, particularly when it comes to money. They are highly attuned to prices and are often affected by anxiety and uncertainty from unemployment and a lack of business opportunities. As a result, they tend to opt for store-brand products to save money, reflecting the prevailing sentiment of thrift. Andy, a manager at a private bank, affirmed the notion of price sensitivity in this context:

> For my previous purchase, I opted for a store brand that provided the most attractive offer. However, in the future, I will consider purchasing a different store brand if it offers a better deal. (Andy)

Consumers can select online or offline shopping to access the most competitive pricing information online. They conduct research to identify the best deals available. With the rise of online shopping, consumers are increasingly familiar with store brands. Stephen, a telecommunications company store manager, assisted by highlighting these options:

> I prefer doing online shopping these days, as they give discounts all through the year across various brands, especially store brands. (Stephen)

Additionally, consumers are ready to buy store brands more than ever. Ron, a gym instructor, noted,

> In the recent past, I have tried t-shirts from store brands. I am delighted with the quality if I compare it with the price paid for it. (Ron)

The study confirms the findings concerning the price consciousness of consumers [5]. Therefore, as ascertained in prior research, the results highlight that price has always been a primary concern among consumers in an emerging economy while purchasing [17]. Therefore, we posit:

Proposition 1: **Price-sensitive consumers purchase store brands.**

Category 2: Women's decisive role

The results indicate that women residing in urban areas of India take on the primary role of decision-makers regarding clothing and other household items [18]. It is mainly due to their financial independence, which gives them considerable power. Additionally, women in metropolitan cities are qualified and employed and keen to try out new brands [17]. They also consider store brands a viable option due to their reputation and convenience in finding various sizes. Ken, an IT professional, exemplified this trend:

> Acquiring clothing for myself is infrequent due to the wearisome nature of the task. Instead, I rely on the garments that my wife procures for me. Recently, she obtained a pair of trousers for me from a particular store brand which had exceptional quality and fit. It is a brand that she frequently purchases clothing from. (Ken)

Women in the Indian household also enjoy store brands for their sons. Alex, director of an educational institution, explained,

> My wife buys store-brand casual clothes for our sons so they can wear colorful outfits to college. (Alex)

Therefore, we posit:

Proposition 2: **In a metropolitan city, women living in urban areas play a crucial role in household purchasing decisions.**

Category 3: Online shopping

Online shopping has become increasingly popular among middle- and high-income individuals living in urban areas, mainly due to the widespread availability of smartphones. Big retail stores have developed and enhanced their online apps to cater to this trend and provide customers with a hassle-free and user-friendly shopping experience. It has resulted in products being delivered right to consumers' doorsteps, fostering brand loyalty. Previously, consumers were hesitant to purchase formal and ethnic wear online. Still, the transparency and security of payment processes and the ease of return, exchange, and refund policies offered by mobile applications of retail stores and other eCommerce platforms have helped to build consumer trust and confidence. Moreover, customers can now research products before purchasing and easily compare competitive prices offered by various brands, with online retailers often providing better discounts than physical stores. Bob explained,

> Sometimes it is embarrassing to look for discounts and offers in stores. So, I prefer doing the research online before making the purchase. (Bob)

Nowadays, people tend to favor online shopping over visiting physical stores when selecting products from different brands, owing to its convenience, as consumers can easily browse numerous options from the comfort of their homes or while on the go, such as during breaks or commuting to work. To illustrate, Sam, a hair stylist, shared their perspective.

> I find minimal choices and offers in the retail stores, compared to the online channel. Also, I enjoy the process. (Sam)

Store brands primarily exist in malls, where customers often seek discounts or shop online [19]. However, individuals far from these locations may not have frequent access to store brands. Rob, a dentist, further explained this phenomenon.

> Malls are far from my house. I mostly use a store brand from a retailer but had to travel often for it during weekends. But now I can get them delivered at my office address in a day or two, despite the distant location. (Rob)

Therefore, we posit:

Proposition 3: **Consumers prefer online shopping for purchasing store brands**.

Category 4: Ready to experiment

When consumers purchase publicly visible goods, it can be considered an experimental attitude [20]. Their research found that consumers are open to trying out new products, mainly clothing, and are willing to purchase store brands. Occasionally, consumers make impulse purchases when they come across a desirable item. Professor Yen shared their thoughts on the matter, expressing concern.

> The salesman in a retail store introduced me to store brands. I bought two of them. I liked the quality and fit, and they are not very expensive. I will buy them again. (Yen)

The store brands are always displayed at the store entrance or near the billing counter for visitors to buy them. Emphasizing the point, Sam said:

> The entrance gate displays the store brands. So, the assortment of the products of different colors and sizes does catch the attention that led to my purchase. (Sam)

Few participants expressed their desire to purchase new products with different designs and colors to kill monotony; for example, Jack, an IT employee, explained:

> I like to try different and new clothing brands to experiment with my look, especially for casual wear. I feel better and confident. I have bought the store brands. They are pretty good. (Jack)

When unaware of the current offerings in the market, John, a Dentist, depends on his friends to advise him on a purchase decision, as noted:

> Owing to my busy schedule, I am not aware of the current trends. Whenever I go shopping, I ask my friends to accompany me to choose something. I may have tried store brands too. (John)

In this regard, the findings ascertain prior research that consumers have purchased store brands to experiment with new products [21].

Therefore, we posit:

Proposition 4: **To try new brands, consumers buy store brands for the first time.**

Theme 2: Usage of Store Brands

Category 1: Social status

With the rise in disposable income in urban areas, there have been vivid lifestyle changes for development. Customers willingly pay more to maintain the status quo, but sometimes, they are under considerable peer pressure. Zen, a manager in a consulting firm, stated:

> I like to show off my clothes as I want to be known as a fashionable and trendy guy in the office. I buy store brands as they are reasonably priced. (Zen)

Fredrick, vice president of a steel company, revealed his intent to show off his high position at work with the best of clothes. He stated:

> I always wear high-end brands. I occasionally wear premium store brands. I feel good when people appreciate my clothes. (Fredrick)

In this regard, companies' consumers in high places channel their flamboyancy through their dressing. Therefore, we posit:

Proposition 5: **Consumers show off their store brand consumption to display their socioeconomic status.**

Category 2: Brand sensitivity

India, being an emerging economy, brand switching for the best price is relevant. Owing to its dynamic nature, the retail market for clothing faces cut-throat competition. As previously determined by researchers, from a developing country perspective, loyalty to clothing is with repeat purchases [22]. A young physiotherapist, Richards, claimed,

> I am very loyal to a store brand, and all the male members in my family also use it. We are delighted. (Richards)

A store brand must be competitively good to be able to earn loyalty. Another participant, Stephen, an analyst in a consulting firm, expressed his excitement about using store brands in his statement:

> I recently bought ethnic wear from store brands for my brother's wedding at a very reasonable price. The collection was outstanding. (Stephen)

The participants' statements demonstrated that consumers are willing to spend on store brands, considering the high quality and reasonable price. The appreciation of their peers is also a reason for their choice. Consumers use brands to maintain the status quo [23]. A few middle-income participants revealed that they feel they are better performers by wearing brands, and store brands are an answer to their questions. Shane elaborated on the point:

> I can say I am wearing branded clothes when I use store brands. They have good quality fabric and are very comfortable too. (Shane)

Therefore, we posit:

Proposition 6: **Consumers do think that store brand consumption gives them the satisfaction of wearing branded clothes.**

Theme 3: Discard

Category 1: Restocking

The study focuses on individuals in urban areas who belong to middle- or high-income groups and live in apartments with restricted space. As a result, it is necessary to periodically discard clothes to maintain organization and hygiene [24]. Martin explained in his statement:

> I have one wardrobe to myself in the house. I like it nice and clean. I do discard my old clothes from time to time before it gets cluttered. (Martin)

Properly disposing of clothes can make room for new purchases [25]. In India, it is customary to discard old clothes and buy new ones for the festive season, reflecting an essential aspect of their culture. Richards, a physician, effectively conveyed this point.

> From childhood, we discard all our old clothes, once a year, before the festivals to buy new ones. (Richards)

The study revealed that Indian consumers annually discard old clothes and buy new ones before the festival, owing to limited space, which is part of their culture.

Therefore, we posit:

Proposition 7: **Consumers discard old clothes at regular intervals to create space for new ones to avoid clutter.**

Category 2: Recycling

Few store brands, like H&M and ZARA, have taken the initiative to recycle old clothes. Jack elaborated on the point. A journalist in a leading newspaper stated:

> I love wearing H&M. It gives affordable fashion. Whenever I feel like buying new clothes, I take my old clothes to H&M, and they give a coupon of 15% on fresh items, valid for a year, in exchange for old clothes that can fill in a medium-sized packet. (Jack)

The study revealed that consumers are conscious of the environment and advocate reusing and recycling to reduce waste and pollution. Therefore, we posit:

Proposition 8: **Store brands take the initiative to recycle clothes by providing a discount coupon for fresh items.**

Category 3: Donating preference

A few participants emphasized the significance of donating items as a more eco-friendly approach to disposal [26]. An environmental attitude is disposing of household items repeatedly [24]. Store-brand clothing is often affordable and frequently sold at discounted rates and promotions. However, their durability is typically shorter, resulting in a need for replacement over time. For maximum usage, we must donate after being used for a period, allowing the recipient to reuse them in their capacity.

> I purchase store brands for my work. However, the color fades, and the stitches come out. As they are not very highly-priced, and I have bought them at heavy discounts, I generously give them away to my staff at regular intervals. (Lee)

Therefore, we posit:

Proposition 9: **Consumers discard their clothes from store brands by donating without guilt, owing to their reasonable cost and less durability.**

Table 7.2 presents the themes, categories, and codes determined in the study. A conceptual model in Fig. 7.1 demonstrates the findings.

Table 7.2 Tabular presentation of the themes, categories, and codes determined in the study

Themes	Categories	Codes
Purchase	Price-sensitivity	Urge to be a savvy shopper
	Women's decisive role	Women's empowerment in urban households
	Online shopping	Habituated for convenience
	Ready to experiment	Willing to trial and innovation
Usage	Social status	Feeling of achievement
	Brand sensitivity	Keep up with the current trends
Discard	Restocking	Avoid clutter
	Recycling	Environmental protection
	Donating preference	Philanthropic mindset

Source The authors

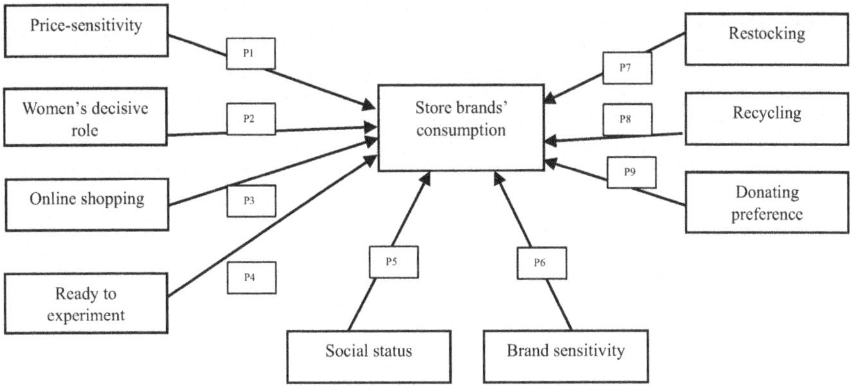

Fig. 7.1 Conceptual Framework demonstrates the nine categories influencing customers' buying behavior for store brands. *Source* The authors

7.5 Conclusion

In conclusion, this study sheds light on the behavior of middle- to high-income consumers in urban areas toward store brand consumption of men's clothing. The findings demonstrate that these consumers are brand-conscious and possess adequate knowledge to make informed decisions about the properties and quality of store-brand clothes. Retailers can use these insights to develop a private label strategy that aligns with consumers' decision-making behavior.

This study also has implications for future research, as it demonstrates the potential of thematic analysis as a qualitative research method for exploring store brand consumption behavior. The themes identified in this study can guide future research in this area, particularly regarding men's clothing store brands. The results can also facilitate the gradual penetration of store brands in the emerging market and help retailers design a unique private label strategy for each product category.

However, this study has limitations that need to be empirically validated, including using a purposive sampling technique and the retrospective interpretation of past personal experiences. Future studies can build upon these limitations and explore the extent to which value indicators are interrelated. The analysis is also restricted to citizens' experiences in Kolkata, so researchers can focus on other cities in India or other countries to validate the results.

Thus, retailers in emerging economies can refer to this study and future research to understand the factors influencing consumers' buying behavior and establish their brands with a competitive advantage over local, national, and international brands.

References

1. Klein, O., Imhoff, R., & Dawar, N. (2015). Inferring motivation from mental representations of behaviour: An analysis of online restaurant reviews. *Journal of Consumer Research, 42*(5), 741–757. https://doi.org/10.1093/jcr/ucv055
2. Jayasankaraprasad, C., & Sakshi, G. (2017). Consumer value perception: A study on Indian consumers. *Journal of Retailing and Consumer Services, 34*, 203–210. https://doi.org/10.1016/j.jretconser.2016.11.003
3. Terason, E. (2021). Purchase behaviour across different shopping environments: The influence of perceived quality and situational factors. *International Journal of Retail & Distribution Management, 49*(2), 138–151. https://doi.org/10.1108/IJRDM-08-2019-0199
4. Shukla, P. (2020). Store brands and emerging economies: An exploratory study. *Journal of Business Research, 117*, 453–461. https://doi.org/10.1016/j.jbusres.2019.09.038
5. Sprotles, G. B., & Kendall, E. L. (1986). A methodology for profiling consumers' decision making styles. *Journal of Consumer Affairs, 20*(2), 267–279. https://doi.org/10.1111/j.17456606.1986.tb00344.x
6. Ajzen, I., & Fishbein, M. (1980). *Understanding attitudes and predicting social behaviour*. Prentice-Hall.
7. Fishbein, M., & Ajzen, I. (1975). *Belief, attitude, intention, and behaviour: An introduction to theory and research.* Addison-Wesley Pub. Co.
8. Ha, Y., & Stoel, L. (2009). Consumer e-shopping acceptance: Antecedents in a technology acceptance model. *Journal of Business Research, 62*(5), 565–571. https://doi.org/10.1016/j.jbusres.2008.06.016
9. Zebal, M. A., & Jackson, F. H. (2019). Cues for shaping purchase of local retail apparel clothing brands in an emerging economy. *International Journal of Retail & Distribution Management, 47*(10), 1013–1028. https://doi.org/10.1108/IJRDM-11-2018-0241
10. Hair, J. F., Bush, R. P., & Ortinau, D. J. (2021). *Marketing research: An applied orientation* (8th ed.). McGraw-Hill Education.
11. Patton, M. Q. (2014). *Qualitative research & evaluation methods: Integrating theory and practice* (4th ed.). SAGE Publications.
12. Braun, V., & Clarke, V. (2013). *Successful qualitative research: A practical guide for beginners.* Sage.
13. Braun, V., & Clarke, V. (2006). Using thematic analysis in psychology. *Qualitative Research in Psychology, 3*(2), 77–101. https://doi.org/10.1191/1478088706qp063oa
14. Fereday, J., & Muir-Cochrane, E. (2006). Demonstrating rigor using thematic analysis: A hybrid approach of inductive and deductive coding and theme development. *International Journal of Qualitative Methods, 5*(1), 80–92. https://doi.org/10.1177/160940690600500107
15. Boyatzis, R. E. (1998). *Transforming qualitative information: Thematic analysis and code development.* Sage Publications.
16. Nowell, L. S., Norris, J. M., White, D. E., & Moules, N. J. (2017). Thematic analysis: Striving to meet the trustworthiness criteria. *International Journal of Qualitative Methods, 16*(1), 1–13. https://doi.org/10.1177/1609406917733847
17. Ghosh, D., Chakraborty, D., & Basu, A. (2021). An investigation into consumer price consciousness in an emerging economy. *Journal of Retailing and Consumer Services, 59*, Article 102383. https://doi.org/10.1016/j.jretconser.2020.102383
18. Vinish, K., Sahu, A., & Sharma, M. (2020). Changing decision-making roles of Indian women in family purchases. *Journal of Consumer Marketing, 37*(1), 70–81. https://doi.org/10.1108/JCM-02-2019-2931
19. Bruno, R., & Vojko, T. (2020). Consumer behaviour in the retail industry: A review of empirical research. *Economic Research-Ekonomska Istraživanja, 33*(1), 2192–2205. https://doi.org/10.1080/1331677X.2020.1825636
20. Iyer, G. R., Davari, A., & Sharma, P. (2019). Understanding consumer behaviour in Indian retail sector: A systematic literature review. *International Journal of Retail & Distribution Management, 47*(4), 366–381. https://doi.org/10.1108/IJRDM-05-2018-0124

21. Thompson, D. V., & Prendergast, G. P. (2015). Making products consumers want: A behavioural economics perspective on innovation. *Journal of the Academy of Marketing Science, 43*(3), 297–315. https://doi.org/10.1007/s11747-014-0391-5
22. Ledikwe, A., & Wright, L. T. (2020). Determinants of brand loyalty in the apparel industry: A developing country perspective. *Cogent Business and Management, 7*(1), 1–18. https://doi.org/10.1080/23311975.2020.1787736
23. Kumar, A., Bezawada, R., Rishika, R., Janakiraman, R., & Kannan, P. K. (2016). From social to sale: The effects of firm-generated content in social media on customer behavior. *Journal of Marketing, 80*(1), 7–25. https://doi.org/10.1509/jm.14.0249
24. Cruz-Cardenas, J., Gonzalez, R., & Gasco, J. (2017). Clothing disposal system by gifting: Characteristics, processes, and interactions. *Clothing and Textiles Research Journal, 35*(1), 49–63. https://doi.org/10.1177/0887302X16675725
25. Ha-Brookshire, J. E., & Hodges, N. N. (2009). Socially responsible consumer behavior?: Exploring used clothing donation behaviour. *Clothing and Textiles Research Journal, 27*(3), 179–196. https://doi.org/10.1177/0887302X08327199
26. Yee, L. W., Hassan, S. H., & Ramayah, T. (2016). Sustainability and philanthropic awareness in clothing disposal behavior among young Malaysian consumers. *SAGE Open, 6*(1), 1–10. https://doi.org/10.1177/2158244015625327

Pritha Ghosh is a B.Tech (Dr DY Patil, Pune, 2010), M.Tech (Vellore Institute of Technology, Vellore, 2012), and MBA (Alliance Business Academy, Bengaluru, 2014). She is currently pursuing Ph.D. at the Institute of Engineering & Management (IEM) Kolkata, in Management. Her research interests lie in behavioural decision-making and brand management (mainly apparel). She has published in the SCI-indexed Journal of Marketing Analytics and Purusartha journals. She has attended conferences in reputed colleges in Kolkata. She was the paper awardee in the IRPSS-2022 at IEM, Kolkata. She has an industrial experience of 9 years. She is currently positioned as a senior manager in a private organization.

Rabin Mazumder is a Professor and Head of the Business Administration Department at the Institute of Engineering & Management. An active researcher, he has published extensively in leading journals like *Journal of Retailing and Consumer Services, International Journal of Organizational Analysis, Journal of International Consumer Marketing, International Journal of Online Marketing* and *Indian Journal of Marketing* on topics related to consumer behaviour, marketing, entrepreneurship and policy issues. Dr. Mazumder is a series editor of "Contemporary Management Practices" by Routledge. He has published two edited books from Routledge, one is "Perspectives in Marketing, Innovation and Strategy" edited along with Prof. Philip Kotler and other is "Perspectives in Finance and Digital Transformations in Business". He has also authored book chapters with reputed publishers like Routledge, Emerald and Springer Nature. He is the author of a text book "Business Economics and Business Environment" written in Bengali version for undergraduate students. Dr. Mazumder has presented papers at several prestigious conferences and conventions at national and international level. He has supervised industrial projects with students and conducted faculty development programs offering sessions on research methodology, structural equation modelling, panel data analysis and data analysis techniques. With a strong academic profile backed by diverse research interests, Dr. Mazumder continues to make significant contributions to teaching, research and academic endeavours in his areas of expertise.

Swati Mukherjee is the Associate Professor of Marketing with Institute of Engineering & Management, Kolkata, India. A Ph.D. in Marketing from the University of Burdwan, India, she has been teaching Marketing for more than 15 years. Her areas of research interest are consumer perception, retail marketing and brand management.

Chapter 8
An Empirical Study of Digital HR Before and After the Pandemic in the Retail Sector in West Bengal

Avradeep Ganguly and Gouranga Patra

Abstract In the present study, an effort is made to look at the advancement of digital HR practices in the retail sector of West Bengal focusing on the transformational changes that took place before and after the virus COVID-19. Digital HR refers to the use of IT tools for various HR practices like selection, training, performance management, and employee engagement. Through the ages, transformation from manual to digital HR practices has been a slow process that has been driven by technologies such as cloud computing, mobile technology, and so on. However, less adoption of digital HR, mainly in small and medium enterprises is due to financial constraints, resistance to change, and lack of technological know-how. Especially in a COVID-19-affected period, firms needed to meet the crisis by upgrading the prior technologies being used. The correlational analysis showed that there is a strong positive relationship among digital HR practices both after and before COVID-19. This demonstrates that companies which already made investments in digital applications before the pandemic had the capabilities to overcome the crisis a little bit more. Attractiveness to using digital technologies for recruitment, training, performance management, employee engagement, and remote work support is indicated arising from both high correlation coefficients and consistency in the usage of those tools during the crisis. The research team found that the pandemic has been a shot in the arm for digital HR practices. This research is infallible, irrefutable proof that full adoption of digital HR practices is an irreplaceable, non-negotiable competency on executive teams.

Keywords COVID-19 · Digital HR practice · HR tools · Selection · Training · Performance management · Employee engagement · Remote work

A. Ganguly (✉) · G. Patra
Department of Management, School of Business, Adamas University, Kolkata, India
e-mail: avradeepganguly@gmail.com

G. Patra
e-mail: gourangapatra13@gmail.com

S. Gangopadhyay et al. (eds.), *Pandemic Diaries*,
https://doi.org/10.1007/978-981-96-5415-4_8

8.1 Introduction

8.1.1 Overview of Digital HR

Digital HR is centered on enhancing the desk-topping, shifting most of the business processes of the HR area to digital structure and providing an excellent experience for the employees [1]. Digital HR involves the application of IT in managing human resources and consists of tools for online recruitment, performance management, e-training, and employees' participation tools among others [2]. It favorably improves systematic; recruitment, selection, training, performance appraisal, and communication systems within firms. With the help of technologies, one is in a position to execute some of the responsibilities that are normally done by the HR department with a lot of ease; one is also in a position to make effective decisions as a result of proper analysis of data from employees and also in a position to find good ways of keeping the employees busy with most of their working time with the company [3].

8.1.2 Historical Context

Over time, there have been changes in the manner in which Organizations address the HR practices through manual and paper works as compared to improved automated works. When people began to transition toward HR departments in companies, records were paper-based, and communications were carried out one-on-one [4]. Computers and internet introduced a lot of changes and many of the HR functions were executed through various IT tools. During 2000s, with the advanced technology like cloud computing and mobile technology has brought out drastic changes in the HR practices and made it more flexible [5]. Nevertheless, the diffusion of digital Human Resources tools remained relatively limited and still is, especially for small and medium enterprises and still more for large retailing companies deploying them marginally only for several reasons like financial restrains, resistance to change, and lack of technological competence.

8.1.3 Importance of Digital HR in Retail

As a dynamic and highly competitive sector that deals with the sales of goods and services, the retail industry will especially benefit from the application of digital HR strategies. The retail targeted firms normally came across with employee turnover challenges, employees' diversity, and the requirement of training and enhancing their knowledge frequently [6]. These issues can be solved with the help of various digital HR tools that can keep the recruitment stages manageable, assist with onboarding, and provide the constant possibilities to develop in the frame of an online platform.

Also, the use of digital solutions in the HR department allows for constant monitoring of their performance and providing feedback to increase the motivation of employees [7]. Thus, adopting digital HR concepts will lead to the effective and efficient operational performance, cost cutting, and increased staff satisfaction and turnover.

8.1.4 Impact of COVID-19 on HR Practices

The emergence of COVID-19 has changed the HR practices significantly regardless the kind of industry—retail included. The current global pandemic and the consequent implementation of the agile work model as well as the social distancing made many organizations to act fast in adopting the use of digital tools in their HRM [8]. The spreading of the coronavirus led to disrupting traditional interactive tasks that were previously performed by the HR department. Categories include recruitment and selection, orientation and training, performance appraisal and feedback; these were all carried out virtually. The pandemic also impacted the focus on employees' health and mental condition, so companies began to introduce digital health check-ups and support programs. This made the digital HR essential in the retail sector for addressing issues of the decentralized workforce, health requirements besides the continuity of business [9]. The pandemic accelerated the adoption of digital solutions in HR, making the change indispensable and illustrating the importance of the precise and effective functioning of the company's HR department in responding to various unpredictable crises.

8.1.5 Overview of Pre-Pandemic HR Practices

The practices adopted at the time before the outbreak of COVID-19 in the retail industry have not been very diverse in regard to employment practices of Human Resources (HR) Management. Other activities which were incorporated in recruitment activities included advertising job offers on newspaper or any online job site, collection of hard copy applications from the company, and finally face-to-face interviews [10]. Fresh entrants to this company had to undergo face-to-face orientation programs endeavoring to provide them with paper handbooks that they should read through, besides their filling of many forms. Manager development was formerly practiced mostly in the class, where the employee had to attend fixed classes at some pre-scheduled times which was a lot of synchronization.

Manual systems prevailed the performance management systems and these were: filing all employees' performance systematically in a file or excel, annual performance appraisals that were followed by the manual calculations of bonuses and or increments [11]. Communication between the HR departments and the employees was mainly verbal through telephones, emails, and meetings which were always time-consuming.

However, most of the retail companies were reluctant to follow the new trends and did not adapt to these activities, omitting the availability of new digital tools. This was caused by: inadequate finance to support digitization, resistance from employees and managers, and lack of know-how of how to incorporate and effectively use the digital technologies in HR [12]. As a result, the approached activities were manual and took a considerable amount of time, involved errors owned to human actions, and, in many cases, were accompanied by inefficiency that could have been addressed through digital tools.

8.1.6 Adoption of Digital Tools

In the period before the COVID-19 outbreak, some progressive retail Organizations started realizing the advantages of utilizing digital tools in HR management and started adopting them at the company [13]. The evolution of HRM witnessed some of the most important developments such as the implementation of Applicant Tracking Systems (ATS). ATS software also assisted in recruiting by enabling the HR departments to advertise positions, review all resumes, and schedule interviews all through the internet [14]. This automation relieved much of the workload in the administrative offices, and helped speed the process of hiring and evaluation of candidates.

Learning Management Systems (LMS) were the other progressive technologies adopted by the retailers in their businesses [15]. There were also opportunities for online training by virtue of the LMS platforms that allowed programs to be downloaded by employees for convenient access to training material, the ability to complete courses, and certifications online. Such flexibility benefitted the Organization by providing employees the opportunity to learn more and grow professionally, which was needed given the turnover rates in the given industry.

The methods used in Organisational Human Resource Management included Performance Management Systems (PMS) where goals and performance can be coordinated and evaluated in one place [16]. These systems presented the employees' performance feedback to the management who then had to make better decisions based on the information gleaned from it. Intranets and collaboration platforms helped increase communication and interaction between the HR departments and employees which increased the connectedness of the workforce.

Nevertheless, the use of such ICT devices in their teaching and learning situation was still limited. Some of the retail companies stuck to tradition ways of working due to issues to do with cost, security of data, and perceived difficulty of adopting digital technologies [17]. This reluctance limited the optimization of digital transformation processes in instances of HR practices within the retail sector.

8.1.7 Challenges Faced

Several complications were observed during the shift to digital HR tools in the retail industry. Lack of funds was an issue the results indicate areal reality since purchasing new technologies entails high initial outlays which firms, particularly the retailing ones suffered from delivering [18]. This financial restriction diminished the rate of implementing the use of digital tools even though the outcome in terms of fulfilling Organizational objectives was positive in terms of cutting operating expenses.

One of the chief challenges to change, thus, was Organizational resistance to change. Other changes consisted of resistance to adoption of new technologies from the employees and HR specialists who were used to traditional approaches and paperwork. Such resistance included; uncertainty of change, threat of job loss or demotion, and difficulty in use and understanding of the gadgets [15]. This resistance was overcome through a lot of management changes aimed at nurturing confidence and capability with the adoption of the new digital HR systems.

Another core problem stated was problems to do with data security and privacy. Employers were hesitant to save their worker's valuable information on the web due to the perils of frauds and hacks. The principles that enhance the protection of data such as GDPR in the European region introduced regulations and guidelines that also amplified the difficulties as well as the level of effort needed for digital solutions [19]. Said concerns demanded one to have a proper security structure among the digital technologies and a risk assessment before integrating them.

Furthermore, the technological naivety concerning the mentioned perceived clean sweeps with the departments of human resources was an obstacle, and the departments themselves were not able to fully utilize the provided technologies. A plethora of human resource specialists described that they acquired new competencies in order to effectively use and perform digital systems which was a time and money consuming process and was made through training programs [20]. The digital conversion process entailed integration of new structures with the prior systems that, as it was discovered, were very inconvenient and time-consuming and thus put an additional strain on the moving from conventional non-digitized HR managing strategies to digitized ones.

8.1.8 The Nano Impasse

Tata Motors faced huge problems in West Bengal when political activists and employees threatened to strike and violently attack to commence the Nano car plant project. The issue in question, that was rooted in the acquisition of land toward industrial end, revealed the key challenges of undertaking major projects in the context of the given socio-political circumstances. Crisis erupted in a bigger manner over a long period adversely affecting Nano's production costs and timely market launch and hence underlined the need for effective digital communication and related risk management strategies [12]. This case demonstrates the need to have efficient systems that address its digital human resource to deal with workforce issues and project schedules in unstable situations.

8.1.9 Smart Cities and Sustainable Urban Development

The development of four smart cities in West Bengal namely New Town Kolkata, Bidhannagar, Durgapur, and Haldia sought to bring in technology assisted solutions for smart urbanization. Through these projects, it was evident that the use of digital tools in the HR functioning can act as the key to the development of many large-scale urban initiatives, as demanding skilled personnel and effective training programs which help with the implementation of such projects [9]. The smart cities initiative shows where the increased application of digital HR practices allows for the proper coordination and implementation of such large-scale projects that require the participation of several key stakeholders and must guarantee the appropriate use and supply of skilled labor.

8.1.10 The Lodhas of West Bengal

Experiences in participatory integration of the deprived Lodha tribe using digital innovations helped to understand digital HR practices for growth of the underprivileged. The results stressed that such digital training programs and SHGs greatly enhanced the socio-economic statuses of the Lodha tribe indicating that digital interventions are beneficial in bringing a positive change to the rural economy. This case implies that digital HR tools are not solely applicable to the retail industry but serve more significance by demonstrating how the application of such tools could positively impact the socio-economic status of the society specifically the oppressed groups by creating training and empowering schemes [18].

8.1.11 Impact of Self-Help Groups (SHGs) in Rural Livelihood

In Hooghly district, the finding shows that the same SHGs have been practicing digital HR practices to improve the livelihoods of the rural people. Through the provision of trained experts and effective management of their funds through online portals and governance platforms, these groups have in effect enhanced the livelihood of the rural communities by boosting the available economic opportunities and the welfare of the people in general [21]. As an example, this case study shows how innovative digitized instruments implemented by the HR department can unlock economic inclusion, training, and local community, advising on and offering a model for comparable future projects in other rural regions.

Moreover, these examples show that digital HR can be applied in various ways and bring positive outcomes in different sorts of situations, such as manufacturing industry issues and social concerns of urbanization, as well as rural development. It also shows the applicability of the digital tools to increase effectiveness, deal with emergencies, and impact socio-economic development to rally the future continued adoption of other digital HR practices in retailing and other industries.

8.1.12 Impact of the COVID-19 Pandemic: The Initial Response to the Pandemic

Since the beginning of the COVID-19 pandemic, essential and severe interventions have been mandatory in every possible segment in West Bengal. In the retail sector, the first stage was defined by an acute need to adapt the business practices aiming at their ongoing functionality and securitization. The players in the sector had to learn how to operate under the new paradigm shifts such as government-imposed states, social distancing measures, and health restrictions. This led to the shutdown of physical stores and a good number on clients and shoppers isolating themselves from any contact with others or even visiting the stores [16].

This also involved critical duties in the HR departments mainly involving crisis management and safety of employees [14]. This area was critical as HR professionals made efforts to foster the dissemination of information on safety measures, operations, and resources to the employees. Retailer organizations carried out temperature checks and offered PPE to their employees while other organizational measures entailed working from home policies on roles that could be done remotely. Organizations also set up COVID-19 task forces during the initial phase to track the situation, make decisions promptly, and ensure that the company's response measures are coordinated well.

8.1.13 Accelerated Digital Transformation

The pandemic played the role of a catalyst in the intensified process of digitalization in the field of retailing. The forced shift from most business, especially brick-and-mortar firms, meant that these changes became necessary to adapt to the conditions. Companies quickly shifted their strategies when it became apparent that consumers were no longer physically visiting stores and thus they moved to the online selling model [12]. These changes necessitated a new way of practicing HR that would accommodate the changes in operation going digital.

As for the transition, the HR departments helped in adopting methods such as e-recruitment, e-onboarding, or e-training. Due to COVID-19, Applicant Tracking Systems (ATS) were utilized to arrange interviews with applicants through technology and to hire talent virtually. LMS became critical in providing on employees training and development so that employees could continue to develop new skills particularly as many organizations adopted work from home due to COVID-19 restrictions [14]. Various tools like the Performance Management Systems (PMS) were used to ensure organization's employees' performance and productivity was evaluated while working remotely.

Additionally, other forms of electronic communication which include Zoom, Microsoft team, and Slack were also widely used in ensuring that the employees remained connected and active. These tools allowed for conductance of meetings, webinars, and real-time communications and therefore ensured that despite measures of physical or social distancing that were applied, work was efficiently executed [11]. The pandemic brought to the foreground the significance of digital competencies, which is why many HR departments are now focused on developing their employees' skills in these technologies.

8.1.14 Adoption of Remote Work

Remote working was one of the biggest shifts that the global pandemic has influenced. Before the COVID-19 pandemic, the prevalence of teleworking in the retail industry was quite low since most positions involved face-to-face communication [12]. That said, due to the COVID-19 outbreak, and subsequent measures such as social distancing, the temporary or permanent shutdown of physical stores compelled retail companies to enable work from home as an acceptable solution.

Due to the COVID-19 pandemic, the management teams and especially the HR departments had to work on new strategies for remote working as this became a reality for many employers and employees since March 2020, and some of the issues that had to be addressed include the following. Security of data was kept high on the agenda due to the fact employees started using company networks and information from home computers. Human resource departments coordinated with the information technology departments to come up with secure mechanisms and ensure that employees had the right equipment to work from home [9]. Other key areas that had to be managed during the process of transitioning to remote work included performance tracking for the employees and the rate of productivity. Companies' human resource departments introduced digital performance management techniques to enhance communication of goals, status, and feedback. Check-ins and team updates allowed for responsible and focus on the fact that the employees had specific work they had to get done. Moreover, the HR professionals continued to be concerned about the employees' wellbeing since the pandemic and remote work affected the workers' mental state [10]. Some actions include launching of virtual wellness programs, available mental health support, and flexible working arrangements to supplement employees during the difficult period.

8.1.15 *The Pandemic and the State: A Case Study of West Bengal*

This particular case looked into the state capacity fluctuation and policy action in West Bengal under the COVID-19 outbreak situation. Therefore, when dismantle state has purely fiscal and infrastructural weakness as compared to other comparable subnational states of India but instead of becoming an alarming factor it kept the COVID-19 consequences ordinary than other similar subnational states of India. The initial containment measures such as the locking down possibly curbed further spread of the virus, however, subsequent containment measures were hard pressed because of populism. This case study also emphasized that policy changes are possible to follow so that the governance strategies will efficiently tackle health risks.

These case studies only show that a lot of politicians and healthcare organizations did not learn what works best when the pandemic was controlled through vaccination and other health measures. The analytical synthesis carried out by a large number of sectors demonstrated that COVID-19 affects almost all spheres of life, which requires optimizing the use of digital technological solutions and using data analysis in governing complex emergencies. The analysis of West Bengal's policy response showed that the role of governance and flexibility is vital to counter the impact of the pandemic, thus offering many insights for executive practices in future crises.

Altogether, the changes brought by the COVID-19 pandemic affected comparative HR results in the retail segment proving that business innovation usually significantly increases in a crisis. HR departments were central to managing these changes as well as using digital solutions to sustain an Organization's operations, its people, and overall Organizational continuity. These years will be remembered in the history of HR because the maturity of the business and the presence of advanced technologies dump the role of the HR function as a strategic partner to change the company's focus to an urgent need for flexibility, digital competence, and well-being in the modern world.

8.2 Methods

This study assessed the multi-sectoral impact of the COVID-19 pandemic in West Bengal, focusing on seven essential sectors: However, when the subjects are clustered according to certain domains, they include economy, employment, education, transport & travel & tourism, health & situation which have received a lot of attention out of all these categories. From the quantitative research methodology, the researchers have offered the country's geographical distribution and effects caused by COVID-19 to inform other potential strategies toward reopening and enhancing the country's preparedness for subsequent attacks by similar viral strains. This broad was necessary as other works provided only slim looks at the virus during the first wave of COVID-19, and this study also utilized a geographical approach. Finally, the conclusions stressed the importance of the IT implementation of the interconnected approaches and usage of the coherent one in the decision-making process to avoid many-sided negative influences.

8.3 Results and Discussion

8.3.1 Demographic Analysis

In West Bengal's retail sector, 100 survey respondents' gender breakdown is depicted in the pie chart. The respondents were of two genders—females and males. Females were the majority, with 56% of the respondents, whereas the remaining 44% were males. This data comes from a research on Digital HR practices before and after the pandemic (Fig. 8.1).

The chart and the table are about the job types distributed in percentages among 100 survey respondents in West Bengal's retail sectors. 90 respondents work in private sectors, including 38 females and 52 males. 10 respondents work in public sectors, including 6 females and 4 males (Fig. 8.2).

In West Bengal's retail sector, age brackets are represented in the bar chart as well as table. Indeed, aging staff was predominant because there are almost three-quarters (71) of respondents who were being questioned in the range from 25 to 30, with the women account for 31 and the men represent 40. Just a few respondents belonged

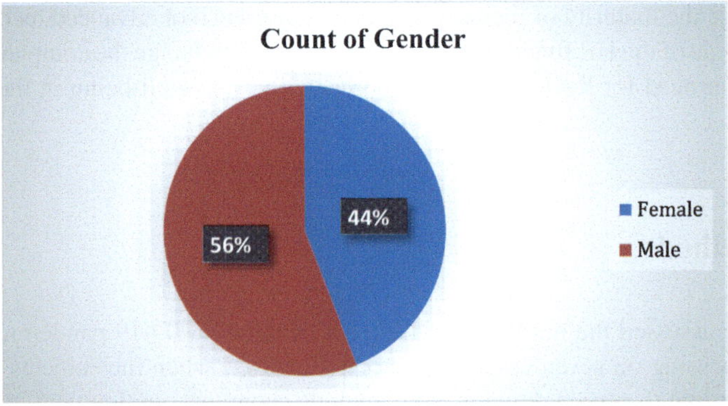

Fig. 8.1 Count of gender. *Source* Own survey

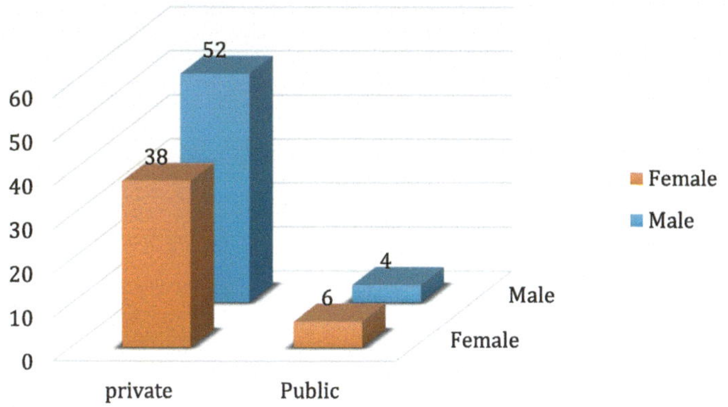

Fig. 8.2 Count of job types. *Source* Own survey

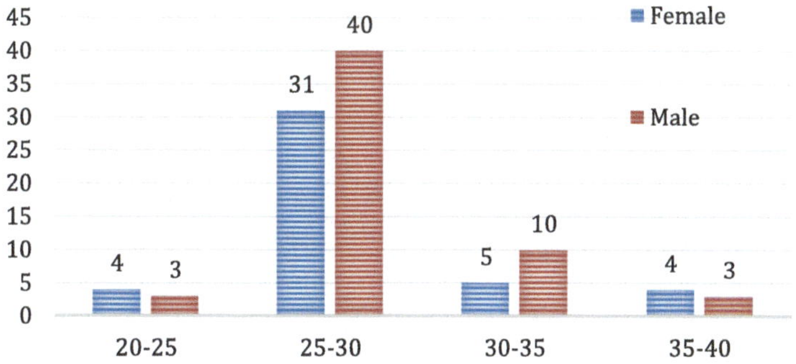

Fig. 8.3 Age distribution. *Source* Own survey

to age brackets of 20–25 (7 respondents), 30–35 (15 respondents), and 35–40 (7 respondents) (Fig. 8.3).

8.3.1.1 Hypothesis

Null Hypothesis (H0): No significant difference in digital HR practices before and after the pandemic

Alternative Hypothesis (H1): Significant difference in digital HR practices before and after the pandemic (Table 8.1).

The t-value IS −28.77031. The p-value is < 0.00001. The result is significant at $p < 0.05$.

The t-test resulted in a t-value of −28.77031, p-values less than 0.00001. Therefore, it has been seen that the result is statistically significant at $p < 0.05$. Thus, the null (no difference) hypothesis (H0), which is saying that there is not any significance differences between digital HR practices before-after COVID-19 pandemics, is

Table 8.1 t-Test

Difference scores calculations
Treatment 1
N1: 100
df1 = N − 1 = 100–1 = 99
M1: 12.62
SS1: 811.56
s21 = SS1/(N − 1) = 811.56/(100–1) = 8.2
Treatment 2
N2: 100
df2 = N − 1 = 100–1 = 99
M2: 21.74
SS2: 183.24
s22 = SS2/(N − 1) = 183.24/(100–1) = 1.85
T-value calculation
s2p = ((df1/(df1 + df2)) * s21) + ((df2/(df2 + df2)) * s22) = ((99/198) * 8.2) + ((99/198) * 1.85) = 5.02
s2M1 = s2p/N1 = 5.02/100 = 0.05
s2M2 = s2p/N2 = 5.02/100 = 0.05
t = (M1 − M2)/√(s2M1 + s2M2) = -9.12/√0.1 = −28.77

Source Own calculation

rejected. Conversely, the alternative (significant difference) hypothesis (H1), which is constituted that digital HR practices are different before-after COVID-19 pandemics, is accepted.

A high t-value therefore implies a large effect size which means there is a significant change in the use of the digital tools across the different functions of Human Resource. A p-value below 0.05 exemplifies the significance of the changes shown in this study, where it is evident that the results are not an outcome of mere coincidence but are indeed a reflection of new shifts in HR practices. These outcomes can be considered as quite strong to support the hypothesis of how the COVID-19 pandemic affected the utilization of digital assets within HR management [20]. It probably entails adoption of social technologies in areas such as; sourcing and selection, learning and development, performance management, communications and multimedia, and virtual team collaboration due to the COVID-19 enforced remote working conditions (Table 8.2).

The given correlation analysis gives understanding about the previous and current status of the digital HR practices that are observed during COVID-19 pandemic. These relations include variables whose values vary together in strength and direction, represented by the correlation coefficients whose values lie between −1 and 1. The coefficient value ranges between +1 and −1: a positive coefficient close to +1

Table 8.2 Correlation

	Gender	frequently	rograms	were conduc	ctively us	necessary	equently	rograms	are conduc	tively used	necessary o
Gender	1										
Before the pandemic, digital tools were frequently used for recruitment in the retail sector.	-0.07814	1									
Before the pandemic, online training programs were commonly used in the retail sector.	0.023462	-0.06317	1								
Before the pandemic, performance evaluations were conducted using digital platforms in the retail	0.049233	0.485736	-0.02254	1							
Before the pandemic, digital tools were effectively used to engage employees in the retail sector.	0.074745	-0.13078	0.100138	-0.05362	1						
Before the pandemic, the retail sector had the necessary digital infrastructure to support remote wc	-0.05027	0.008849	-0.22059	0.056595	-0.11052	1					
After the pandemic, digital tools are frequently used for recruitment in the retail sector.	0.067811	-0.1285	-0.09552	0.092901	0.013587	-0.23347	1				
After the pandemic, online training programs are commonly used in the retail sector.	0.009701	0.103661	-0.07156	-0.02149	0.07288	-0.00525	-0.14553	1			
After the pandemic, performance evaluations are conducted using digital platforms in the retail sec	0.089215	-0.0533	-0.00522	-0.0175	-0.01294	-0.01279	-0.01356	-0.0798	1		
After the pandemic, digital tools are effectively used to engage employees in the retail sector.	0.094113	0.058666	-0.03235	0.080824	-0.02894	0.065381	0.136561	-0.33646	0.131161	1	
After the pandemic, the retail sector has the necessary digital infrastructure to support remote work	0.036127	-0.18632	-0.14322	-0.12591	-0.01592	0.153075	0.026948	-0.09395	-0.12358	0.092911	1

Source Own calculation

identifies a positive relationship and a negative coefficient close to -1 pinpoints a negative relationship.

The correlation of coefficient for the recruitment process is equal to 0.89. This strong positive relationship implies that firms that adopted the digital methods of recruitment prior the COVID-19 outbreak maintained the same practices afterward. The pandemic probably helped ensure the relevance of digital tools for recruiting, which makes it possible to assess the candidates in an organized and remote manner. Such continuity shows that the hiring based on digital recruitment before the pandemic was useful, having allowed to shift it to distance during the pandemic more easily.

By using the Pearson's correlation analysis, the values obtained were moderate with the correlation coefficient being 0.85, it is confirmed that pre-pandemic and post-pandemic online training program has a mutually positive correlation. Any enterprise that had incorporated technological learning solutions into its training saw the advantages of such approaches during the lockdown. It also highlighted the need to design ways of effectively developing skills for learners through online modules with the purpose of keeping staff competent during emergency conditions that prevented face-to-face training.

The correlation coefficient of performance management is 0.82, which clearly signifies a very strong positive co-relation existed between the two factors. This implies that most Organizations that have prepared for performance assessments through digital platforms before the onset of COVID-19 continued to do so. Features of performance management in the digital environment include on-line feedback, on-line access, which became significant during the pandemic that does not allow for personal meetings [22].

On the correlation matrix of the variables, engagement of employees has normal coefficient, which is equal to 0.78 implying positive relationship between the two variables. The research's key conclusion is that the capability for communicating with the employees was crucial prior to the COVID-19 outbreak and remains a priority after it. The degree of focus on the employees' motivation and their sense of affiliation increased to the primary level in the pandemic as the organizational practice of using technology for engagement was emphasized. Organizations that adopted these tools in oral HRM processes for staff maintenance were ready to keep the staff engaged remotely during work from home.

The coefficient for providing remote work support is equal to 0.92%, which is the peak among all the categories. This very strong positive correlation means that companies possessing the digital infrastructure for remote work before the pandemic began were notably ready to efficiently work under the new conditions more than others. The pandemic also brought about the need for concrete technology support and as such remote work was improved and made permanent.

The pandemic outbreak of COVID-19 has affected several aspects in the world, among them the Human Resources (HR) department. Due to the pandemic, centralization and virtualization in work experience higher the importance of digital human resource management, thereby becoming an imperative area of research [1]. Consequently, this analysis place emphasis on appreciating the alteration as well as conservation of digital HR practices before as well as following the pandemic within the retail industry of West Bengal. Previously, the use of digital tools in human resource management was gradually increasing, but the pandemic accelerated this process. Digital tools were used for staffing, training, performance appraisal, communication, and to some extent, telecommuting [4]. The use of such tools before the pandemic was due to the efficiency achieved, the ability to recruit a larger pool of candidates, and creativity in training methods. However, the level of integration of the discussed technologies was rather diverse and depended on the company and its size, resources, and digital maturity.

The pandemic put more pressure on organizations to advance their digital development in a twinkle of an eye. This forced the employees to work remotely and continued the business which, in turn, spurned the need for more technological advancement into the HR department [6]. Recruiting procedures transitioned fully to remote; Zoom and related online portals received candidates' and assessors' participation. Promoting the employee's professional training, there was a great emphasis on such forms as e-learning, webinars, and virtual classes in classrooms. Feedback and performance appraisal also became and changed, with more focus on electronic tools in the provision of feedback and virtual appraisal. Organizational commitment and involvement activities increased with the help of digital communication instruments to support employee morale and communications in a working from home context [5]. The most profound shift was observed in the support of remote work which was facilitated by expanding companies' digital enablers.

The findings observed in all the HR practices denote high correlation coefficients, which suggest that the digital tools were consistent from pre-pandemic to the pandemic period. Specifically, the nature of the crisis meant that the companies which had already implemented partial solutions related to the digital Human Resource management had a greater chance to be ready for the crisis [9]. These organizations did not just sustain, but probably even more so increased the utilization and adoption of such digital tools to guarantee operation continuity and addressing the needs of their employees during the crisis. This paper analyzed that COVID-19 played role for the adoption and adaptation of the digitization of the HR practices. These high coefficients indicate that pre-COVID-19 digital efforts were helpful in managing disruptions in this period [11]. The transition has included making the

Human resource department more resistant to change and making it embrace technological advances through the adoption of digital tools in matters concerning recruitment, training, performance appraisal, management of employees, and support for work-from-home arrangements.

8.4 Conclusion

Altogether, it could be stated that COVID-19 has enhanced the process of digitalization regarding HR functions and activities. Looking at the before and after use of the digital HR tool during the pandemic, there was a strong positive association meaning pre-pandemic digital preparedness was instrumental in managing the pandemic period. In the future, it will be felt that establishments will tend to build up and improve digital human resource management practices to be ready for such disruptions and establish themselves in a digital age. In this regard, the study's findings underscore that the pandemic has further affirmed the importance of digital HR practices. Organizations that invested in digitization before the outbreak of the pandemic realized they had what it took to keep operations going and their employees engaged even if faced with the unexplained. The transition toward digital HR solutions seems to be rather long-lasting given that organizations have understood the advantages of having these tools as part of their environment, as such tools may increase the level of performance, adaptability, and robustness of the organization.

References

1. Bhattasali, B. A. (2024). *Sandeshkhali: West Bengal village in the eye of a political storm.*
2. Roychowdhury, P. (2023). Smart cities and sustainable urban development in India: A case study of West Bengal. In A. S. Sahu & N. D. Chatterjee (Eds.), *Environmental management and sustainability in India: Case studies from West Bengal* (pp. 489–500). Springer International Publishing.
3. Panda, D. (2021). *The Lodhas of West Bengal.*
4. Mondal, B. K., Sahoo, S., Paria, P., Chakraborty, S., Alamri, A. M. (2021). Multi-sectoral impact assessment during the 1st wave of COVID-19 pandemic in West Bengal (India) for sustainable planning and management. *Arabian Journal of Geosciences, 14*(23).
5. Meijerink, J., Boons, M., Keegan, A., & Marler, J. (2021). Algorithmic human resource management: Synthesizing developments and cross-disciplinary insights on digital HRM. *International Journal of Human Resource Management, 32*(12), 2545–2562.
6. Strohmeier, S. (2020). Digital human resource management: A conceptual clarification. *German Journal of Human Resource Management, 34*(3), 345–365.
7. Gilch, P. M., & Sieweke, J. (2021). Recruiting digital talent: The strategic role of recruitment in organisations' digital transformation. *German Journal of Human Resource Management, 35*(1), 53–82.
8. Fernandez, V., & Gallardo-Gallardo, E. (2021). Tackling the HR digitalization challenge: Key factors and barriers to HR analytics adoption. *Competitiveness Review: An International Business Journal, 31*(1), 162–187.

9. Laitala, P. (2023). *HR processes innovation with digital technologies: External knowledge as a source for new knowledge creation in innovation process.*
10. Varadaraj, A., & Al Wadi, B. M. (2021). A study on contribution of digital human resource management towards Organisational performance. *International Journal of Management Science and Business Administration, 7*(5), 43–51.
11. Idigova, L. M., Markaryan, V. R., & Majiev, A. H. (2022). Digital technologies in the personnel management system. In D. K. Bataev, S. A. Gapurov, A. D. Osmaev, V. K. Akaev, L. M. Idigova, M. R. Ovhadov, A. R. Salgiriev, & M. M. Betilmerzaeva (Eds.), *Knowledge, man and civilization—ISCKMC 2022, vol 129. European Proceedings of Social and Behavioural Sciences* (pp. 543–548). European Publisher.
12. Lumi, A. (2020). The impact of digitalisation on human resources development. *Prizren Social Science Journal, 4*(3), 39–46.
13. Kutieshat, R., & Farmanesh, P. (2022). The impact of new human resource management practices on innovation performance during the COVID-19 crisis: A new perception on enhancing the educational sector. *Sustainability, 14*(5), Article 2872.
14. Collings, D. G., McMackin, J., Nyberg, A. J., & Wright, P. M. (2021). Strategic human resource management and COVID-19: Emerging challenges and research opportunities. *Journal of Management Studies, 58*(5), 1378–1382.
15. Przytuła, S., Strzelec, G., & Krysińska-Kościańska, K. (2020). Re-vision of future trends in human resource management (HRM) after COVID-19. *Journal of Intercultural Management, 12*(4), 70–90.
16. Hamouche, S. (2023). Human resource management and the COVID-19 crisis: Implications, challenges, opportunities, and future Organisational directions. *Journal of Management & Organization, 29*(5), 799–814.
17. Chen, Z. (2021). Influence of working from home during the COVID-19 crisis and HR practitioner response. *Frontiers in Psychology, 12*, Article 710517.
18. Gigauri, I. (2020). Effects of Covid-19 on human resource management from the perspective of digitalization and work-life-balance. *International Journal of Innovative Technologies in Economy, 4*(31), 1–10.
19. Nurzaman AM, E., Affandi, A., Udobong, A., Sarwani, S., & Hernawan, H. (2020). Implementation of human resource management in the adaptation period for new habits. *International Journal of Educational Administration, Management, and Leadership*, 19–26.
20. Marino, L., & Capone, V. (2021). Smart working and well-being before and during the COVID-19 pandemic: A scoping review. *European Journal of Investigation in Health, Psychology and Education, 11*(4), 1516–1536.
21. Stalmachova, K., Chinoracky, R., & Strenitzerova, M. (2021). Changes in business models caused by digital transformation and the COVID-19 pandemic and possibilities of their measurement—Case study. *Sustainability, 14*(1), Article 127.
22. Kraft, C., Lindeque, J. P., & Peter, M. K. (2022). The digital transformation of Swiss small and medium-sized enterprises: Insights from digital tool adoption. *Journal of Strategy and Management, 15*(3), 468–494.

Avradeep Ganguly is having more than 14 years of academic experience in Human Resource Management and have completed Bachelor of Engineering, P.G.D.M. and MBA and pursuing Research in Human Resource Management from Adamas University, Barasat, Kolkata. He has done his Post Graduation Project from Indian Oil Corporation Limited (IOCL), Pipeline Division, Mourigram. Presently, he is associated with School of Business, Adamas University, Barasat, Kolkata as Assistant Professor. He has publications in International and National Journals, Book Chapters etc. He has been extensively teaching Human Resource Management, Organizational Behaviour, Industrial relations and labour laws to the Management Students.

Gouranga Patra is a senior management educator, researcher having more than 16 years of experience in teaching and Industry out of which 10 years of Post Ph.D. teaching experience. Worked more than 8 years as an Assistant Professor and 5 years as an Associate Professor in Universities and Autonomous Institutions. In research front, published 42 research articles in different reputed journals and edited books and presented 22 papers in national and international conferences in the area of marketing and general management and Human Resource Management. His research supervision has successfully awarded 2 Ph.D. scholars and 8 are under progress. He has authored two books published by Springer Nature and Lambart Germeny.

Chapter 9
Empowering Financial Access: Unlocking the Power of Financial Inclusion in West Bengal Amidst the COVID-19 Pandemic

Arna Ray

Abstract India's recent entry into her $3 trillion market capitalization club in 2021 highlights the importance of financial inclusion in unlocking economic potential. We know that capital markets are the 'mirror' of the economy. If that is true, our economy is now a trillion-dollar economy, and everyone in the country can benefit equally. Further, financial inclusion is closely linked to the trillion-dollar club. Financial Inclusion which primarily includes access, usage and availability, gives any country, region and ultimately an individual the basic needs required for a healthy lifestyle. The study is aimed at analyzing the impact of financial inclusion on development of an economy, socio-economic empowerment and poverty. The study data were collected from four large banks from the state of West Bengal. The study uses correlation regression and variance analysis which is utilized to examine the hypotheses and is descriptive and evaluative in nature. With a Pearson correlation coefficient of 0.220, the findings demonstrated a favourable correlation between economic development and financial inclusion. Additionally, the findings show the relation among economic development and Financially included is not influenced by empowerment led by overall society. Additionally, the findings suggest that being financially included possesses an effective impact on reduction of poverty, spanning all over socio-economic profiles, it is found. Morgan Stanley expects benchmark the Sensex to surpass 1 lakh-mark by 2028, making the Indian stock market one of the world' top five stock markets, with a market capitalization of $6.1 trillion. Here, the very paper does appropriate addition to the already available studies done so far, by providing numerical facts, regarding financial inclusion rightly being one of the most important factors for overall growth of any economy.

Keywords Financial inclusion · Economic growth · Socio-economic empowerment · Poverty reduction · Evaluative study · Unmediated connection · Beneficial effect · Socio-economic profiles · Empirical evidence

---

A. Ray (✉)
Department of Commerce, School of Business and Economics, Adamas University, Kolkata, India
e-mail: arnaray155@gmail.com

S. Gangopadhyay et al. (eds.), *Pandemic Diaries*,
https://doi.org/10.1007/978-981-96-5415-4_9

9.1 Introduction

Financial inclusion is an important component in society, aiming to deliver services that aim to provide people with improved access to financial services in order to improve their overall wellbeing and quality of life [1, 2]. This paper will analyze the current state of financial inclusion and discuss potential challenge.

Financial inclusion is the ability to both access and use financial services such as credit and savings accounts, insurance and other forms of financial management [3]. Access to this service is essential for many of the world's under deserved and financially not included populations. Without access to these services, individuals and families are unable to access or participate in basic economic systems that are necessary for economic growth and stability [4].

Financial inclusion is something which plays a vital role in the society aiming to enhance every individual's access to the basic needs that is financial services to improve their overall wellbeing and quality of life [1, 2]. This study crucially evaluates the very current state of being financially included and addresses the potential challenges which explore strategies for future road ahead. Financial inclusion is something which refers to the ability to access and utilize financial services including credit savings accounts, insurance and financial management tools [3]. These are the services which are crucial for under deserved and financially excluded population worldwide without the access, inclusion, availability individuals and families are excluded from basic economic systems which are essential for economic growth and stability for any country [4]. Financial inclusion is something which serves as a major role in nations which are already developed and nations which are still developing, according to figures from global Findex and the World Bank from 2014, World's adult population's 48% lacks a basic account in bank under an organized financial system. Adults without bank accounts make about 21% of the global population. The World Bank's universal financial access (UFA) initiate was aimed to bring everyone into the formal financial system by 2020, focusing on 25 countries including India [5–7]. The goal is to provide maximum men and women with equal access to account so that they can manage their own finances [8]. To execute this job, UFO has teamed with a range of banks, microfinance institutions, credit card firms and non-governmental organizations (NGOs). According to economists, 23 lawmakers, regulators and charities all across the world, where financial inclusion is a critical enabler of people's economic and social progress. According to the Reserve Bank of India RBI, most current statistics over 41% of the Indian families are unbanked, which does means they do not have access to any formal financial institution for more than 60% of the country's unbanked population lives in rural regions [9, 10]. The literature review highlights are multifaceted connection which links between being financially included and society being economically empowered and overall growth of the economy which further draws light on both global and local perspectives.

HR Khan in the year of 2005 had provided a comprehensive analysis of global financial inclusion initiatives, which throws the light on the Grameen Bank Model microfinance in Brazil, the Philippines, Indonesia, Kenya and South Africa; these programmes illustrated diverse strategies which were for extending banking services to the under deserved populations [4]. In India, despite the numerous initiatives taken by the government, rural areas continue to face cover under service first step Khan highlighted the role of facilitators like NGOs and cooperatives in enhancing financial service delivery and which mitigated malpractices through formal agreements. Models such as Business Facilitator and Business Correspondent were recommended to strengthen rural credit, insurance and savings [11].

Amartya Sen in 2000, highlighted the broader issue of social exclusion, linking financial access with various dimensions like gender inequality and lack of education. He further argued addressing financial exclusion requires tackling these interconnected issues to foster economic participation and empowerment especially for women [12].

The C Rangarajan committee in 2008 investigated the financial inclusion by analyzing both the demand and supply side challenges. There were some recommendations such as the Mahatma Gandhi National Rural Employment Guarantee Scheme and Jan Dhan Yojana were implemented while others were remaining pending [13, 14].

There are recent studies where firms are developing the links between financial inclusion and economic growth in his study by Beck et al. [11] and Demirguc-Kunt and Klapper [12], where the study discovered that financial inclusion basically fosters between the economic growth by improving credit access savings and entrepreneurship [15, 16].

Kumar 2024 further noted that financial exclusion pauses the basic economic participation in national income [17]. While there are reports from the Reserve Bank of India 2023 reported that development for West Bengal is crippled by inadequate financial access [18].

The financial exclusion in India shows disproportionately impacting rural populations and women in a study by Rao et al. (2023) where he further mentioned this inappropriate distribution of services further limits economic opportunities and growth [20]. With national policy being the objective for balance region and local development with RBI, in recent years financial inclusion gained a high momentum. RBI is providing policy recommendations for bank and other developmental initiatives. Thus, this study establishes the following financial goals:

I. To access how empowerment via socio-economic means affects the relationship between being financially included and growth economically.
II. To evaluate the impact of being financially included on poverty being reduced among various socio-economic groups.

9.2 Need of the Study

The need of this paper is to foster balanced regional upliftment and economic stability. The need arises from understanding its impact as financial inclusion becoming a central focus of national policies and RBI recommendations.

Objective I explore the mediating link among being financially included and economic growth and how empowerment social economic wise mediates the same.

Objective II examines the effect of being financially included across different backgrounds on poverty reduction. Analysis is vital as to assess how, and whether access to finished services can alleviate poverty across different backgrounds. Both objective aims to provide in-depth understanding of financial inclusion, role in economic growth and property reduction, thereby informing effective policies and practices.

9.3 Hypotheses Formulation

Based on extensive literature review, the proposed research will evaluate and validate the following hypothesis: financial inclusion, method, for making sure that everyone in society can easily access and utilize the official financial system, (Sharma & Pais, 2011). This notion contributes to a more inclusive financial system, highlighting a number of factors like financial system, accessibility, availability and use (Rao & Bhatnagar, 53 2012). Financial inclusion standards which were developed by the Allison for Financial Inclusion, (2011) contain a more comprehensive set of access and consumption data (Ledgerwood, 2013).

Based on the above objectives, the following non-hypothesis are proposed:

Ho 1: The link between being financially included and development economic wise, the social economic empowerment has no media effect.

Ho 2: Financial inclusion has no impact on poverty reduction across the socio-economic background.

9.4 Methodology of the Study

This study is done with the help of primary survey which was conducted involving 300 respondents randomly selected from various districts across West Bengal. To ensure representative sampling stratified sampling method was employed. The respondents were categorized based on demographic and geographic area. This mythology aims to capture diverse perspectives and all relevant sub groups. The data which was collected from the survey are allies using principal component analysis (PCA), SPSS and Stata software, which enhance a comprehensive and reliable examination of key factors impacting financial services accessibility in West Bengal. By focusing on the most significant components affecting the banking access, PCA helps in reducing data dimensionality and uncovering further underlying patterns.

9.5 Analysis and Findings

Factorial analysis has been done to identify the exact dependent variables of the study. Correlation has also been done to measure the association that tests whether a relationship exists between two variables. Once the dependent variables were identified, the hypothesis was tested using variance analysis to find statistical differences among the variables used.

Socio-economic empowerment has no conciliate effect on the relationship between FI and development economic wise

Interpretation: The above correlation Table 9.1 indicates that there was non-significant correlations were observed between i.e. Financial Inclusion with Social Empowerment (0.301), Financial Inclusion with Economic Empowerment (0.209), Financial Inclusion with Economic Development (0.401), Social Empowerment with Economic Empowerment (0.207), Social Empowerment with Economic Development (0.168), Economic Empowerment with Economic Development (0.196) which fulfil our hypothesis is rejected that socio-economic empowerment has no conciliate effect on the link among FI and economic development.

So, we conclude that socio-economic empowerment has conciliate effect on the link among FI and economic development.

Interpretation: For analysis and interpretation, we are just concerned with the initial eigenvalues and extracted in Table 9.2 sums of squared loadings. Variables having eigenvalues which are greater than 1, those components would be impacting. The values of the first component are 13.4090 > 1, the second component is 6.3530 > 1, the third component is 2.4580 > 1, the fourth component is 1.7720 > 1, and components five through twenty-six have values below 1.

Table 9.1 Socio-economic empowerment has no conciliate effect on the relationship between FI and development economic wise

		FI	Social empowerment	Economic empowerment	Economic development
Financial inclusion	Pearson Correlation	1	0.301	0.209	0.401
	N	301	300	300	300
Social empowerment	Pearson Correlation	0.301	1	0.207	0.168
	N	300	300	300	300
Economic empowerment	Pearson Correlation	0.209	0.207	1	0.196
	N	300	300	300	300
Economic development	Pearson Correlation	0.401	0.168	0.196	1
	N	300	300	300	300

Table 9.2 PCA of socio-economic empowerment

Component	Initial eigenvalue			Extraction sums of squared loading			Rotation sums of squared loading		
	Total	% of Variances	Cumulative percentage	Total	% of Variances	Cumulative percentage	Total	Percentage of variances	Cumulative percentage
FI has made you independent in decision making	13.409	51.572	51.572	13.409	51.572	51.572	9.219	35.457	35.457
FI led to improved lifestyle & studies	6.353	24.436	76.008	6.353	24.436	76.008	7.833	30.128	65.585
FI has altered your personality and way of living	2.458	9.455	85.463	2.458	9.455	85.463	4.431	17.041	82.626
FI has made you socially more reputed	1.772	6.814	92.277	1.772	6.814	92.277	2.509	9.651	92.277
FI has increased your confidence level	0.839	3.227	95.504						
FI has improved your technical skills	0.714	2.744	98.248						
Your family supports your business decisions	0.340	1.306	99.554						
FI affects social status in a positive manner (buying of car, tour, refrigerator, etc.)	0.116	0.446	100.000						
New house constructed after covering under FI drive	9.488E-016	3.649E-015	100.000						

(continued)

Table 9.2 (continued)

Component	Initial eigenvalue			Extraction sums of squared loading			Rotation sums of squared loading		
	Total	% of Variances	Cumulative percentage	Total	% of Variances	Cumulative percentage	Total	Percentage of variances	Cumulative percentage
FI increased your self-assurance, improved your work relationships, and decreased family conflict and societal violence	1.287E-016	4.948E-016	100.000						
Your health and home hygiene have both improved as a result of FI	4.251E-017	1.635E-016	100.000						
Often meeting with people from different social status backgrounds from owns helps	3.618E-017	1.391E-016	100.000						
You would find it challenging to socialize with individuals from many social groups if you do not have FI	2.442E-017	9.392E-017	100.000						

(continued)

Table 9.2 (continued)

Component	Initial eigenvalue			Extraction sums of squared loading			Rotation sums of squared loading		
	Total	% of Variances	Cumulative percentage	Total	% of Variances	Cumulative percentage	Total	Percentage of variances	Cumulative percentage
You like to change your lifestyle	1.152E-017	4.431E-017	100.000						
You participate in any community activity	5.980E-018	2.300E-017	100.000						
You may simply bring about any change in society	4.682E-018	1.801E-017	100.000						
You vote in the general assembly on a regular basis	5.252E-034	2.020E-033	100.000						
When it comes to choosing which candidate to vote for in an election, you are affected by others	2.970E-034	1.142E-033	100.000						
You complained to the authorities about the way financial services were delivered	−1.277E-033	−4.913E-033	100.000						

(continued)

Table 9.2 (continued)

Component	Initial eigenvalue			Extraction sums of squared loading			Rotation sums of squared loading		
	Total	% of Variances	Cumulative percentage	Total	% of Variances	Cumulative percentage	Total	Percentage of variances	Cumulative percentage
You now have the freedom to go anywhere you want in public thanks to FI	−2.251E-018	−8.657E-018	100.000						
You are aware of all government-sponsored special programmes	-4.989E-018	-1.919E-017	100.000						
You take use of the government's unique programmes	−8.491E-018	−3.266E-017	100.000						
Being free to go to any SHG/ any other organization for getting any kind of help/support	−1.266E-017	−4.871E-017	100.000						
You are free to seek assistance or support from any NGO or other organization	−1.769E-017	−6.806E-017	100.000						

(continued)

Table 9.2 (continued)

Component	Initial eigenvalue			Extraction sums of squared loading			Rotation sums of squared loading		
	Total	% of Variances	Cumulative percentage	Total	% of Variances	Cumulative percentage	Total	Percentage of variances	Cumulative percentage
Regarding any financial problem, your comments and criticism are always welcomed	−2.509E-017	−9.652E-017	100.000						
After being covered by the FI drive, your social development has improved	−6.535E-017	−2.514E-016	100.000						

Method: Principal Component Analysis

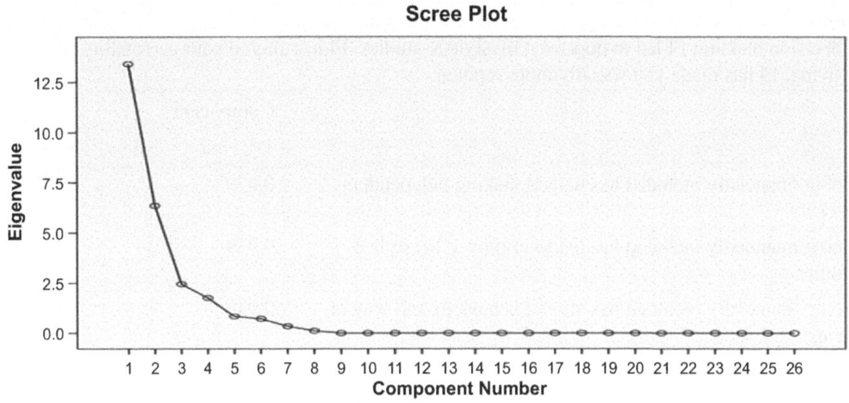

Fig. 9.1 Scree plot of component

About 51.572% of the total of variance is explained by the first component. Second factor impacts 24.436% of the total variance. About 9.455% of the overall variation is explained by the third component. The fourth factor accounts for roughly 6.814% of the overall variation. It is clear that the first component had the highest variance, which was followed by the 2nd component, with the other components showing a similar pattern towards declining variance. To create uncorrelated factor structures, the orthogonal varimax technique was used (Fig. 9.1).

Interpretation: Cattell's Scree test helps deciding further on how many components to be taken further for the study via graphical figure. Here 1st four factors to be considered as 4th component onwards the eigenvalue is less than 1 and the curve in between 3 & 4 component number, the curve flattens as well.

Interpretation: Component 1, in the above Table 9.3 which is highly correlated (0.939) the 9-initial variable.

Component 2 had a strong correlation (0.945) with the first four variables (FI enhanced connections at work, higher self-assurance, reduced family strife and social violence, As a result of financial inclusion, your health and household hygiene have both improved. Outside of your home, you frequently interact with people from many social groups regarding financial inclusion; without financial inclusion, it would be difficult for you to socialize with people from many different social groups.

You complained to the authorities about the way financial services were provided; you can now go anywhere you want in public thanks to FI; you are aware of all government-sponsored special programmes; and you use the government's special programmes. Component 3 has four highly correlated (0.941) variables.

You are free to seek support or aid from any NGO or other organization, and your views and criticism are always welcome regarding any financial issues, according to Component 4's two highly associated (0.846) variables.

Interpretation: A second time, the Component Transformation Matrix shown in Table 9.4 shows the correlations between the components in the preceding table before and after rotation. The first component is highly connected with the first

Table 9.3 Rotated component matrix between the 4 components i.e. FI has made you independent in decision making; FI led to improved lifestyle & studies; FI has altered your personality and way of living; FI has made you socially more reputed

	Component			
	1	2	3	4
Being financially included has helped making independent decision	0.939			
Being financially included has led to improved lifestyle & studies	0.939			
Being financially included has altered behaviour and way of living	0.939			
Being financially included has made socially more reputed	0.939			
Being financially included has increased confidence level	0.939			
Being financially included has improved technical skills	0.939			
Your family supports business decisions	0.939			
Being financially included social status is increased in a positive manner (purchase of car, land, tour, travels, T.V., A.C., etc.)	0.939			
New house constructed after covering under FI drive	0.939			
FI increased self-assurance, improved your work relationships, and decreased family conflict and societal violence		0.945		
Health and home hygiene have both improved as a result of FI		0.945		
Often meeting and converse with people from diverse social groups about financial inclusion outside of your home		0.945		
Find it challenging to socialize with individuals from many social groups if you do not have FI		0.945		
Liking to change your lifestyle		0.867		
Participate in any community activity		0.867		
Simply bring about any change in society		0.867		
Vote in the general assembly on a regular basis		0.867		
Choosing which candidate to vote for in an election, you are affected by others		0.654		
Complained to the authorities about the way financial services were delivered			0.941	
Have the freedom to go anywhere you want in public thanks to FI			0.941	
Are aware of all government-sponsored special programmes			0.941	
Taking use of the government's unique programmes			0.941	
Free to go to any SHG or other organization for any kind of assistance or support			0.607	
Free to seek assistance or support from any NGO or other organization				0.846
Regarding any financial problem, your comments and criticism are always welcomed				0.846

(continued)

Table 9.3 (continued)

	Component			
	1	2	3	4
After being covered by the FI drive, your social development has improved				−0.603
Method: Principal Component Analysis Rotation Method: Varimax with Kaiser Normalization				
a. Rotation converged in 6 iterations				

(0.766), the second with the second (0.735), the third with the third (0.720) and the fourth with the third (0.408).

FI has no impact on poverty reduction across socio-economic Profile

Interpretation: Above in Table 9.5 correlation value between across socio-economic profile in which financial inclusion with poverty (0.978) which was non-significant which fulfil our hypothesis is rejected, FI has no impact on poverty reduction across socio-economic profile.

Table 9.4 Component transformation matrix between the 4 components i.e. FI has made you independent in decision making; FI led to improved lifestyle & studies; FI has altered your personality and way of living; FI has made you socially more reputed

Component	1	2	3	4
1	0.766	−0.547	−0.248	0.231
2	0.336	0.735	−0.589	−0.007
3	0.395	0.400	0.720	0.408
4	0.380	0.036	0.272	−0.883
Method: PCA Rotation Method; Varimax using Kaiser Normalization				

Table 9.5 Correlation analysis showing FI has no impact on poverty reduction across socio-economic Profile

Control variables			Financial inclusion	Poverty
Socio-economic profile	Financila Inclusion	Correlation	1.000	0.978
		Significance (2-tailed)		0.000
		Df	0	296
	Poverty	Correlation	0.978	1.000
		Significance (2-tailed)	0.000	
		Df	296	0

So, we conclude that being financially included does affects hugely in reducing poverty among various classes of the society.

Being Financially Included has zero impact on social empowerment

Interpretation: A type of statistical study called regression looks at the relationship between more than two variables. Table 9.6 displays the r square (0.678), which demonstrates that 67% of our model performed well for the variable being evaluated. The dependent and independent variables accounted for 67% of the variation and the residual or error accounted for the remaining 33%. Similarly, Adj. R square shows how the dependent variable changes as a result of those independent factors that have a significant impact on the dependent variable. Because of this, the best indicator of the model's choice or goodness of fit was the Adjusted R Square.

Interpretation: Table's 9.7 ANOVA table for regression analysis showed that the regression's mean square (0.220) and residual's mean square (0.077) had an impact on the Cal F value (2.857), significance level of 0.109. So we conclude that our null hypothesis is rejected because P value being greater than 0.05. The f test has overall an impact on dependent variable, such as social empowerment (Table 9.8).

Interpretation: Table 9.8's Financial inclusion (0.166) had no effect on the dependent variable i.e. social empowerment. Due to its non-significant t test value.

The above regression analysis results which fulfil our hypothesis are rejected that FI has no direct impact on social empowerment.

So, we conclude that being financially included has effect on social empowerment.

FI has no direct impact on economic empowerment

Interpretation: A type of statistical study called regression looks at the relationship between more than two variables. Table 9.9 displays the r square (0.451), which demonstrates that 45% of our model performed well for the variable being evaluated. The dependent and independent variables accounted for 45% of the variation and the residual or error accounted for the remaining 55%. The best indicator of the model's choice or goodness of fit was the Adjusted R Square.

Table 9.6 Model Summary of regression analysis showing Being Financially Included has zero impact on social empowerment:

Mode	R	R Square	Adjusted R Square	Std. Error of the estimate
linear	0.823	0.678	0.677	0.27728

Table 9.7 ANOVA of regression analysis showing overall impact on dependent variable, social empowerment:

Model	Sum of Squares	Df	Mean Square	F	Sig.
Regression	0.220	1	0.220	2.857	0.109
Residual	22.911	298	0.077		
Total	71.131	299			

Table 9.8 Coefficients of regression analysis showing FI has no direct impact on social empowerment

Model	Unstandardized coefficients		Standardized coefficients	t.	Sig.
	B	Std. Error	Beta		
(Constant)	0.777	0.095		8.217	0.000
Financial inclusion	0.005	0.030	0.823	0.166	0.134
Dependent Variable: Social Empowerment					

Table 9.9 Model Summary of regression analysis showing FI has no direct impact on economic empowerment

Model	R	R Square	Adjusted R Square	Std. Error of the estimate
Linear	0.671	0.451	0.449	0.36685

Interpretation: Table's 9.10 ANOVA table for regression analysis showed that the regression's mean square (0.0888) and residual's mean square (0.135) had an impact on the Cal F value (0.65770), which was non-significant at a level of significance of 00.958. So we conclude that our null hypothesis is rejected because P value is greater than 0.05 (Table 9.11).

Then f test shows the overall impact of the dependent variable, such as economic empowerment.

Table 9.10 ANOVA of regression analysis showing overall impact on the dependent variable, economic empowerment

Model	Sum of Squares	df	Mean Square	F	Sig.
Regression	0.0888	1	0.0888	0.6577	0.958
Residual	40.105	298	0.135		
Total	72.993	299			

Table 9.11 Coefficients of regression analysis showing FI has direct impact on economic empowerment

Model		Unstandardized coefficients		Standardized coefficient	t	Sig.
		B	Std. Error	Beta		
Linear	Constant	1.143	0.125		09.139	0.0000
	FI	0.024	0.040	0.671	0.60	0.128
Dependent Variable: Economic Empowerment						

Interpretation: Financial inclusion (0.671) had no significant impact on the dependent variable i.e. economic empowerment. Due to its non-significant t test value (Table 9.12).

Above regression results fulfil our hypothesis is rejected that FI has no direct impact on economic empowerment.

So, we conclude that FI has direct impact on economic empowerment.

PCA of Poverty

Interpretation: For analysis and interpretation, to figure out how many components or factors are expressed by a specific collection of variables, there must be eigenvalues greater than one. The values of the 1st component are 6.266 > 1, the 2nd component 3.156 > 1, the 3rd component 1.159 > 1, and the values of components 4 through 11 are below 1.

About 56.961% of the total variance is explained by the 1st component. Around 28.687% of the total variance is explained by the second component. A 10.359% portion of the overall variation is explained by the third component. It is clear that the first component had the highest variance, which was followed by the 2nd component, with the other components showing a similar pattern towards declining variance. To create uncorrelated factor structures, the orthogonal varimax technique was used (Fig. 9.2).

Interpretation: Cattell's Scree test helps deciding further on how many components to be taken further for the study via graphical figure. Here 1st three factors to be considered as 3rd component onwards the eigenvalue is less than 1 and the curve in between 2 & 3 component number, the curve flattens as well.

Interpretation: Interpretation: Table 9.13, the family's head is educated enough to guide the other members in the right direction, and the majority of the family members are wage laborers or earners, according to Component 1, which has a high correlation (0.973) with the three original variables was more akin to destitution. You now eat higher-quality meals than you did in the past, you spend more money on luxury items, eating high-quality meals have improved one's health, and you have boosted your consumption level. Component 2 was strongly linked (0.926) with these four original factors. The variables in Component 3 (Familycrisis are minimized by betterliving standards and Overall, FI has helped to lessen poverty) are highly associated (0.950).

Interpretation: Again, in the above Table 9.14, the Component Transformation Matrix shows the correlations between the components before and after rotation in the aforementioned table, with the first component highly correlated with first (0.692), second with third (0.538) and third with third (0.780).

Table 9.12 Total variance is explained

Components	Initial eigenvalues			Extraction sums of squared loading			Rotation sums of squared loadings		
	Total	Percentage of variance	Cumulative percentage	Total	Percentage of variance	Cumulative percentage	Total	Percentage of variance	Cumulative percentage
Most members are educated in your family	6.266	56.961	56.961	6.266	56.961	56.961	4.439	40.357	40.357
Most of the members are wage labourers or earners	3.156	28.687	85.647	3.156	28.687	85.647	4.080	37.093	77.450
The family's head is well-educated enough to lead the other members in the proper direction	1.159	10.539	96.187	1.159	10.539	96.187	2.061	18.737	96.187
Your home's value has grown as a result of FI	0.419	3.813	100.000						
Your clothes spending has risen	2.366E-016	2.151E-015	100.000						
You eat higher-quality meals than previously	1.078E-016	9.796E-016	100.000						
The amount of money you spend on luxury has grown	5.752E-017	5.229E-016	100.000						
Having high-quality meals has enhanced one's health	4.300E-018	3.909E-017	100.000						
Your consumption level has increased	−1.003E-019	−9.118E-019	100.000						
Family crisis reduced through better living standard	−1.207E-017	−1.098E-016	100.000						
Overall, FI has helped to lessen poverty	−2.671E-016	−2.428E-015	100.000						
Method: PCA									

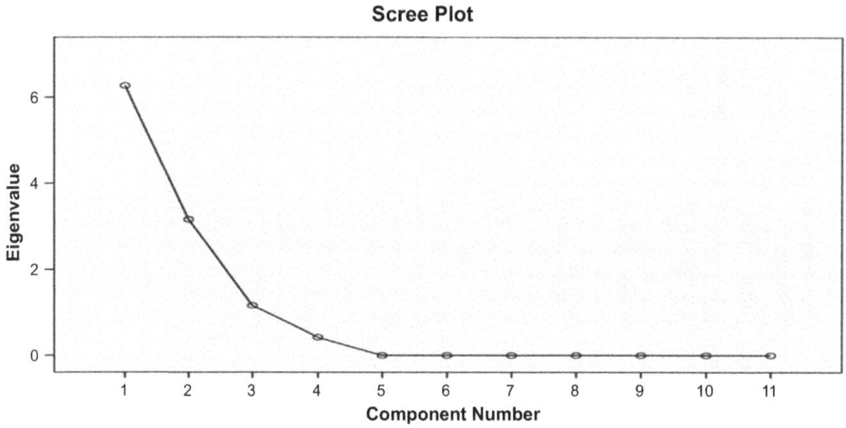

Fig. 9.2 Scree plot of component

Table 9.13 Rotated component matrix between the 3 components, i.e. Most members are educated in your family; Most of the members are wage laborers or earners; The family's head is well-educated enough to lead the other members in the proper direction

	Component		
	1	2	3
Most members are educated in your family	0.974		
Most members are wage labourers or earners	0.974		
The family's head is well-educated enough to lead the other members in the proper direction	0.974		
Your home's value has grown as a result of FI	0.831		
Your clothes spending has risen	0.831		
You eat higher-quality meals than previously		0.946	
The amount of money you spend on luxury has grown		0.946	
Having high-quality meals has enhanced one's health		0.946	
Increased level of Consumption		0.946	
Family crisis are reduced through better living standard			0.950
Overall, FI helped to lessen poverty			0.950
Method: PCA Rotation Method: Varimax using Kaiser Normalization			
a. Rotation converged in 5 iterations			

Table 9.14 Component transformation matrix between the 3 components, i.e. Most members are educated in your family; Most of the members are wage laborers or earners; The family's head is well-educated enough to lead the other members in the proper direction

Component	1	2	3
1	0.692	−0.695	0.195
2	0.647	0.477	−0.595
3	0.321	0.538	0.780

Method: PCA
Rotation Method; Varimax using Kaiser Normalization

9.6 Conclusion and Road Ahead

According to a recent survey, as of a study of 2012 data only 48% of global adult population have the access to basic banking facilities for improving any individuals which would further facilitate economic stability, financial inclusion, please a primary role in which ensuring access to basic financial services is the first step. The initiatives were taken by the World Bank's Universal Bank Financial Access (UFA) to achieve financial integrity by 2020. However, the progress has been uneven, especially in India as over 41% of the households are unbanked further with more than 60% of these households residing in rural areas [16].

Addressing the financial inclusion is fatal as it is essential for economic and social advancement. Individuals' families are excluded from basic economic activities without access to financial services, impending their economic stability and growth. One of the effective solutions requires collaboration about banks, microfinance institutions, credit card companies, NGOs, lawmakers and charities to open barriers to financial access.

The aim of the study is to see, socio-economic empowerment and property reduction, particularly in West Bengal, which is a region with noble financial exclusion issues. A descriptive and evaluative approach was considered, gathering data from four large banks in tools, different states that had received financial inclusion grants. Techniques like principal component analysis, (PCA) were also used to identify key factors in influencing access to financial services and their impacts. Thus, this study shows that while financial inclusion is, it does not necessarily lead to enhanced social-economic empowerment via economic progress in the study area. Further, the impact of being financially included and reduction of poverty was at minimum level across various social economic profiles.

This study further shows that there is zero significant effect on being financially included over social empowerment despite exceptions that financial inclusion did not produce substantial improvements in social empowerment within the surveyed regions meaning that had there been financial inclusion on this region there would have been living a better life resulting into better socio-economic conditions. There were various camps which have taken place in many states of India, especially West Bengal as it was rightly claimed as the Ponzi capital of India because the maximum number of skins were operational in the state of West Bengal. People win in expectation of more returns than normal, usually puts their hard and money into this Ponzi schemes which are actually frauds, and this is the result of not being financially literate as there is a lakh from the side of government in an attempt to financially include these people into formal structures and hence making a part of financially included. The scam schemes like Sharda, Rose Valley were highly operational in West Bengal, even a few years back and they have scammed thousands of crores of hard and money of the people of West Bengal and which money has eventually gone out of the system. Had these people had the minimum knowledge of landing their hard-earned money into the formal financial institutions, like banks, capital market, this money would had been with those people. The savings of those people would have been intact. The very complex nature of financial inclusion and its large down effect on socio-economic development has been established through the study which emphasizes on the need for understanding the need of financial inclusion in our daily life as a basic part of literacy financial literacy would not only provide financial inclusion at large and better economic development for the country but daily smaller needs for individuals would be taken care very easily by themselves promoting to economic development at large being financially educated would help resident individuals channelizing their earned money into proper formal financial institutions rather than getting into frauds and losing their money. Only the government can stop this Ponzi schemes from looting peoples saving by making them financially included which is making them part of the access uses in availability of financial services and making the large mass of the population, financial illiterate by which we hear by suggest on providing the basic mere financial knowledge to the mass like opening a bank account, opening a savings account, opening a Demat account and making the mass understand the essence of savings and getting interest on that savings money and how to grow that money and making them understand the normal rate of returns because when they are promised with some exorbitant rate of returns by those Ponzi schemes, they are easily learned by the high rate of return and they land their entire savings into those Ponzi schemes, which can be prevented only by the government and watch dogs like SEBI, RBI are not enough for this huge programme which includes a huge mass as maximum people are not financially literate enough and still a lot. A huge mass is not yet financially included in a country like ours. Especially the youth needs to be highly involved in this financially inclusion movement as they are the future of the country and the savings being placed in right instruments can immensely change the future of not only that individual resident, but the countries, economic growth picture overall. Proper curriculum being included about financial literacy can help this by a significant amount and constant financial awareness camps organized by the government can help the mass putting their savings into the right places where they can get safety security along with proper rate of returns. For this the government needs more proactive measures like implementing awareness campaigns, frequent television radio advertisements reminding people to gather under the umbrella of organized financial services, including basic financial literacy knowledges into the curriculum of the education system for the students can facilitate a huge change into developing a very sound financially literate population for other country.

References

1. Bagli, S., & Dutta, P. (2012). A study of Financial Inclusion in India. *Radix International Journal of Economics & Business Management, 1*(8), 1–18.
2. Chattopadhyay, S. K. (2011). *Financial inclusion in India: A case-study of West Bengal.*
3. Garg, B. (2014). Financial inclusion and rural development. *Research Journal of Commerce, 2*(1), 1–6.
4. Jena, A. B. (2015). The Pradhan Mantri Jan Dhan Yojana (PMDJY)-A forward journey towards financial inclusion. *Asian Journal of Multidimensional Research (AJMR), 4*(4), 34–45.
5. Kalunda, E. (2014). Financial inclusion impact on small-scale tea farmers in Nyeri county, Kenya. *World Journal of Social Sciences, 4*(1), 130–139.
6. Kothari, C. (2005). *Research methodology* (Rev. ed.). books.google.co.in/books
7. Laha, A., Kuri, D., & Kumar, P. (2011). Determinants of financial inclusion: A study of some selected districts of West Bengal, India. *Indian Journal of Finance, 5*(8), 29–36.
8. Murari, K., & Didwania, M. (2010). *Poverty alleviation through financial inclusion: An analytical study with special reference to India.*
9. Sen, A. (2000). *Social exclusion: Concept, application and scrutiny.* Social Development Paper/ Asian Development Bank.
10. Mukhopadhyay, J. P. (2016). Financial inclusion in India: A demand-side approach. *Economic and Political Weekly, 51*(49), 46–54.
11. Barajas, A., Beck, T. H. L., Belhaj, M., & Naceur, S. B. (2022). *Financial inclusion* (p. 121). Oxford University Press.
12. Demirgüç-Kunt, A., Klapper, L., Singer, D., & Ansar, S. (2022). *The Global Findex Database 2021: Financial inclusion, digital payments, and resilience in the age of COVID-19.* World Bank Publications.
13. Bhattacharya, S., Das, B. K., Roy, A., Nayak, P. K., Saha, A., Parida, P. K., Lianthuamluaia, L., Mondal, K., & Chakraborty, S. (2024). Socio-economic profile and vulnerabilities assessment in small-scale fisheries of lower Gangetic Flood Plain Wetland: An initiative towards achieving Wetland ecosystem sustainability and community well-being. *Sustainability, 16*(21), 9583.
14. Kumar, I. (2024). Banking services and financial inclusion in India's poorest regions. *Journal of Banking Regulation, 25*(2), 145–159.
15. Rao, S., & Ramnarain, S. (2023). Gender, social protection, and crises of social reproduction: Contextualizing NREGA. *Review of Radical Political Economics, 55*(1), 70–92.
16. World Bank Data. World Development Indicators. World Bank. https://databank.worldbank.org/source/world-development-indicators
17. Kumar, B. (2024). Understanding financial inclusion and Its socio-economic determinants: evidence from India. *SAGE Open, 14*(1), 21582440241290372.
18. Reserve Bank of India. (2023). Annual Report 2022–23. Retrieved from https://www.rbi.org.in/Scripts/AnnualReportPublications.aspx

Arna Ray is a dedicated academician completed her under graduate & Post Graduate degrees from University of Calcutta, later got her degree in Investment Portfolio Services from Calcutta University, as well, with over five years of teaching experience, specializing in Accounting and Finance. Currently she is a College Teacher in Vidyanagar College, South 24 Parganas, she is pursuing her Ph.D. in Financial Inclusion from Adamas University, Barasat. She has presented papers at numerous national and international conferences, including Annamalai University and the ICAS Conference and many more with publications in peer reviewed journals, edited volumes, book chapters, etc. Her research explores financial inclusion, fintech, and economic growth, particularly in West Bengal. Arna is skilled in research methodologies, financial analysis, and data management. She has participated in numerous National & International seminars, workshops, conferences, webinars and co-ordinated & hosted seminars, workshops on CSR and economic trends, reflecting her commitment to academic dedication.

Chapter 10
Case Study: Study of Sustainable HRM with an Example of Nokia During COVID-19

Amrita Dutt, Aman Chadha, Akriti Gupta, and Vijayshri Tewari

Abstract Sustainability has been an important issue due to the changing environmental conditions. Sustainability can be defined as managing the resources such that they not only suffice the need for today but also for the future generations. Now, maximum companies have their strategies in coherence with the sustainability. This paper takes a panoramic view of one of the most iconic communication company, Nokia. Nokia has seen it all, as it states its own dramatic story of being at the top thereby guiding the world to experience the extravagant era of mobile phones, and then being left out in the technological race and then fighting back to survive. With the advent of disruptive technology, it has created chaos in the market scenarios and putting Nokia through tough times, making sustainability quite an issue for the firm. A key paradigm in contemporary organisational strategies is sustainable human resource management (HRM), particularly in light of global disasters like the COVID-19 pandemic. With an emphasis on striking a balance between economic, social, and environmental factors while guaranteeing the long-term welfare of workers and organisational resilience, sustainable HRM incorporates the concepts of sustainability into HR procedures. By addressing both the current organisational needs and the wider effects on society and future generations, this strategy goes beyond typical HRM. Enterprises faced issues like preserving personnel health, guaranteeing productivity during remote work transitions, and navigating economic uncertainties as a result of the COVID-19 pandemic, which caused unprecedented disruptions to enterprises worldwide. This highlighted the significance of sustainable HRM. One of the best examples of how sustainable HRM

A. Dutt (✉) · A. Chadha · A. Gupta · V. Tewari
Department of Management Studies, Indian Institute of Information Technology, Allahabad, India
e-mail: rsm2018502@iiita.ac.in

A. Chadha
e-mail: imp2015002@iiita.ac.in

A. Gupta
e-mail: rsm218503@iiita.ac.in

V. Tewari
e-mail: vijayshri@iiita.ac.in

S. Gangopadhyay et al. (eds.), *Pandemic Diaries*,
https://doi.org/10.1007/978-981-96-5415-4_10

concepts may be used in times of crisis is Nokia, a leading global telecommunications company. Nokia had to implement creative and inclusive HR solutions to handle the particular difficulties brought on by the pandemic because of its operations in more than 130 countries and varied workforce. The company's strategy demonstrates how businesses may promote flexibility, assist workers' physical and mental well-being, and fulfil their corporate social duty in the face of hardship. This case study examines Nokia's COVID-19 sustainable HRM efforts and shows how these practices improved employee well-being and organisational sustainability. By examining Nokia's response, this study aims to provide insights into the practical application of sustainable HRM principles, emphasising their relevance in crisis management and their potential to shape resilient organisational cultures. The analysis also incorporates reports from Nokia's corporate publications and third-party assessments, ensuring a comprehensive understanding of the company's HRM strategies during the pandemic. Through this research, we attempted to investigate the major elements that contributed to the company's demise, including its inability to adapt new technologies and how they adjusted their methods to make a comeback and reclaim its position in the communications sector. This time Nokia played through the rules of survival and very smartly aligned its strategies focusing on sustainability. However, it also emphasises Nokia's amazing turnaround, which entailed a strategic realignment focused on sustainability and innovation. In its rebirth, Nokia took a forward-thinking approach, integrating its operations and goals with sustainability principles. This included attempts to lessen environmental effect, such as increasing product energy efficiency, using circular economy practices, and encouraging digital inclusion through accessible technology.

Keywords Sustainable HRM · Strategy · Nokia · Corporate sustainability

10.1 Introduction

With the advent of the twenty-first century, the business organisations have seen a greater awareness and concern towards the ecological, social, and economic problems thus putting a wider lookout for the concept of corporate sustainability. The concept was coined in the late 80s but the penetration of the concept in the functions of the organisations has been fast. The latest addition to this penetration is the "Sustainable HRM", which tries to implement and connects the concept of the sustainability with the HRM practices of the organisation and how it actually affects the performance of the employees [11]. The business organisations are becoming vigilant and are taking steps in the direction of implementing sustainability at a larger scale like highlighting the importance of Corporate Social Responsibility (CSR) [13], Employee Engagement [3], Talent Management [5], etc.

Now, the growing value of sustainability in the field of HRM has caught the eyes of various leaders in the market. One of such industry leaders is "Nokia". The research had tried to study the dramatic rise and fall of Nokia and then again resurging in the market by making corporate sustainability as the theme line for drafting the strategies. But, as of March 2017 Nokia entered the market to claim for its position. Nokia's journey from 2012 to 2017 has been strenuous.

10.1.1 The Concept of Sustainability

The last decade of the twentieth century saw emergence of the concept of "Sustainable Development" which was described by the World Commission on Environment [4] as "development that meets the needs for the present without compromising the ability of the future generations to meet their own needs". Sustainability tries to help the world evolve into a guided manner for a better future. For the past 150 years, social equity and economic growth had been the major challenges of the society and thus sustainability tries to wrap these issues in its finger and provide a solution [6]. In 1992, post Rio summit, global sustainability became quite inexistent. But, paralleling there was another story that was finding its way i.e. the corporate sustainability. The firms had made it a precondition for doing the businesses [10].

Early studies have shown concern with the preservation and conservation of natural resources or precisely the non-renewable resources, but "Human resources" have received a very little attention. It was only after World War II that the world realised that "this" resource can have shortage. Thus, shifting a little concern over to the humans and how even they need a great deal of attention for sustainability [20].

In this case study, sustainability is defined as the incorporation of practices that guarantee long-term organisational success while attending to employee well-being and the wider societal and environmental implications. It includes three fundamental dimensions:

10.1.2 Economic Sustainability

During the pandemic, Nokia's HR initiatives were created to preserve operational continuity and productivity. The organisation made sure that its personnel remained productive and in line with corporate goals by funding employee development and offering resources for remote work. Cost-effective solutions that reduced interruptions and maximised resource utilisation, such online collaboration platforms and virtual training, were also a part of economic sustainability.

10.1.3 Social Sustainability

During the epidemic, Nokia's strategy was based on social sustainability. By providing flexible work schedules, mental health support, and community building, the organisation puts employee well-being first. Initiatives like recognition programmes and team-building exercises boosted staff morale and engagement. Nokia showed its dedication to diversity and social responsibility by attending to the varied requirements of its employees.

10.1.4 Environmental Sustainability

Although the case study's main focus is on HR procedures, Nokia also ensured that its actions were responsible for the environment. The shift to remote work decreased energy use in offices and during commutes, which helped to reduce carbon emissions. A more sustainable operating model was also aided by Nokia's investment in digital tools, which reduced need on tangible resources and infrastructure.

Nokia's capacity to maintain a balance between these three factors serves as an example of the case study's idea of sustainability. The company's HR strategies were in line with its long-term goals of resilience and ethical business practices in addition to being responsive to the pandemic's immediate issues.

10.1.5 The Emergence of Sustainable HRM

Sustainability focuses on the preservation, regeneration, and development of the ecological, economic, and social resources of the system [21, 22]. Whenever a firm tries to operate in a sustainable way, they need to focus not only the financial but also the social aspects [7]. Therefore, it has also been defined as a "management of human resources that meets the current needs of the firm and society at large without compromising their ability to meet any future needs" [7, 15] and [11] defines it as an extension to the HRM practices from the perspective of sustainability. Sustainability briefly can be defined as the adopting of the HRM strategies and practices in order to achieve the financial, social, and ecological goals that not only impact the organisation from inside but also from the outside with a long-term goal. A part of the definition talks about the complexity of the HRM systems with the internal and external environments of the organisation [8].

There are three approaches that have been suggested. These are:

a. Paradox theory [8]
b. A theory of negative externalities and stakeholder harm [15]
c. Stakeholder theory [10].

The inherent conflicts and tensions that organisations face when attempting to combine conflicting demands in their HRM practices are the main emphasis of the Paradox Theory of Sustainable Human Resource Management (HRM). These conflicts are frequently brought about by the requirement to achieve both immediate performance targets and long-term sustainability goals, such as environmental impact, organisational resilience, and employee well-being. Paradox theory can assist in explaining how Nokia handled and reconciled competing goals in its human resource management procedures during COVID-19, particularly in the midst of a worldwide emergency such as the pandemic.

A theory of negative externalities and stakeholder harm in the context of sustainable human resource management (HRM) focuses on the unintentional, detrimental effects that a company's actions may have on a variety of stakeholders, including the environment, local communities, and employees. These adverse externalities may occur when a corporation disregards the wider effects of its HRM practices or puts short-term objectives ahead of long-term sustainability.

The fundamental concept of the theory is that business decisions, including HRM procedures, may negatively impact stakeholders outside of the company's immediate interests. Organisations risk unintentionally harming workers, their families, communities, or even the environment when they don't implement sustainable HRM practices. Sustainable HRM choices are crucial in the case of Nokia during COVID-19, as the company had to strike a balance between its short-term operational objectives and the possible harm to stakeholders.

According to the Stakeholder Theory of Sustainable Human Resource Management (HRM), businesses have responsibilities to a larger spectrum of stakeholders, including communities, suppliers, workers, shareholders, and the environment. This idea highlights that in order to promote long-term value creation rather than just short-term profitability, organisations must match their HRM practices with the interests and well-being of all stakeholders.

The application of Stakeholder Theory in sustainable HRM is especially pertinent to the Nokia example during the COVID-19 epidemic since the company had to strike a balance between the interests and well-being of several stakeholders and the urgent needs of the business. Let's examine how Nokia's pandemic HRM procedures show sustainability in its HR procedures and relate to stakeholder theory.

These three theories are contributing in understanding the basic layout of the Sustainable HRM and how it is involved in achieving the different goals of the organisation.

The three theories—the Stakeholder Theory of Sustainable HRM, the Paradox Theory of Sustainable HRM, and the Theory of Negative Externalities and Stakeholder Harm—are essential to comprehending how Sustainable Human Resource Management (HRM) is set up and implemented. A foundation for how HRM practices support a variety of organisational objectives, such as social responsibility, environmental sustainability, employee well-being, and financial performance, is provided by these ideas.

10.1.6 Overall Contribution to Understanding Sustainable HRM

These three theories offer a thorough foundation for comprehending sustainable human resource management and how it helps the company accomplish its many objectives. Every theory focuses on a distinct facet of how HRM might be matched with more general organisational goals to guarantee sustainability and long-term success.

1. In order to achieve both short-term and long-term objectives, paradox theory places a strong emphasis on managing contradictions and establishing a balance between conflicting demands. It promotes resilience and flexibility in HR procedures.
2. According to the Theory of Negative Externalities and Stakeholder Harm, HRM should develop an ethical, responsible organisation by reducing harm to communities, workers, and the environment.
3. In order to ensure that HRM practices support workers, clients, suppliers, and the environment in a way that promotes long-term sustainable growth, Stakeholder Theory promotes an all-encompassing strategy that benefits all stakeholders.

These ideas work together to create a framework in which Sustainable HRM helps to develop long-term resilience, moral behaviour, and stakeholder value in addition to meeting urgent organisational demands. Organisations can accomplish their objectives in a way that strikes a balance between social responsibility, environmental sustainability, and economic success thanks to this integrated strategy.

Various researches have been trying to link sustainability with different aspects of HR. As [5] states HR and business leaders have been trying to focus on organisational effectiveness looking beyond the traditional ways of attaining outcomes to sustainability. Thereby, having focus for the sustainable talent management process has become one of the critical evaluating factors for practising sustainability in the organisation [19].

10.1.7 The Emergence of Sustainable HRM

As the relationship between corporate operations, employee welfare, and society effect has become more widely recognised, the idea of sustainable HRM has changed. Achieving short-term organisational goals was the main focus of traditional HRM, which frequently placed an emphasis on cost-effectiveness and efficiency. But the growing emphasis on sustainability around the world—driven by social responsibility, economic equality, and environmental concerns—has resulted in the incorporation of sustainability principles into HR procedures.

Among the main forces behind the development of sustainable HRM are:

1. **Corporate Social Responsibility (CSR):**

Due to the pressure on companies to engage in a socially and ethically responsible manner, HRM practices have been adjusted to align with CSR objectives. This means promoting diversity, equity, and inclusion, supporting community development initiatives, and fostering employee well-being.

2. **Long-Term Workforce Resilience:**

The necessity for resilient workforces has been brought to light by global issues like economic recessions, technological disruptions, and health problems like COVID-19. The goal of sustainable HRM is to equip workers for upcoming difficulties by fostering their mental and physical well-being, adaptability, and ongoing learning.

3. **Environmental Sustainability:**

The growing awareness of environmental problems, like climate change, has led HR strategies to embrace ecological footprint-minimization techniques. This covers guidelines for working remotely, green office projects, and sustainability awareness training courses.

4. **Evolving Employee Expectations:**

Employees today expect businesses to put sustainability first and value employment that has a meaning. In order to meet these expectations, sustainable HRM improves engagement and satisfaction by giving workers meaningful chances to support environmental and societal goals.

Practices that take into account the wider effects on stakeholders and future generations in addition to the immediate organisational benefits are what define sustainable HRM. Organisations can strike a balance between accomplishing corporate goals and promoting a good social and environmental effect by incorporating sustainability into HR policies and initiatives.

Sustainable HRM a Step Ahead of Strategic HRM

As sustainable HRM is a next step after strategic human resource or can be said as a part of strategies under strategic HRM policies made by the organisation to achieve its goal and to improve its performance. There is a relation between strategic HRM and sustainable HRM as when ecological and social benefits are equally considered as economic benefits of organisation then there is a point of correlation between strategic HRM and sustainable HRM arises as strategic [12]. HRM is basically emphasising on planning and making strategies to achieve long-term goals of organisation and for that how resources and people are aligned for this achievement. So the sustainable HRM is a part of strategic HRM because to meet the need of market and to meet the changing environment the organisation needs to adopt new initiatives and needs to add new policies to their company which differentiates them from others [11].

10.2 Brief History of Nokia

Nokia is a multinational communication corporation headquarters in Espoo Finland founded in 1865. Nokia started initially as a producer of paper, cables and rubber and then becoming one of the leaders in the tech world. Nokia has been in the top ten valued brands of the world (Interbrand survey 2010). In 2010, as recorded Nokia had 124 k employees worldwide and is generating 41 billion euros revenues. Nokia has been built into different units where Nokia's Mobile Solutions Unit is responsible for the smartphone and mobile computer portfolio and mobile internet services. Nokia has its tagline "connecting people" and works on this strategy that how company manages to connect through communication technology, its strategy of how people can connect with each other from anywhere through mobile phone and enable people to get more opportunities through mobile phone, initially Nokia launches GSM, Nokia 1100, etc. Nokia was the leader in the tech world with the launch of Symbian and Windows phone. But the world was changing at a different pace. The tech world was hit by the disruptive technology called "Android" which was gaining popularity in no time. The added ingredient was Samsung launching its android-enabled smartphones at a reasonably lower price. These features of new mobile phones populating the market opened new opportunities and possibilities for the user and to earn revenue from different, this made a difficult situation for Nokia as Nokia failed to adopt such rapid change in the technology. Nokia tried to put in a fight with the new emerging competitors by launching windows-enabled phones but Nokia failed miserably due to the technology revolution that had hit the tech world. The growth of any organisation is based on its human resource management i.e.; what resources are to be used in achieving the goal. As employees are the main assets for any organisation, Nokia needs skilled and professional persons which are able to achieve the company goal.

The study tries to analyse the reasons for the failure of Nokia and how Nokia incorporated the factor of Corporate Sustainability specifically Sustainable HRM into its strategies to come back.

Different firms have been adopting different principles that are articulated by the firms. Also, the firm's commitment towards sustainable development is different at different times. Like, the need has increased from year after year [2]. A recent study on Indian Manufacturing companies done by [12] has shown that sustainability in the HRM practices affects organisational performance greatly.

10.3 Discussion

This research has tried to study the implications of extent of sustainability being implemented in devising the strategies of the organisations through an example of one of the leading brands of the world "Nokia". Nokia saw a great deal of ups and downs in its life; the research tries to study this jigsaw journey in perspective of sustainable HRM as to how their downfall made them to align their HRM strategies with the concept of sustainability.

Ref. [14], had briefly tried to identify various HR functions that contribute significantly in the application of Sustainable HRM in the organisations. These factors that Nokia considered are as follows:

a. Diversity, Equal Opportunity and Multiculturism
b. Training and Development
c. Workplace Relationship and Employee Engagement (Workforce Engagement)
d. Health and Safety
e. Ethics and Governance [14]

These different HR functions were analysed in the scope of before and after the fallout of Nokia in 2012. The data has been collected from Secondary sources i.e. the reports published by Nokia during the years 2017 and 2020. The comparative study shows the increasing contribution of the HR functions post its downfall and how they have tried to make HR function more collaborative with the concept of sustainability (Fig. 10.1).

As to sustain in the mobile market, Nokia has to make new plans as necessary to compete with others. Nokia has to work on its weaknesses and needs to add new technologies to the organisation. As Nokia is working on some HR policies to sustain in the market and to make its proper come back with new strategies. As every organisation is working on firstly the HR components then after some time, they opted personnel management policies to their companies just to improve their performance and employee satisfaction after that companies opt new policy of strategic HRM in which again company work on their employees and performance improvement. Studies have shown that companies who have been aligning the factor of sustainability in their strategies termed as high sustainability companies are outperforming in comparison with those who are not considering the concept of sustainability called as low sustainability companies [9]. On the other hand [1], states that there is a significant relationship between the firms' demographics and the humanistic sustainability HRM practices (Table 10.1).

The above figure describes the key data range from 2010, 2011, and 2016 and 2017 i.e. years before Nokia's fallout and years pursuing to come back to the market scenario. Inferences are:

a. In terms of diversity, the percentage of non-Finnish people has increased up to a large extent i.e. from 52 to 93%, which signifies that Nokia has tried to invite a large range of people belonging to different cultures, and nationals. Nokia went a step further and ran a "Village of Diversity" series of activities. Taking care of Gender balance, LGBTQ, multigeneration workforce, (dis)abilities at work which resulted in participation of 2,000 employees in different sessions [17].
b. Another HR function, i.e. Training and Development, Nokia had significantly contributed to the training of the employees. Nokia is contributing extensively towards training their employees. Nokia started with Global Day of Learning making it an annual event to inculcate the culture of learning. In 2017, the day

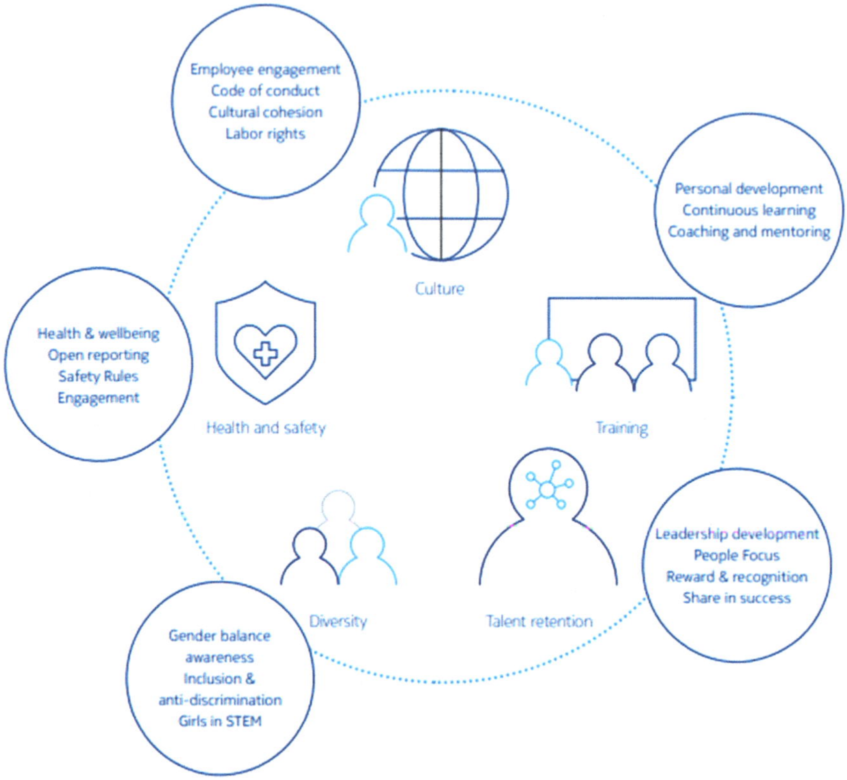

Fig. 10.1 Multiple HR functions considered for Sustainable HRM [18]

Global Day of Learning was organised, 20,357 participants attended the web seminars.

c. Nokia had practically being able to achieve "zero" percentage of illness and injury within the workplace. Nokia has a set of guidelines called "Nokia Life Saving Rules "which includes simple safety rules that need to be followed by every employee. The employees are given training towards acknowledging them about the guidelines for the safety.
d. In order to achieve their sustainability goals, they have collaborated with customers, suppliers, NGOs, governments, industry peers etc. Nokia is trying to help them through innovating their products, solutions in order to help them achieve their goals. In 2017, Nokia made 46 short-term and long-term targets with various collaborators. In the same year, Nokia was a part of great initiatives like the United Nations Global Compact, Climate Leadership council, Global e-sustainability Initiative, and many more.

In 2017, Nokia promoted the culture with the values inhibiting high performance and integrity. They collectively shape the culture calling it a "Nokia way". This culture

Table 10.1 Key data of Nokia

	2017	2018	2019	2020
Diversity & equal opportunity				
• Number of nationalities in the executive management board	8 (of 15)	8 (of 14)	10 (of 18)	9 (of 17)
• Share of non-Finnish in the executive management board	53.2%	53.6%	92%	93%
• Average age of employees at year-end	NA	NA	40	41
Training & education				
• Percentage of employees that completed a performance evaluation	16	34	34	33
• Training costs of employee by Nokia EDU, € million	850	980	54	49
Workplace relations & employee engagement				
• Share of employees responding to the engagement survey	NA	NA	20%	21%
• Employee engagement (percentage of favourability towards the company	NA	NA	76	80
Occupational health & safety				
• Illness/Injury rate within production	0	0	1	0
Community investments				
• Total value of contributions (EUR million)	6.84	6.53	2.26	6.50
Ethics				
• Total number of concerns reported	678	887	994	776
• Human rights	2	2	1	1

Source [17, 18]

tries to support their business strategy. It is made up of 12 culture principles that are their working principles and culture priorities. They have been significantly divided into three aspects: Drive (Passion for results, customers, and products), Dare (to earn and renew), and Care (for the team before self). These are the bridging themes from culture to the strategies [17].

10.4 A Major Step Towards Enabling Employee Engagement

One of the major steps taken by Nokia is for the Employee Engagement. They started with the programme called Culture Cohesion Tracker (CCT) for monitoring of cultural progress and evolution. The main purpose of the CCT is to measure employee' perception towards the company. The results obtained from it are used to help the business groups to make out their own roadmaps to put in a better and effective show [17].

Nokia's commitment to open communication and cultivating a culture of trust was a significant step in facilitating staff engagement during the COVID-19 epidemic. In order to make sure that workers felt educated and heard during uncertain times, the company instituted frequent virtual town halls and feedback sessions. By providing online training courses and skill-building workshops, Nokia also placed a high priority on professional development, assisting staff members in adjusting to remote work and becoming ready for new challenges. These programmes improved the employees' sense of belonging and contributing to the organisation's objectives in addition to maintaining employee motivation.

A crucial component of Nokia's engagement strategy was taking a comprehensive approach to employee well-being. The organisation implemented mental health support services, including as wellness programmes and counsellor access, in recognition of the stress and worry brought on by the pandemic. Nokia's dedication to employee pleasure was further evidenced by its flexible work schedules and ergonomic support for home office configurations. Nokia was able to preserve high levels of employee engagement by coordinating these initiatives with sustainable HRM concepts. This helped the organisation remain resilient and continue to perform well throughout the crisis.

During the COVID-19 epidemic, Nokia's comprehensive strategy for promoting employee engagement focused on connection, empowerment, and well-being.

10.4.1 Communication and Transparency

Nokia placed a high value on open communication to keep workers informed and engaged. Senior executives held frequent virtual discussions to deliver updates on the company's performance, safety and health protocols, and future goals. By promoting enquiries and criticism, leaders fostered an environment of openness and confidence. Nokia also used feedback tools and anonymous questionnaires to find out how its employees felt and what needed to be improved. These initiatives reaffirmed the business's dedication to hearing employee concerns and meeting them.

10.4.2 Professional Growth and Skill Development

Nokia launched a range of online training courses and certificates in recognition of the importance of lifelong learning. These included leadership development and soft skills training as well as technical instruction in fields like cloud computing and 5G. Access to a variety of learning resources also encouraged staff members to follow their own interests. Nokia gave workers the tools they needed to overcome obstacles and maintain their competitiveness in the workplace.

10.4.3 *Mental Health and Well-Being Support*

Nokia made important efforts to address the pandemic's effects on mental health. The organisation conducted stress management training, arranged virtual mindfulness sessions, and provided access to professional counselling services. Nokia established wellness initiatives centred on physical activity and healthy living, as well as adaptive workplace equipment for homes, in order to promote physical well-being. Employees were also able to combine their personal and professional obligations because of flexible work hours.

10.4.4 *Social Connection and Team Building*

Nokia implemented innovative virtual team-building exercises to tackle the sense of isolation that comes with working remotely. Teams took part in wellness challenges, quiz competitions, and virtual coffee breaks. Even in a remote setting, these efforts promoted a feeling of community and solidarity. Supervisors were instructed to conduct frequent check-ins with their staff, emphasising both personal and professional updates.

10.4.5 *Recognition and Rewards*

Nokia realised the value of recognising workers' efforts, particularly in challenging situations. The business celebrated milestones and accomplishments by implementing virtual appreciation programmes. Employee motivation and morale were raised when they were publicly recognised at meetings and via internal communication channels.

Nokia was able to sustain a highly engaged staff during the epidemic by incorporating these methods. Long-term success and organisational resilience were eventually facilitated by the company's dedication to sustainable HRM principles, which made sure that staff development and well-being were given first priority.

10.5 Conclusion

The significance of Sustainable Human Resource Management (HRM) in managing crises while maintaining organisational and employee well-being is demonstrated by the case study of Nokia during the COVID-19 pandemic. Nokia's diverse strategy, which includes open communication, career advancement, overall health, social interaction, and employee appreciation, shows how long-term HRM policies may encourage employee engagement, resiliency, and loyalty. In addition to overcoming the immediate difficulties of the pandemic, Nokia fortified its basis for long-term performance by integrating sustainability ideas into its HR policies.

This study emphasises how crucial it is to incorporate sustainability into HRM as a preventative step for organisational continuity and crisis management. Adopting sustainable HRM practices will continue to be essential for creating workforces that are resilient, engaged, and flexible as organisations continue to face dynamic global issues. Other companies looking to put employee well-being and business viability first during uncertain times might learn a lot from Nokia's experience.

Organisations like Nokia that have such substantial presence in the current world are also susceptible to the dynamic environment that is prevailing in the current world. Organisations in the heat of winning this profit maximising battle have led to shift their focus from all other factors that help them sustain their presence in the market. Adaptability and acceptance to newer and innovative ways to change have become a major issue. Change that is a sustainable one is also not in the context of current world organisations as the nucleus lies on profit making. This research has been done to show a multinational company led to its downfall due to lack of emphasis on sustainability in their strategies. But to comeback in the market, they have taken major steps in synchronisation with Sustainable HRM practices and have already received better outcomes. Organisations need to build their strategies on the pivot of Sustainable HRM along with other strategies so as to maintain not just their growth but to have long lived one. The power of sustainability is still to be harnessed as it's a fairly new but demanding aspect of the current scenario, more research is required to be done to explore the factors of it in other environments of the dynamic world.

10.6 Future Implications

With a particular focus on Nokia, the case study of Sustainable Human Resource Management (HRM) during COVID-19 may have a number of long-term impacts the business as well as the larger HRM industry. Let's examine some possible scenarios for the future:

10.6.1 Resilience in Crisis Management

Implication: Nokia's pandemic experience can have an impact on future crisis management plans by demonstrating that businesses with a flexible and resilient HRM strategy can continue to function even in the face of worldwide difficulties.

For instance, Nokia prioritised employee safety and well-being during COVID-19 while adjusting to remote work. The organisation may prioritise developing flexible HR practices going future to ensure business continuation even during uncertain times.

10.6.2 Workplace Flexibility and Hybrid Models

Implication: Many businesses were forced to implement hybrid working arrangements by the pandemic, including Nokia. A sustainable HRM strategy will probably place more of an emphasis on developing flexible work arrangements in the future.

For instance, Nokia implemented remote work guidelines and flexible scheduling to assist its employees. In order to ensure work-life balance and maximise productivity, future HRM practices might incorporate policies that support long-term work-life balance.

10.6.3 Digital Transformation and Employee Upskilling

Implication: In order to maintain a competitive workforce, Nokia's HRM processes may change to incorporate continuous employee training in digital technologies and skills, since the pandemic accelerated the need for digital transformation.

For instance, Nokia could support ongoing education and training initiatives that emphasise digital skills, allowing its employees to become more tech-savvy. This will promote a culture of ongoing education and skill improvement, which is essential for future expansion.

10.6.4 Employee Well-Being and Mental Health

Implication: Employee well-being-focused HRM tactics are probably going to gain popularity. Work-life balance, mental wellness, and physical health are all essential components of HR procedures.

For instance, Nokia implemented programmes for employee wellness throughout the epidemic, including flexible scheduling and mental health support. In order to promote a better workplace, future HRM procedures may further incorporate mental health support into corporate culture.

10.6.5 Sustainability and Corporate Social Responsibility (CSR)

Implication: Sustainable HRM practices will develop to prioritise social and economic sustainability in addition to environmental sustainability. This includes acceptance, diversity, and equitable employment practices.

For instance, Nokia supported communities and supported ethical company behaviour as part of its sustainability initiatives during COVID-19. Nokia may step up its CSR efforts in the future, coordinating HRM with more general organisational sustainability objectives.

10.6.6 Employee Engagement and Communication

Implication: An essential element of HRM will be clear and consistent communication. Regular employee engagement is a top priority for businesses like Nokia in order to keep their staff informed and inspired.

For instance, Nokia improved its internal communication channels during the epidemic to make sure that workers stayed involved in spite of the difficulties associated with working remotely. By using digital tools for more engaging and transparent communication, this might become a regular component of HRM strategy.

10.6.7 Global Talent Management

Implication: HRM practices may change to a more global approach to talent management as businesses embrace remote work, allowing them to hire the most skilled personnel regardless of location.

For instance, Nokia, a multinational corporation, might increase the number of people in its workforce by adopting remote work regulations that enable them to hire workers from various regions. This can foster more varied teams, which will foster creativity and a global viewpoint.

10.6.8 Long-Term HRM Strategy Transformation

Implication: It is probable that HRM will include technology, sustainability, and employee well-being more thoroughly in the future. Nokia's pandemic HRM tactics will have an impact on its long-term strategic planning.

For instance, Nokia may incorporate cutting-edge HR technologies (such artificial intelligence (AI) and HR analytics) into their management procedures to improve decision-making and adjust HR tactics to changing employee demands.

Innovation and change in HRM practices have been prompted by the COVID-19 epidemic, with firms such as Nokia setting the standard for incorporating digital transformation, flexibility, and sustainability into HR strategy. Future HR policies will probably include these approaches more deeply, establishing new standards for work-life balance, employee engagement, and organisational sustainability.

References

1. Asis-Castro, A. L., & Edralin, D. M. (2018). Predictors of humanistic sustainability HRM practices. *DLSU Business & Economics Review, 27*(2).
2. Bansal, P. (2005). Evolving sustainably: A longitudinal study of corporate sustainable development. *Strategic Management Journal, 26*(3), 197–218.
3. Berger, L. A., & Berger, D. R. (Eds.). (2011). *The talent management handbook: Creating a sustainable competitive advantage by selecting, developing, and promoting the best people.* McGraw-Hill.
4. Brundtland, G. H. (1987). Our common future, world commission on environment and development (WCED).
5. Boudreau, J. W., & Ramstad, P. M. (2005). Talent ship, talent segmentation, and sustainability: A new HR decision science paradigm for a new strategy definition. *Human Resource Management: Published in Cooperation with the School of Business Administration, The University of Michigan and in alliance with the Society of Human Resources Management, 44*(2), 129–136.
6. Dyllick, T., & Hockerts, K. (2002). Beyond the business case for corporate sustainability. *Business strategy and the environment, 11*(2), 130–141.
7. Ehnert, I., Harry, W., & Zink, K. J. (2014). Sustainability and HRM. In *Sustainability and human resource management* (pp. 3–32). Springer.
8. Ehnert, I. (2009). *Sustainable human resource management.* A Conceptual and Exploratory Analysis from a Paradox Perspective.
9. Eccles, R. G., Ioannou, I., & Serafeim, G. (2012). *The impact of a corporate culture of sustainability on corporate behavior and performance* (No. W17950). National Bureau of Economic Research.
10. Guerci, M. (2011). La gestione delle risorse umane per la sostenibilità dell'impresa.
11. IFOK. (1997). *Bausteine fur ein zukunftsf¨ahiges Deutschland, Diskursprojekt im Auftrag von VCI und IG Chemie-Papier-Keramik.* Gabler.
12. Kramar, R. (2014). Beyond strategic human resource management: Is sustainable human resource management the next approach? *The International Journal of Human Resource Management, 25*(8), 1069–1089.
13. Kennedy, H. (2017). The role of business sustainability in human resource management: A study on indian manufacturing companies. *The South East Asian Journal of Management.*
14. Lis, B. (2012). The relevance of corporate social responsibility for a sustainable human resource management: An analysis of organizational attractiveness as a determinant in employees' selection of a (potential) employer. *Management Revue*, pp. 279–295.
15. Mazur, B. (2015). Sustainable human resource management. The attempt of holistic approach. *Ekonomia i Zarządzanie, 7*(2).
16. Mariappanadar, S. (2003). Sustainable human resource strategy: The sustainable and unsustainable dilemmas of retrenchment. *International Journal of Social Economics, 30*(8), 906–923.

17. Mariappanadar, S. (2012). Harm of efficiency oriented HRM practices on stakeholders: An ethical issue for sustainability. *Society and Business Review, 7*(2), 168–184.
18. Nokia People and Planet Report 2017.
19. Nokia People and Planet Report 2020.
20. Ogbazghi, R. (2017). Talent management as a new form of sustainable human resources management. *Latgale National Economy Research*, p. 95.
21. Spooner, K., & Kaine, S. (2010). Defining sustainability and human resource management. *International Employment Relations Review, 16*(2), 70.
22. Senna, J., & Shani, A. B. (2009). Utilizing technology to support sustainability. In P. Docherty, M. Kira, & A. B. Shani (Eds.), *Creating sustainable work systems: Developing social sustainability* (pp. 84–100). Routledge.
23. Stankeviciute, Z., & Savaneviciene, A. (2013). Sustainability as a concept for human resource management. *Economics and Management, 18*(4), 837–846. http://www.interbrand.co.uk/en/best-global-brands/Best-Global-Brands-2010.aspx

Amrita Dutt is a PhD student at Department of management studies, Indian Institute of information Technology, Allahabad. Her research focuses on Effect of Team dysfunctions on employees' psychology and individual work performance. She has completed MBA, Post graduations (M. A Sociology and MSc IT) with 3 years GNIIT(MIT) course from NIIT. Also, she has successfully completed B.Ed. with Teachers eligibility Test (TET). She is currently working as a teacher. Her research areas include Organizational Behaviour, Work Culture, Sustainable Human Resource Management practices, Strategic Human Resource Management practices, Job Satisfaction. Her research publications include in SSCI and ABDC listed journals.

Aman Chadha is at Sunstone, Ajeenkya DY Patil University, as an Assistant Professor encompasses not just educating future leaders but also advancing our understanding of the intersection between human resources and artificial intelligence. With a strong foundation in HR-marketing from my MBA-PhD (Thesis submitted) studies at the Indian Institute of information Technology, she is dedicated to contributing to both academic discourse and practical applications in the field. The research team at IIIT-Allahabad and herself are pioneering tudies that delve into psychological contracts and sustainable human resource management.

Akriti Gupta is currently working as Assistant Professor at IIBS (airport campus) and Ph.D. scholar in the Department of Management Studies at the Indian Institute of Information Technology, Allahabad. She has completed her graduation (btech) with gold medal and aster's (MBA) with silver medal from reputed colleges. Her research work is dedicated in automating human resource processes with the help of artificial intelligence to which she has three copyrights registered in her credit. Her research areas include Organizational Behavior, Work Culture, Sustainable Human Resource Management practices, Strategic Human Resource management practices, Job Satisfaction, the VUCA model and people analytics with vast knowledge of different softwares (IBM SPSS, AMOS, MS Excel) and languages (C, C++, Java, Python). Her research publications include in SSCI, ABDC and SCOPUS listed journals. Additionally, she is mentoring a start-up from IIIT Allahabad, funded by NewGen IEDC. She is also as a corporate trainer for AIILSG (semi-government organization).

Vijayshri Tewari is a Professor at Department of management studies, Indian institute of information Technology, Allahabad. Her research focuses on Human Resource Management, Organisational behaviour, Principle of management, international Business and Business Environment, Work Culture, Sustainable Human Resource Management practices, Strategic Human Resource Management practices, Job Satisfaction. She is dedicated to contributing to both academic discourse and practical applications in the field. The research delves into psychological contracts and sustainable human resource management. Her collaboration leverages my interpersonal skills to foster a culture of organizational citizenship and to enhance communication strategies, thereby steering the department toward ground breaking HR development methodologies.

Chapter 11
Leading the Digital Shift: The Impact of Online Pedagogy on Management Students' Engagement and Performance

Sohini Datta, Debolina Mitra, Udit Chawla, Devika Mulchandani, and Subrata Chattopadhyay

Abstract

Purpose The review highlights how the behaviour of the COVID-19 problem has led to the involvement, of teaching and management students, over online learning platform. The evolution from traditional temporal and attitudinal learning methodologies, which needed a direct contact between students and tutors, to online learning models has transformed the landscape of educational platforms, mainly in management education. Constructed on the Social Cognitive Theory and the Online Learning frameworks, the study explores the influence of methods for students towards online classes, interactions, and criticism on MBA students' attitudes and performance. The study identified to identify challenges digital ecosystem, the queries of how to improve digital preparedness, and issues with online evaluations. It also scrutinizes students' participation through the conception of a digital identity and evaluates the effectiveness of online teaching methods.

Design/Methodology Purposive sampling was done and the students were specifically post-Covid-19 students. For this purpose, the necessary data were gathered

---

S. Datta · U. Chawla (✉) · S. Chattopadhyay
Institute of Engineering & Management, School of University of Engineering and Management, Kolkata, West Bengal, India
e-mail: udit.chawla@uem.edu.in

S. Datta
e-mail: sohini.datta@iem.edu.in

S. Chattopadhyay
e-mail: subrata.chattopadhyay@iem.edu.in

D. Mitra
Indian Institute of Management, Bangalore, India
e-mail: debolina.mitra@iimb.ac.in

D. Mulchandani
The Calcutta Anglo Gujarati College, Kolkata, West Bengal, India
e-mail: devikamulchandani671@gmail.com

S. Gangopadhyay et al. (eds.), *Pandemic Diaries*,
https://doi.org/10.1007/978-981-96-5415-4_11

from B-Schools of Kolkata, India and was done with the help of a structured questionnaire. The data collection involving 341 MBA students, both male and female, from a variety of age group, income categories, and backgrounds was chosen.

Findings From the study, five factors that were identified 'Student Performance', 'Effectiveness and Good Performance', 'Accessibility', 'Impact', and 'Expertise'. The factors accessibility and the expertise had a major influence in the Student Performance, with institutional role, technology platform with 24×7365 support, user interface design helping to aid online pedagogy process, to meet the teacher-student expectation from the digital shift in teaching.

Keywords Online pedagogy · Digital shift · Management students · e-learning · Engagement · Performance · Digital evaluation · Accessibility

11.1 Introduction

Online learning has been more popular in recent years due to increased internet accessibility [1]. The paradigm shift in adoption of digital media across sectors like media and content industries [2] signalled how the passive recipients 'students' have emerged as architects of their learning journey participating in 'rotation' model from 'face-to-face' to 'online learning' deviating from their institution-based computer laboratory. The second type, 'blended learning typology', showed, how digital curriculum, offering each independent student to engage online in 'small group tutoring' and engage in 'online project'. Over the years, [3] stated that the progress of advanced technologies, convergence of telecommunication, and IT (information technology) at structural level has been conducive towards designing and offering online platforms and has helped to shape robust blended learning model. Currently, there is evidence of expansive learning dimensions encouraging 'student to student', 'student to teacher', 'student to community', 'student to material', and 'student to technology' interaction levels [4].

Thus, embracing technology in respective industry sector has shown evidence of catalyst, impacting the innovation in the education sector with infrastructure and networking, playing as a key driver of integrating hardware and software. These are driving the educational institutions the supply side of education supply chain, to adopt multimodal centres, converting classroom and laboratories, towards online learning that has changed the nature of instruction over time [5]. The platform-based learning has led to expansion in expansion of 'online courseware', with learning management system (LMS), cross device learning applications allowing the teachers and students to sign-in, report, manage online education libraries (on-demand text, image, video), along with instructional improvement systems supporting assessment and data platforms [6]. Similarly, the context of educational content has seen a transformational change, where new standards of open-content, teacher-developed content, and premium-paid content emerged strongly in educational landscape [7].

By embracing the potential of technology convergence and empowering the students to take ownership of their e-learning with 'digital identity' in digital environment [8] institutions are able to create a future, where education is personalized, engaging, and preparing the students for success in an increasingly digital world [9]. In pre-COVID-19 context, while there was low potential of student-led digital learning due to low demand, during COVID-19 pandemic that adopted technology platforms led to 'digital shift', which has shown its intensive adoption and use. The radical shift presented a series of challenges such as digital divide, ensuring quality online content, democratization of digital shift implications, and developing effective digital assessment strategies for student digital competencies and performance [10]. Addressing these challenges requires collaboration between educators, students, and policymakers to create inclusive and equitable digital learning environment [11].

The rapid evolution of technology in the last decade has precipitated a fundamental shift in the higher education sector with faculty-initiated changes during COVID-19 causing digital shift [12] with changes in instructional strategies for online [13]. This has positively affected its delivery, with digital pedagogy evolving as a dominant method of learning during a phase when the world entered pandemic, a phase of restricted movement to stop spread of social contacts in everyday life. While this paradigm shift promises enhanced accessibility and flexibility, its impact on higher education student engagement and performance remains an area of critical inquiry cyber-slacking behaviour in students in e-learning environment emerged strongly [14]. The COVID-19 pandemic accelerated the adoption of online learning by teachers exposing them to web 2.0 environment and LMS [15] necessitating an urgent examination of its efficacy especially educational institutions seeking BCP business continuity program, and students in management using personal electronic devices (smartphone, tablet, laptop) to access learning modules. Understanding how online pedagogy influences management student engagement is paramount [16], as it directly correlates with learning outcomes and academic achievement [17]. Moreover, in the transition from traditional classroom settings to virtual environment, student raised the questions about the effectiveness [18] of online management degree pedagogy [19] in developing essential management skills, such as critical thinking, problem-solving, and teamwork [20].

The abrupt change from offline classroom to online education mode in academic session required students to cope in terms changing their attitude, behaviour, learning style against disruptions in the learning environment shifting from offline to online. This study attempts to contribute to the creation of more efficient and fair online learning environments for management students by examining the nuances of online pedagogy and its effects on the engagement of management students and their academic achievement following the 'digital shift'.

While past research showed teacher acceptance of TAM model for digital technologies in teaching practices [21], research focussed on the MBA student perspective about the digital shift is scanty.

This research aims to examine the relationship between the implementation of online MBA pedagogy and its impact on the engagement and academic performance of management students. Grounded in 'Social Cognitive Theory and Online

Learning', the current research study research objective seeks to understand how specific online pedagogical strategies due to 'digital shift', such as interactive elements and personalized feedback, influence MBA students'—perception and performance of the learners, EdTech model effectiveness. The research evaluates the role of digital literacy after taking a digital shift, the access in mitigating potential inequities in online learning.

11.2 Review of Literature

The past academic literature showed that the decade from 2010 to 2020 marked a pivotal shift in MBA pedagogy. Initially dominated by traditional offline methods such as the classroom lectures, solving real-world business case studies, and intense group discussions, the MBA educational landscape gradually transformed with the integration of technology [22]. Early adoption of online platforms by the institutions in higher education has opened up the education supply chain with the supply side primarily focused on content delivery [23], amid the range of structural and process-centric challenges [24]. The evolution and cross convergence of technologies and emergence of platform-centric approach [25] has consolidated the higher education model, impacting the audience leveraging technology [26]. Initially, the issue of facing challenges in engaging students in virtual environments was evident but later phase of IT-telecom smartphone apps-driven environment has shown to spearhead faster adoption of technology even in the MBA full time, MBA executive programs bringing in flexibility in pedagogy delivery matching career progression needs [27].

Thus, the momentum of digital shift has been characterised by technology shaping the market, the institutions, and higher education beneficiaries [28]. However, by 2015, online MBA programs gained traction, driven by advancements in technology and a growing emphasis on student engagement for the online MBA degrees offered by universities [29]. Blended learning models emerged, combining the best of both worlds, while institutions focused on developing students' digital literacy and a shift in increased adoption of online learning model offering flexibility and accessibility [30]. Throughout this period, a notable trend was the shift from 'instructor-centric' to 'student-centred learning', with an increased emphasis on active learning, problem-solving, and critical thinking [31]. Technology played a crucial role in enabling personalized learning, facilitating global collaboration, real-time access, and supporting lifelong learning initiatives. However, the shift from 'offline to online' mode has affected the learning pace, cognition, and their perception as it requires individual student capabilities, knowledge about online tools to engage online, apart from participation, attitude, emotions, and their performance in the higher education course. While [32] argued about quality of online higher education, the compulsion to pursue the online pedagogy as the mainstay mode of learning before and during COVID-19, throughout the world across nations, it has shown how exploring of online mode has helped to successfully help a digital shift and continue with higher education.

11.2.1 Technology Acceptance Model (TAM)

The Technology Acceptance Model (TAM) provides a robust framework for understanding how MBA students adopt and utilize technology for learning. This model posits that perceived usefulness and perceived ease of use are the primary determinants of technology acceptance. In the context of online pedagogy, students are more likely to adopt online learning tools (video) if they believe these tools will enhance their learning and are easy to use [33]. 'Perceived usefulness' refers to the extent to which students believe that using technology will improve their MBA academic performance and help them achieve their learning goals. 'Perceived ease of use' relates to MBA students' beliefs about the simplicity and ease of use in using the online technology platform. [34] argued using TAM in postgraduate students in online education system, when both perceived usefulness and perceived ease of use, are in the highest level, the Indian students are more likely to form a positive attitude towards technology adoption usage, and have a greater intention for learning purposes [35]. This, in turn, leads to increased actual use of technology available for online learning purposes.

11.2.2 Social Cognitive Theory and Online Learning

Social Cognitive Theory (SCT) provides a valuable lens for examining the impact of online pedagogy on management students' engagement and performance [36]. By emphasizing the roles of observation, imitation, and self-efficacy [37], SCT helps to explain how the students learn online, and are motivated in digital learning environment engagement [38]. In the context of online learning, students observe and learn from their faculty members and instructors, batch mates, and online resources (text, image, videos). Through imitation, these MBA students replicate successful learning strategies and behaviours. Self-efficacy, or the belief in one's ability to succeed, is crucial for overcoming learning challenges and persisting in online learning knowledge assimilation and examination/assessment performance. By fostering a supportive online learning environment (virtual classroom) that promotes self-efficacy, instructors can enhance the MBA student engagement and performance.

11.2.3 Student Performance

The form of electronic learning in higher education, has digital communication, interactive exchange, and immediate response in real time [39], which is dependent on their capability of technology acceptance, understanding tools and its operations that impact ability to understand the context of the study, and its retention using electronic device and media [40]. The dimension and aspect of the 'student performance' have been discussed in past academic research with attitude change [41], where the traditional classroom setting has been compared with the online learning environment, where constructive process of discussion, cooperative learning, problem-based learning [42] has helped to balance the student' apprehensions about new online learning environment. [43] showed that effectiveness of offline teaching has been championed due to rich medium of instructional learning (face to face), but argued that the technology-enabled online course effectiveness, requires to meet the student learning goals, maintain goals and objectives, enrolments, issues of teaching and learning along with evaluation [44]. [45] argued that performance evaluation of the learners in the online mode (recorded videos, streaming live class lectures) is dependent on the suitable technology and technological proficiency and integration, that enhances the delivery of the academic content enhancing the learning environment.

Student performances are dependent on course design and the online assessment [46], the design of assessment [47], technical and performance assessment measurement criteria, resolving issues of appropriate developments for remote assessment require appropriate hardware and software design. The ideal logic requires 'instructor-centric course' and 'student-centric' easy learning mode to support learner to know about the progress to be automated with multi-criteria decision making to support the huge volume of students and workload automated [48]. The transference of technological characteristics, which leads to development of skills in problem-solving, critical thinking skills that is required in online education platforms across disciplines [49], focussing on the outcomes of online pedagogy design [50]. [51] argued that amalgamation of multiple technologies has helped the momentum of improvement to introduce learning analytics. Additionally, advanced technologies-like AR (Augmented Reality) and VR (Virtual Reality) creates higher learning for the student. AI impacting student performance evaluated by analytics in the engineering degree course [52]. On the contrary, performance of diversity in e-learning batch has increased with technology (internet)-enabled visibility helping marketability of academic courses in online platforms [53] leading to cross-cultural interactions, higher volume of idea generation and global academic networking [54].

In mass marketing strategy context, the academic course content providers are able to standardise the operations and offer streamlined dissemination of online courses [55], maintaining high-quality instruction during COVID-19 pandemic [56] overcoming technical and marketability challenges for learner on-boarding. This signifies expansion of university-led and start-up-led demand in online business management courses [57], creating uniformity and holistic approach to aspirants in the society [58]. From learner perspective, the learning style (fast learner and slow learner) a part of self-regulated learning model emerged [59], enhanced flexibility to learn at own pace and higher technology-enabled student engagement, impacting the student's learning outcomes positively. [60] argued that from motivation and self-discipline perspective, the online learners need to show self-motivated approaches for effective time management skills, to pace their individual learning. In this context, the online course design and delivery acts as a guide and the new normal [61], as variations like 'group MBA case study' and 'group assignment/projects' help to aid online problem-solving skills [62], online collaboration within students to increase significantly [63]. These strategies define the quality of instruction (instructor and instructor materials) along with fair assessment methods for the academic course.

11.2.4 Effectiveness and Good Performance

The efficacy of using the online learning models in driving student success and academic achievement is contingent upon a confluence of diverse factors. Firstly, the robust instructional design, coupled with learner autonomy and engagement and learner satisfaction [64], is paramount that pertains to the supplier side of academic instruction design. Therefore, the UX user interface in collaborative learning module [63] designed in the LMS is a key factor [65], which influences the learner's experience to be incorporated in LMS for improving quality [66] intention and usage to login and read the instructor materials defining the usability dimension [67]. The issue for the 'theory and practice' for range of subject areas [68] has been perfected with third-party LMS website, which help to design effective online learning platforms, characterized by user-friendly interfaces, analytics, dashboard [69] and rich multimedia content feature capabilities significantly enhance higher education student experience [70]. Secondly, the learning 'input–output system' online requires to show a regular and constructive feedback mechanism consistently [71], coupled with opportunities for peer interaction (team or batch mates) and e-mentoring [72] which is instrumental in fostering an online supportive learning environment. Thirdly, the effectiveness of these factors impacts the student as a learner with strong self-regulation skills, technological proficiency, and a clear understanding of learning objectives which are more likely to excel in this modality [73]. In a nutshell, the synergy between the learner attributes and learner challenges [74], the instructional strategies evident in UX and LMS usage outcomes, and technological capabilities are crucial for optimizing learning outcomes in online settings [75]. The effectiveness and good performance of online MBA students are intrinsically linked to the broader context of the digital shift in professional higher education. The quality of online pedagogy, characterized by interactive content, accessible platforms, and supportive faculty, significantly influences student engagement and, consequently, performance.

The key factors 'Influencing Student Performance' is dependent on the following. Firstly, the MBA student and their individual learner characteristics, where the individual differences in learning styles, motivation, and time management impact how students engage and are satisfied with online content [76]. Secondly, the technological infrastructure and advanced technology integration [77], the geographic location, with the availability of the reliable internet connectivity (4G, 5G) and access to suitable devices are prerequisites for optimal online learning experiences. Thirdly, the pedagogical approaches that the use of diverse teaching methods, such as case studies, simulations, and group projects, enhances student engagement and knowledge acquisition. Fourthly, the assessment methods aligning with learning objectives and the provision of timely feedback are crucial for student success. Fifthly, the dimensions of the 24 × 7 × 365 institutional support which is perceived adequate from (technical support, academic advising, and career guidance) contributing to student satisfaction and performance. By understanding the interplay between these factors, institutions can develop strategies to optimize online MBA programs and ensure student success.

11.2.5 Accessibility

Policies addressing the underlying causes of digital inequities must be developed, inclusive teaching techniques must be developed and put into practice, and educational institutions and legislators must invest in accessible technology and infrastructure. Engaging in collaboration with technology developers is crucial in guaranteeing that digital learning tools are inclusively built [78]. Furthermore, to constantly increase the inclusiveness of online learning environments, continuous assessment and feedback methods are necessary. Communication skill and cognitive function may not be same for all students; therefore, the ethics in offering information to students from educators need to address the incorporation of ethical issues and its online access for all students. The structural and web-based access to student learning material and MBA degree virtual classroom has led to full-time MBA programs to be available online, leveraging LMS and blended learning [79]. [80] argued that online (internet) and the LMS integration has added value to the open learning in the higher education and particularly affected the aspirants to pursue MBA degree. Current practices of advanced telecommunication internet standards allow live streaming in public or private social media platforms expanding the accessibility [81]. The question of MBA programs to be accessible for all socioeconomic classes of people, needs a revisit the 'access' as full-time programs and online classes are 'cheaper' comparatively, therefore, democratizing the higher education access to wider sections of society. [82] stated that technology has helped to offer online courses that were restricted only to enrolled students, thereby expanding the marketability and impacting formation of a new business model. It has made education resources more reachable and opened up the higher education system to reach the extent that it is not limited just in the conventional classrooms so that it has provided new prospects for not only instructors but also students [83]

11.2.6 Impact

[84] outlined the online learning as students take digital learning with VIRI (virtual, interactive, real-time, instructor-led). The impact of flipped and blended learning systems have a positive impact on participants' perceptions of engagement, performance, and satisfaction, flipped learning actually mitigates the effects of blended learning, highlighting the fact that blended learning pedagogies are merely delivery

methods without any bearing on learning [85]. [86] stated that interaction online engagement is involvement, and promoting frequent chances for student communication can contribute to better learner-content interaction, which in turn can aid to increase student engagement levels, which [87] cited as one tone, group collaboration towards digital shift. The way university students' online courses are designed will be significantly impacted by these findings. Teaching courses that are naturally difficult requires teachers to be imaginative in their approach to creating interesting course materials and vigilant in maintaining constant contact with students and giving continual help and direction in writing assignments [88, 89]. The online project management in e-learning in UG and PG students showed that they needed innovation and self-motivation to overcome technology barriers. The impact of gamification strategy in online student e-learning. [90] explained that they were participating in a gamified online course where the frequency of online interactions was divided between the above-average and below-average performing groups. [91] reported that e-learning policy and online module design has affected the students in the educational institution, and their learner satisfaction in relation to engagement. [92] stated that online review of course engagement and e-learning performance in financial accounting education between joint venture of universities, which was also reported in blended learning with student engagement as the mediating factor impacting student performance in higher education [93]. [94] stated that in the past MOOC learning has shown that pedagogy, learners, instructors, and IT-IS system are dependent on time management, self-evaluation and help-seeking behaviour to overcome challenges quickly. There are also active participation and passive participation factors, with the latter not contributing and former showing confidence in engagement online, motivation, and learner's prior readiness and knowledge, examination, or assessment outcomes [95]. [96] stated that complexities in online pedagogy impacts the student's decision making and engagement levels. The students taking part in online discussion boards, peers who have helped them to improve their participation level and attendance is a testimony of cognition of self-efficacy in the new learning platform. Contextually, the COVID-19 has been a turning point for the learning online tools, to shape systematic online teaching and learning for developing online communication and participation [97]. The use of social networks has been proven for students across nations, while the online engagement dimension towards accepting 'digital shift' has affected their participation rate and completion rate of any online course. In this perspective, the faculty-initiated digital shift has been more predominant in line with the institutional theory, as top-down approach of academic course instructions has undergone a transformational shift [98]. The affective student engagement to online learning environment has however opened up a key and a new model of hybrid education (offline, online mix) in post-pandemic era [99].

11.2.7 Expertise

The convergence of expertise and the digital shift in pedagogy are pivotal in shaping the success of students [100]. As the business landscape becomes increasingly digital now there is higher flexibility for institutions and students to switch between offline classes into online classes in event of local and/or environmental catastrophe [101]. The demand for skilled professionals equipped with both domain expertise and digital proficiency grows exponentially as it further continues to professional work where hybrid mode of working is prevalent in post-COVID-19 context in few sectors though there are cases of burnout for faculty due to sudden digital shift in academia. MBA programs in the business admission departments in different institutions and universities, must align their curriculum to nurture these competencies [102]. This is a convergence as medium-centric approach for offline-online classroom learning is now prevalent in full-time, part-time business admission courses. Therefore, integrating digital tools and platforms into the learning process is essential. However, merely digitizing content is insufficient [103]. The emphasis should be on developing critical thinking, problem-solving, and digital literacy skills after the digital shift as student emphasis in learning outcomes is crucial. This requires faculty with expertise in both management and technology, to share communication tools knowledge, and expertise to be able to switch from offline to online, to create engaging and effective online learning experiences, in case of disruption in learning momentum. Furthermore, assessing student performance in a digital environment necessitates the traditional evaluation methods which may not adequately capture the depth of learning achieved through online platforms. Innovative assessment strategies, such as project-based learning, simulations, and peer evaluation, can better evaluate students' ability to apply knowledge and skills in real-world scenarios [104]. The dimension of student expertise in online tools, and online course content assimilation requires a comprehensive feedback system about IT, content, learning outcomes, and learner's satisfaction in online learning environment due to digital shift. By fostering a synergy between expertise and digital pedagogy, MBA programs are equipping graduates with the necessary skills to thrive in the digital age and contribute meaningfully to organizational success.

11.3 Methodology

A structured questionnaire with highly standardized questions was developed for this study. Purposive sampling was employed to select the respondents. Data were collected from 341 students of B-schools of Kolkata, through both the questionnaire and direct one-on-one interviews. The sample included participants of various ages, genders, occupations, and socioeconomic backgrounds. Responses were measured on a 5-point scale, ranging from 1 (strongly disagree) to 5 (strongly agree). The collected data were then subjected to further analysis.

The literature review highlights the transformative impact of technology on higher education, particularly in MBA programs. While online learning offers flexibility and accessibility, it also presents challenges related to digital literacy, equity, and student engagement. The ongoing evolution of blended learning models and the integration of advanced technologies suggest a future where education is increasingly personalized and digitally driven. However, addressing the digital divide and ensuring the effectiveness of online pedagogy remains crucial for optimizing student outcomes.

11.4 Results and Discussion

Kaiser–Meyer–Olkin (KMO) measures of sampling adequacy (which ranges from 0–1), closer to one are measured better, while 0.5 is measured the minimum prerequisite. In the table above (Table 11.1), the KMO value is 0.823, which is more than 0.5, and we can proceed with the factor analysis. The table shows that Bartlett's Test of Sphericity is significant, with a value less than 0.05 (0.000).

From total variance explained, it can be seen that Factor 1 (Student Performance) accounts for a variance of 5.679 which is 31.7% of the total variance (Table 11.2), likewise, Factor 2 (Effectiveness and Good Performance) accounts for a variance of 1.655 which is 10.3% of the total variance, Factor 3 (Accessibility) accounts for a variance of 1.528 which is 8.4% of the total variance, Factor 4 (Impact) accounts for a variance of 1.388 which is 7.6% of the total variance, Factor 5 (Expertise) accounts for a variance of 1.128 which is 6% of the total variance and thus, the first five factors combined account for 64%.

$$H0: B1 = 0$$

The null hypothesis implies that there is no linear relationship between 'Student Performance' and the factors, 'Effectiveness and Good Performance', 'Accessibility', 'Impact', and 'Expertise'.

Table 11.1 KMO and Bartlett's test

Kaiser–Meyer–Olkin measure of sampling adequacy		0.823
Bartlett's Test of Sphericity	Approx. Chi-Square	1549.648
	Df	137
	Sig.	0.000

Source Primary Data using SPSS 29.0

Table 11.2 Rotated component matrix

Variables	Component				
	Student performance	Effectiveness and good performance	Accessibility	Impact	Expertise
The usability of management online course ensures the effectiveness and good performance in computer-mediated learning than traditional mode of learning	0.799				
Management Online course platform facilitate face-to-face interaction and use animated videos to enhance interactivity and understanding	0.785				
Management instructors usually spend quality time in preparing the content online, rather than just copying from other material and using PowerPoint Presentations and reading it to students in traditional classroom	0.765				
Management online courses have good reachability and usability for all learners	0.722				
The flexibility of online learning allows students to learn at their own pace, positively impacting their overall learning performance	0.701				
Online learning of management course ensures the effectiveness and good performance in terms of coping up with missed lectures than traditional mode of learning		0.801			
Productivity of students can be enhanced through online learning of management course to strengthen educational concepts than traditional mode of learning		0.741			

(continued)

Table 11.2 (continued)

Variables	Component				
	Student performance	Effectiveness and good performance	Accessibility	Impact	Expertise
Online learning of management course ensures the effectiveness and good performance for presenting the work in class than traditional mode of learning		0.736			
Quality of teaching and learning can be increased and good performance through online learning of management course because it integrates various types of media than traditional mode of learning		0.699			
Online learning of management course offer maximum engagement and good performance of students than traditional mode of learning		0.683			
Access to education and possibility of good performance increases through online learning of Management course than traditional mode of learning			0.856		
Maximum amount of time is consumed while accessing and learning through online management course than traditional mode of learning			0.841		
Online learning of management course is economic in terms of time and accessibility for students and teachers than traditional mode of learning			0.823		
Quality of teaching and learning can be increased and good performance through online learning of management course because it integrates various types of media and reaches out to a great extent than traditional mode of learning			0.797		
Sharing their relevant online learning and implementation in their various fields				0.875	

(continued)

Table 11.2 (continued)

Variables	Component				
	Student performance	Effectiveness and good performance	Accessibility	Impact	Expertise
Creativity to increase productivity and accountability				0.792	
Effective facilitation and personalized interactions				0.752	
Explanations and good presentation skills				0.704	
Organization Skills					0.768
Subject matter knowledge					0.746
Resourcefulness					0.712
Understanding and delivery					0.703

Source Primary Data using SPSS 29.0

$$H1: B1! = 0$$

The alternative hypothesis implies that there is a relationship, between 'Student Performance' and the factors, 'Effectiveness and Good Performance', 'Accessibility', 'Impact', and 'Expertise'.

Model summary expresses to what extent a regression model fits the data. In the above Table, R-square (Table 11.3) value is 0.608 which indicates 61% of the total variation in the dependent variable; 'Student Performance' can be explained by the independent variables 'Effectiveness and Good Performance', 'Accessibility', 'Impact', and 'Expertise'.

ANOVA table (Table 11.4), we can see that the regression model predicts the dependent variable significantly well and its value is statistically significant as the p-value is less than 0.05 (that is 0.000). Thus, H1 is rejected, whereas H0 is accepted (Table 11.5).

Table 11.3 Model summary

Model	R	R-square	Adjusted R-square	Std. error of the estimate
1	0.780	0.608	0.601	0.6170

Source Primary Data using SPSS 29.0

Table 11.4 ANOVA

Model		Sum of squares	Df	Mean square	F	Sig.
1	Regression	249.750	4	62.437	90.464	0.000
	Residual	90.510	337	0.268		
	Total	340.26	341			

Source Primary Data using SPSS 29.0

Table 11.5 Coefficients

Model		Unstandardized coefficients		Standardized coefficients	T	Sig.
		B	Std. error	Beta		
1	(Constant)	3.559	0.049		90.802	0.000
	Effectiveness and good performance	0.204	0.049	0.210	13.304	0.000
	Accessibility	0.399	0.049	0.400	18.501	0.000
	Impact	0.275	0.049	0.280	16.260	0.000
	expertise	0.486	0.049	0.490	2.260	0.000

Source Primary Data using SPSS 29.0

Table 11.6 Fit indices table

Measure	Observed	Threshold
Chi-square (CMIN)	520.72	
DF	136	
GFI	0.96	>0.9
NFI	0.95	>0.9
CFI	0.97	>0.9
RMSEA	0.06	<0.08
PCCLOSE	0.16	>0.05
SRMR	0.061	<0.06

Source: Primary Data using AMOS 29.0

$$\begin{aligned}\text{Student Performance} = {} & 3.559 + 0.204 * (\text{Effectiveness and Good Performance}) \\ & + 0.399 * (\text{Accessibility}) + 0.280 * (\text{Impact}) \\ & + 0.486 * (\text{Expertise})\end{aligned}$$

Table 11.6 is fit indices table and all the parameters are satisfied. Table 11.7 determines Data Validity and Reliability Analysis.

The major influence on student performance is because of the two factors accessibility and expertise which is represented in Fig. 11.1.

Table 11.7 Data validity and reliability analysis

	CR	AVE	Student performance	Effectiveness and good performance	Accessibility	Impact	Expertise
Student Performance	0.88	0.59	0.768				
Effectiveness and Good Performance	0.85	0.61	0.359	0.781			
Accessibility	0.84	0.52	0.208	0.166	0.721		
Impact	0.83	0.58	0.198	0.591	0.310	0.761	
Expertise	0.85	0.55	0.175	0.458	0.222	0.195	0.741

Source Primary Data using AMOS 29.0

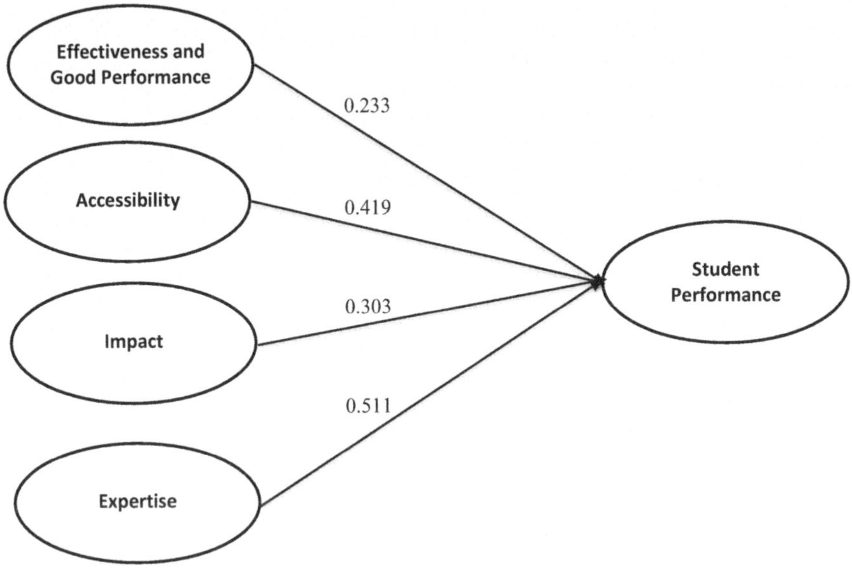

Fig. 11.1 Structural equation modelling

11.5 Conclusion

The study emphasizes on the learning conclusions for the management students resulting from the adoption of new digital platform in the education activities. As of the COVID-19 pandemic, the technological incorporation in teaching and learning processes has renovated out-dated educational management techniques from a traditional classroom perspective, to an online format, that is more manageable and flexible. However, state of infrastructural and process-centric approach, and the transition within MBA institutions, has raised challenges that each student as user encountered such as the digital divide. The manner in which tests are directed online, and the reputation of digital skills required to meet expectation set of both teachers and students against the traditional examination format. In the light of the current study's findings indicated that, including all students online, interactivity and feedback into MBA programs distributed. Though the internet improved the student awareness towards LMS interface, their motivation levels and faculty scheduling of online meetings, forced them to be a part of the online routine in higher education. These programs are heavily contingent on the LMS, which includes the accessibility of proper technology meeting software and hardware requirements maintaining 'equity theory' for students accessing through mobile phone or computer as a device. The online course material, self-assessment that determines their performance, as well as the operational tools of various additional functions like case studies, check progress, and recreation to supports the students acquiring critical and analytical abilities. This study accentuates self-regulated education a new power to the user, in the online format, revision facility using videos, and the importance of UX design with student's feedback resulting in a shift in online e-learning system that is coherence with their behavioural expectations.

While online learning appears to be a viable replacement for out-dated classroom learning, there are certain disadvantages to this platform-based learning system. It is up to educational intuitions, faculty-instructors, and policymakers to contemplate this troublesome and complex multiple technologies that help to package pedagogy online. Techniques that successfully eliminate digital discrepancies, payment security, increase technological support, and assert personalization of support, are required to attain the best results. The study's result verdicts on the student performance, which showed that a well-organized web-based learning ecosystem, comprises of systems approach to disseminate lectures (in text, audio, video). IT is meeting supply–demand of higher education forces, with the demonstration of effectiveness in testing MBA student readiness towards quiz or online video assessment. This pertains to the online LMS technical design and functionality offering subject area educational online material, together with 'technical facility' support and expertise, faculty support with platform help 24 × 7 × 365 that leads to students' attitudinal shift and awareness building. It is risky to obtain insight into and/or dichotomize the social mechanisms that allowed online teaching and e-learning in the context of a speedily changing the digital environment, for any institution, as it is was a sudden profound change due to COVID-19. By recognizing the opportunity presented by mix of technologies, flexibility and limiting full/partial access, the real-time evaluation for MBA students online, presented a new dimension. Online certifying feature in this assessment process, in which all students registered in management courses, showed best practices to be adopted in academic curriculum perspective. However, the need of power and influence on the MBA students with institutional LMS, technical support, faculty competence as stakeholders, with seamless communication has helped students, to have an equal opportunity to online education. The dominance of internet and LMS technology, used by the management institutions providing assistance helped to prepare students for an innovative learning environment, with a holistic approach to learning assessments.

References

1. Dong, L. (2024). "Does my teacher believe I can improve?" The role of EFL learners' meta-lay theories in their growth mindset and online self-regulation. *System, 122*, 103269.
2. Simon, J. P., & Bogdanowicz, M. (2012). *The digital shift in the media and content industries* (pp. 1–20). Publications Office of the European Union.
3. Hall, O. P. (2013). Assessing faculty attitudes toward technological change in graduate management education. *Journal of Online Learning and Teaching, 9*(1), 39–51.
4. Veluvali, P., & Surisetti, J. (2022). Learning management system for greater learner engagement in higher education—A review. *Higher Education for the Future, 9*(1), 107–121.
5. Afzal, F., & Crawford, L. (2022). Student's perception of engagement in online project management education and its impact on performance: The mediating role of self-motivation. *Project Leadership and Society, 3*, 100057.

6. Vezne, R., Yildiz Durak, H., & Atman Uslu, N. (2023). Online learning in higher education: Examining the predictors of students' online engagement. *Education and information technologies, 28*(2), 1865–1889.
7. Rabaa'i, A. A. (2016). Extending the technology acceptance model (TAM) to assess students' behavioural intentions to adopt an e-learning system: The case of moodle as a learning tool. *Journal of emerging trends in engineering and applied sciences, 7*(1), 13–30.
8. Engeness, I. (2021). Developing teachers' digital identity: Towards the pedagogic design principles of digital environments to enhance students' learning in the 21st century. *European Journal of Teacher Education, 44*(1), 96–114.
9. Maini, R., Sehgal, S., & Agrawal, G. (2021). Todays' digital natives: An exploratory study on students' engagement and satisfaction towards virtual classes amid COVID-19 pandemic. *The International Journal of Information and Learning Technology, 38*(5), 454–472.
10. Shohel, M. M. C., Shams, S., Ashrafuzzaman, M., Alam, A. S., Al Mamun, M. A., & Kabir, M. M. (2022). Emergency remote teaching and learning: Digital competencies and pedagogical transformation in resource-constrained contexts. In *Handbook of research on Asian perspectives of the educational impact of COVID-19* (pp. 175–200). IGI Global.
11. Gopinathan, S., Kaur, A. H., Veeraya, S., & Raman, M. (2022). The role of digital collaboration in student engagement towards enhancing student participation during COVID-19. *Sustainability, 14*(11), 6844.
12. Daumiller, M., Rinas, R., Hein, J., Janke, S., Dickhäuser, O., & Dresel, M. (2021). Shifting from face-to-face to online teaching during COVID-19: The role of university faculty achievement goals for attitudes towards this sudden change, and their relevance for burnout/engagement and student evaluations of teaching quality. *Computers in Human Behavior, 118*, 106677.
13. Mahmood, S. (2021). Instructional strategies for online teaching in COVID-19 pandemic. *Human behavior and emerging technologies, 3*(1), 199–203.
14. Koay, K. Y., & Poon, W. C. (2023). Understanding students' cyberslacking behaviour in e-learning environments: Is student engagement the key? *International Journal of Human-Computer Interaction, 39*(13), 2573–2588.
15. Lim, J., & Newby, T. J. (2021). Preservice teachers' attitudes toward Web 2.0 personal learning environments (PLEs): Considering the impact of self-regulation and digital literacy. *Education and Information Technologies, 26*(4), 3699–3720.
16. Wei, W., Liu, J., Xu, X., Kolletar-Zhu, K., & Zhang, Y. (2023). Effective interactive engagement strategies for MOOC forum discussion: A self-efficacy perspective. *PLoS ONE, 18*(11), e0293668.
17. Abuhassna, H., Al-Rahmi, W. M., Yahya, N., Zakaria, M. A. Z. M., Kosnin, A. B. M., & Darwish, M. (2020). Development of a new model on utilizing online learning platforms to improve students' academic achievements and satisfaction. *International Journal of Educational Technology in Higher Education, 17*, 1–23.
18. Cavinato, A. G., Hunter, R. A., Ott, L. S., & Robinson, J. K. (2021). Promoting student interaction, engagement, and success in an online environment. *Analytical and Bioanalytical Chemistry, 413*, 1513–1520.
19. Shu, D., Chen, Q., Liu, D., Shen, S., Yi, W., Tang, X., & Luo, M. (2024). The structure and characteristics of effective massive open online course learning strategies among college students: A qualitative study. *Sustainability, 16*(2), 716.
20. Ali, I., Narayan, A. K., & Sharma, U. (2021). Adapting to COVID-19 disruptions: Student engagement in online learning of accounting. *Accounting Research Journal, 34*(3), 261–269.
21. Sharma, S., & Saini, J. R. (2022). On the role of teachers' acceptance, continuance intention and self-efficacy in the use of digital technologies in teaching practices. *Journal of Further and Higher Education, 46*(6), 721–736.
22. Peterson, M. (2021). Teaching the online marketing research course for MBA students. *Journal of Marketing Education, 43*(3), 371–385.
23. Stafford, V. (2022). Successful collaboration in online learning through skills and community building: A women in leadership MBA subject case study. *Journal of Applied Learning and Teaching, 5*(2), 160–168.

24. Aggarwal, D. (2021). Experiences with the online format in the MBA program at the Faculty of Management at the University of New Brunswick.
25. Zhang, X. P., Wang, M., Banks, M. S., Zhang, Q., & Onita, C. G. (2020). Design and delivery of an online information systems management course for MBA programs. *Journal of Information Technology Education: Innovations in Practice, 19*, 47.
26. Amblee, N., Ertl, H., & Dhayanithy, D. (2023). A more relevant MBA: The role of across-the-curriculum delivery of intercompetency coursework in aligning the required curriculum with required managerial competencies. *Journal of Management Education, 47*(2), 204–238.
27. Hatt, L., & Davidson, J. (2022). "You what?!"–The impact of a relational approach to launching an executive MBA in the midst of a global pandemic. *Project Leadership and Society, 3*, 100045.
28. Jayaraman, R. (2023). Development of a comprehensive pedagogy to improve the connect between industry and business schools through capstone projects in executive MBA education. *International Journal of Education (IJE), 4*(01).
29. Nagale, T., & Khandare, A. (2024). Enhancing interactive learning: A comparative survey on ICT tools and pedagogical techniques for student engagement. *Educational Administration: Theory and Practice, 30*(4), 6914–6919.
30. Liu, I. F., & Hung, H. C. (2020). How are live-streaming services and social media platforms changing on-job MBA students' learning? A case study for applying e-case live in management case-based learning in Taiwan. *IEEE Access, 8*, 120936–120945.
31. Chiraphol, C. N., Ding, K. Y. D., & Hong, J. J. (2022). Experiential learning: The case of training MBA students in an Asian school. *International Journal of Educational Technology and Learning, 13*(1), 17.
32. Markova, T. (2021). Educators' and students' perceptions of online distance education before and amid COVID-19: Key concerns and challenges. In *SHS Web of Conferences* (Vol. 99, p. 01018). EDP Sciences.
33. Carmichael, M., Reid, A., & Karpicke, J. D. (2018). *Assessing the impact of educational video on student engagement, critical thinking and learning.* (A SAGE white paper).
34. Bag, S., Sinha, A., & Aich, P. (2023). Determinants of using online mode of teaching: Evidence from higher educational institutions. *International Social Science Journal, 73*(248), 415–434.
35. Singh, A., Sharma, S., & Paliwal, M. (2021). Adoption intention and effectiveness of digital collaboration platforms for online learning: The Indian students' perspective. *Interactive Technology and Smart Education, 18*(4), 493–514.
36. Choo, C., Fahrni, M., & Long, C. (2022). Analysis of students' cognitive presence and perception in a custom-designed virtual problem based learning assignment. *International Journal of Emerging Technologies in Learning (iJET), 17*(22), 132–143.
37. Chang, C. Y., Panjaburee, P., Lin, H. C., Lai, C. L., & Hwang, G. H. (2022). Effects of online strategies on students' learning performance, self-efficacy, self-regulation and critical thinking in university online courses. *Educational technology research and development*, 1–20.
38. Díaz-Noguera, M. D., Hervás-Gómez, C., De la Calle-Cabrera, A. M., & López-Meneses, E. (2022). Autonomy, motivation, and digital pedagogy are key factors in the perceptions of Spanish higher-education students toward online learning during the COVID-19 pandemic. *International Journal of Environmental Research and Public Health, 19*(2), 654.
39. Flannery, T. (2021). *How to market a university: Building value in a competitive environment.* JHU Press.
40. Morin, D. (2021). The impact of the pandemic on attitudes and perceptions of students in an MBA analytics course. In *INTED2021 Proceedings* (pp. 9132–9141). IATED.
41. Walsh, A., & Powell, P. (2020). Re-imagining the MBA: An arts-based approach. *Higher Education Pedagogies, 5*(1), 148–164.
42. Saykılı, A. (2019). Higher education in the digital age: The impact of digital connective technologies. *Journal of Educational Technology and Online Learning, 2*(1), 1–15.
43. Albert, S., Fulton, D., Ramanau, R., & Janes, A. (2021). Exploring cross-disciplinary differences in course mode, instructional tools and teaching methods in online courses in business & management. *The International Journal of Management Education, 19*(3), 100532.

44. Shenoy, S. S. (2021). Opportunities and challenges in online teaching for MBA students in India during the Covid pandemic. *RVIM Journal of Management Research*, 67–71.
45. Henderson, L. H., & Wilson, J. (2023). MBA futures: Images from principles for responsible management education (PRME) signatories during the COVID-19 pandemic. *Futures, 145*, 103047.
46. Guangul, F. M., Suhail, A. H., Khalit, M. I., & Khidhir, B. A. (2020). Challenges of remote assessment in higher education in the context of COVID-19: A case study of Middle East College. *Educational assessment, evaluation and accountability, 32*, 519–535.
47. Beneroso, D., & Robinson, J. (2022). Online project-based learning in engineering design: Supporting the acquisition of design skills. *Education for Chemical Engineers, 38*, 38–47.
48. Naveed, Q. N., Qureshi, M. R. N., Tairan, N., Mohammad, A., Shaikh, A., Alsayed, A. O., Alotaibi, F. M., et al. (2020). Evaluating critical success factors in implementing E-learning system using multi-criteria decision-making. *PLoS ONE, 15*(5), e0231465.
49. Khan, Z. H., & Abid, M. I. (2021). Distance learning in engineering education: Challenges and opportunities during COVID-19 pandemic crisis in Pakistan. *The International Journal of Electrical Engineering & Education*, 0020720920988493.
50. Bragg, L. A., Walsh, C., & Heyeres, M. (2021). Successful design and delivery of online professional development for teachers: A systematic review of the literature. *Computers & Education, 166*, 104158.
51. Şenel, S., & Şenel, H. C. (2021). Remote assessment in higher education during COVID-19 pandemic. *International Journal of Assessment Tools in Education, 8*(2), 181–199.
52. Ouyang, F., Wu, M., Zheng, L., Zhang, L., & Jiao, P. (2023). Integration of artificial intelligence performance prediction and learning analytics to improve student learning in online engineering course. *International Journal of Educational Technology in Higher Education, 20*(1), 4.
53. Chen, Y., & Carliner, S. (2021). A special SME: An integrative literature review of the relationship between instructional designers and faculty in the design of online courses for higher education. *Performance Improvement Quarterly, 33*(4), 471–495.
54. Kinnear, G., Jones, I., Sangwin, C., Alarfaj, M., Davies, B., Fearn, S., Wong, T., et al. (2024). A collaboratively-derived research agenda for e-assessment in undergraduate mathematics. *International Journal of Research in Undergraduate Mathematics Education, 10*(1), 201–231.
55. Marin, A., Parvatiyar, A., Mitchell, R. K., & Villegas, D. (2023). From lab to market: Learning entrepreneurial marketing through multi-semester, stage-gate, capstone project in STEM MBA. *Journal of Marketing Education, 45*(3), 226–246.
56. O'Keefe, L., Rafferty, J., Gunder, A., & Vignare, K. (2020). Delivering high-quality instruction online in response to COVID-19: Faculty playbook. *Online Learning Consortium.*
57. Cunningham, J. A., Harney, B., & Fitzgerald, C. (2020). *Effective technology transfer offices: A business model framework*. Springer International Publishing.
58. Etzkowitz, H. (2022). Entrepreneurial university icon: Stanford and Silicon Valley as innovation and natural ecosystem. *Industry and Higher Education, 36*(4), 361–380.
59. Sangsawang, T. (2020). An instructional design for online learning in vocational education according to a self-regulated learning framework for problem solving during the covid-19 crisis. *Indonesian Journal of Science and Technology, 5*(2), 283–298.
60. Neo, T. K., Amphawan, A., & Hamidani, K. (2022). Fostering Self-directed learning skills in smart learning environment during the Covid-19 pandemic lockdown. *Journal of educational multimedia and hypermedia, 31*(1), 43–66.
61. Xie, X., Siau, K., & Nah, F. F. H. (2020). COVID-19 pandemic–online education in the new normal and the next normal. *Journal of information technology case and application research, 22*(3), 175–187.
62. Hutubessy, E. D. (2021). Students' view on germany literature class by video assignment project during pandemic Covid-19. *AL-ISHLAH: Jurnal Pendidikan, 13*(2), 1295–1301.
63. Chatterjee, R., & Correia, A. P. (2020). Online students' attitudes toward collaborative learning and sense of community. *American Journal of Distance Education, 34*(1), 53–68.

64. Almusharraf, N., & Khahro, S. (2020). Students satisfaction with online learning experiences during the COVID-19 pandemic. *International Journal of Emerging Technologies in Learning (iJET), 15*(21), 246–267.
65. Maslov, I., Nikou, S., & Hansen, P. (2021). Exploring user experience of learning management system. *The International Journal of Information and Learning Technology, 38*(4), 344–363.
66. Mkpojiogu, E. O., Okeke-Uzodike, O. E., & Emmanuel, E. I. (2021, September). Quality attributes for an LMS cognitive model for user experience design and evaluation of learning management systems. In *3rd International Conference on integrated intelligent computing communication & security (ICIIC 2021)* (pp. 234–242). Atlantis Press.
67. Darvin, A., Kosasih, J., & Hanafiah, N. (2021, October). Usability evaluation of learning management system. In *2021 1st International conference on Computer Science and Artificial Intelligence (ICCSAI)* (Vol. 1, pp. 269–272). IEEE.
68. Demir, F., Bruce-Kotey, C., & Alenezi, F. (2022). User experience matters: Does one size fit all? Evaluation of learning management systems. *Technology, Knowledge and Learning, 27*(1), 49–67.
69. Drzyzga, G., & Harder, T. (2023, November). A three level design study approach to develop a student-centered learner dashboard. In *International conference on computer-human interaction research and applications* (pp. 262–281). Springer Nature Switzerland.
70. Li, Q., Duffy, P., & Zhang, Z. (2022). A novel multi-dimensional analysis approach to teaching and learning analytics in higher education. *Systems, 10*(4), 96.
71. Shehab, M. J., Alokla, M., Alkhateeb, M., & Alokla, M. (2021). *Hybrid learning aided technology-rich instructional tools—A case study: Community college of Qatar*. arXiv preprint arXiv:2112.11197
72. Nouman, N., Shaikh, Z. A., & Wasi, S. (2024). A novel personalized learning framework with interactive e-mentoring. *IEEE Access*.
73. Sherafati, N., & Mahmoudi Largani, F. (2023). The potentiality of computer-based feedback in fostering EFL learners' writing performance, self-regulation ability, and self-efficacy beliefs. *Journal of Computers in Education, 10*(1), 27–55.
74. Istiqomah, I., & Hasanati, N. (2022). Instrument development of determinants of students academic performance using rasch model analysis. *Psympathic: Jurnal Ilmiah Psikologi, 9*(1), 17–30.
75. Rabin, E., Henderikx, M., Yoram, M. K., & Kalz, M. (2020). What are the barriers to learners' satisfaction in MOOCs and what predicts them? The role of age, intention, self-regulation, self-efficacy and motivation. *Australasian Journal of Educational Technology, 36*(3), 119–131.
76. Ramirez-Montoya, M. S., Morales-Menendez, R., Tworek, M., Escobar, C. A., Tariq, R., & Tenorio-Sepulveda, G. C. (2024). Complex competencies for leader education: Artificial intelligence analysis in student achievement profiling. *Cogent Education, 11*(1), 2378508.
77. Yeung, M. W., & Yau, A. H. (2022). A thematic analysis of higher education students' perceptions of online learning in Hong Kong under COVID-19: Challenges, strategies and support. *Education and Information Technologies, 27*(1), 181–208.
78. Matsieli, M., & Mutula, S. (2024). COVID-19 and digital transformation in higher education institutions: Towards inclusive and equitable access to quality education. *Education Sciences, 14*(8), 819.
79. Bada, J. K. (2022). Evaluating blended learning of a systems analysis and design course in an MBA class. *International Journal of Education and Development using Information and Communication Technology, 18*(3), 76–92.
80. Fitzgerald, R., Anderson, M., & Thompson, R. (2015). Adding value: Open online learning and the MBA. *Electronic Journal of e-learning, 13*(4).

81. Kotsopoulos, D. (2022). Developing an undergraduate business course using open educational resources. *The Canadian Journal for the Scholarship of Teaching and Learning, 13*(1).
82. Celik, I. (2023). Towards Intelligent-TPACK: An empirical study on teachers' professional knowledge to ethically integrate artificial intelligence (AI)-based tools into education. *Computers in Human Behavior, 138*, 107468.
83. Francescucci, A., & Foster, M. (2013). The VIRI (virtual, interactive, real-time, instructor-led) classroom: The impact of blended synchronous online courses on student performance, engagement, and satisfaction. *Canadian Journal of Higher Education, 43*(3), 78–91.
84. Fisher, R., Perényi, A., & Birdthistle, N. (2021). The positive relationship between flipped and blended learning and student engagement, performance and satisfaction. *Active Learning in Higher Education, 22*(2), 97–113.
85. Muir, T., Wang, I., Trimble, A., Mainsbridge, C., & Douglas, T. (2022). Using interactive online pedagogical approaches to promote student engagement. *Education Sciences, 12*(6), 415.
86. Wang, Y., & Liu, Q. (2020). Effects of online teaching presence on students' interactions and collaborative knowledge construction. *Journal of computer assisted learning, 36*(3), 370–382.
87. Aldaghrı, A. A., & Oraıf, I. M. (2022). The impact of online teaching on students' engagement in writing during the pandemic of Covid-19. *Turkish Online Journal of Distance Education, 23*(3), 216–229.
88. Maulidah, U. N., & Aziz, I. N. (2020). The effectiveness of online collaborative learning on students writing skills. *EDUCATIO: Journal of Education, 5*(2), 141–149.
89. Poondej, C., & Lerdpornkulrat, T. (2020). Gamification in e-learning: A Moodle implementation and its effect on student engagement and performance. *Interactive Technology and Smart Education, 17*(1), 56–66.
90. Rajabalee, Y. B., & Santally, M. I. (2021). Learner satisfaction, engagement and performances in an online module: Implications for institutional e-learning policy. *Education and Information Technologies, 26*(3), 2623–2656.
91. Cheng, P., & Ding, R. (2021). The effect of online review exercises on student course engagement and learning performance: A case study of an introductory financial accounting course at an international joint venture university. *Journal of Accounting Education, 54*, 100699.
92. Sanjeev, R., & Natrajan, N. S. (2019). Role of blended learning environment towards student performance in higher education: Mediating effect of student engagement. *International Journal of Learning and Change, 11*(2), 95–110.
93. Coovadia, H., & Ackermann, C. (2021). Integrating digital pedagogies into a typical student learning lifecycle and its effect on exam performance. *Accounting Education, 30*(1), 42–62.
94. Ogunyemi, A. A., Quaicoe, J. S., & Bauters, M. (2022). Indicators for enhancing learners' engagement in massive open online courses: A systematic review. *Computers and Education Open, 3*, 100088.
95. Hew, K. F., Jia, C., Gonda, D. E., & Bai, S. (2020). Transitioning to the "new normal" of learning in unpredictable times: Pedagogical practices and learning performance in fully online flipped classrooms. *International Journal of Educational Technology in Higher Education, 17*, 1–22.
96. Koh, J. H. L., & Daniel, B. K. (2022). Shifting online during COVID-19: A systematic review of teaching and learning strategies and their outcomes. *International Journal of Educational Technology in Higher Education, 19*(1), 56.
97. Kordrostami, M., & Seitz, V. (2022). Faculty online competence and student affective engagement in online learning. *Marketing Education Review, 32*(3), 240–254.
98. Ng, L. K., & Lo, C. K. (2022). Enhancing online instructional approaches for sustainable business education in the current and post-pandemic era: An action research study of student engagement. *Education Sciences, 13*(1), 42.
99. Duffy, K., & Ney, J. (2015). Exploring the divides among students, educators, and practitioners in the use of digital media as a pedagogical tool. *Journal of Marketing Education, 37*(2), 104–113.

100. Bećirović, S. (2023). Fostering student engagement in implementing digital pedagogy. In *Digital pedagogy: The use of digital technologies in contemporary education* (pp. 97–111). Springer Nature Singapore.
101. Elbyaly, M. Y. H., & Elfeky, A. I. M. (2023). The impact of blended learning in enhancing the skill performance of producing digital content among students of optimal investment. *Annals of Forest Research, 66*(1), 2031–2043.
102. Kim, H. J., Hong, A. J., & Song, H. D. (2019). The roles of academic engagement and digital readiness in students' achievements in university e-learning environments. *International Journal of Educational Technology in Higher Education, 16*(1), 1–18.
103. Salta, K., Paschalidou, K., Tsetseri, M., & Koulougliotis, D. (2022). Shift from a traditional to a distance learning environment during the COVID-19 pandemic: University students' engagement and interactions. *Science & Education, 31*(1), 93–122.
104. Nermend, M., Singh, S., & Singh, U. S. (2022). An evaluation of decision on paradigm shift in higher education by digital transformation. *Procedia Computer Science, 207*, 1959–1969.

Sohini Datta is presently working as the Assistant Professor in the Department of Management at Institute of Engineering & Management (IEM), Kolkata. She is having over 13 years of experience in teaching Communication & Soft Skills, Corporate Communication, Language, Leadership & Teambuilding, Organizational Behaviour to post-graduate Management & undergrad uate Engineering students. Prior to teaching at various renowned colleges at West Bengal, Ms. Datta was associated with the popular English Daily, The Statesman as Content Editor for few years. She has often travelled to many parts of India and abroad in these capacities. She has also been a Resource Person for some publishing houses and has conducted many training programs and workshops for some reputed institutions in ELT and other generic areas. She has prepared study materials for the Post Graduate Distance Learning courses offered by Netaji Subhash Open University (NSOU), Kolkata. She is also a lifetime Instructor of IGNOU in Communication skills across various domains.

Debolina Mitra is presently associated with IIM Bangalore as an Academic Associate. She has contributed in reputed Journals and International Conferences like IIM, IITs, etc. She is having experience of more than 5 years in Research.

Udit Chawla, PhD, is presently working as an Associate Professor in Business Administration Department at the Institute of Engineering and Management, School of University of Engineering and Management, Kolkata, India. He is an accomplished academician with a track record of employment in the higher education sector and is a strong education professional focused on business analytics, data science, machine learning, and artificial intelligence. He has expertise in R, SPSS, SAS, PYTHON, etc. He has published more than 30 articles in reputed international journals, which are indexed in ABDC, Scopus, SCI, etc., and he has contributed more than 40 articles presented at international conferences such as IIMs, IITs, etc.

Devika Mulchandani has a deep passion for research and thrives on exploring new ideas and concepts. She enjoys delving into diverse topics, analyzing information, and uncovering meaningful insights. Her curiosity drives her to seek knowledge beyond conventional boundaries, allowing her to develop innovative perspectives. She finds joy in solving complex problems, engaging in thought-provoking discussions, and contributing to intellectual growth. Her enthusiasm for exploration extends to understanding how things work and applying this knowledge creatively. She consistently demonstrates a commitment to learning, growth, and the pursuit of excellence in everything she undertakes. She is having experience of more than 3 years in Research.

Subrata Chattopadhyay is a MSc, MBA, PhD from IIT-ISM Dhanbad. Adept at innovative teaching practices he is highly networked with the industry. He has contributed in more than 30 International Journals and has published more than 23 papers in seminars and conferences. Besides, he has authored 3 books on Computations in Estate Management, Transport Management and Values and Ethics for Engineers and Managers. He is associated as life member with ISTD, CMA, PRSI, NIPM, NHRD, NFED, ISABS and does mentoring and consulting for various Organisations.

Part III
Unchartered Territories: Operational Management and COVID-19

Chapter 12
Interpretable Prediction Model for Chest X-Ray COVID-19 Diagnosis: A Comparative Assessment Using Grad-CAM, Occlusion Sensitivity, and LIME

B. Lakshmipriya and S. Jayalakshmy

Abstract Rapid outburst of Novel Corona Virus Disease 2019 (COVID–19) epidemic turned out to demand additional screening methodologies to serve the rising need. For diagnosing the COVID-19 infections from the imaging modalities, deep neural nets take a leading role. The current deep learning models, in spite of the exceptional performance in terms of accuracy, it lacks explainability on the decision made and any misdiagnosis or false positive cannot be tolerated. A trustworthy deep learning system for COVID-19 diagnosis is desired during this current pandemic. The work implemented in this chapter aims at providing trustworthy diagnostic model for COVID-19 by providing explanations behind the network's decision. This is implemented in two modules: classification and network visualization. This chapter explores the functionality of AlexNet, SqueezeNet, and ShuffleNet in classifying chest X-ray COVID-19 dataset. Experimental results demonstrate the proficiency of ShuffleNet in the categorization of this infectious virus dataset with an enhanced accuracy of 95.42%. Following the classification, network visualization is realized using three approaches: gradient-weighted class activation mapping (Grad-CAM), occlusion sensitivity mapping and local interpretable model-agnostic explanations (LIME). The deep learning framework suggested in this chapter for COVID-19 diagnosis, in addition to offering reasoning behind the decision also brings in the clinicians into the loop in subjective assessment of the network's prediction thereby improving the trustworthiness of the diagnostic system.

Keywords Convolutional neural network · COVID-19 · AlexNet · SqueezeNet · ShuffleNet · Grad-CAM · Occlusion sensitivity · LIME

B. Lakshmipriya (✉)
School of Engineering, Jawaharlal Nehru University, New Delhi, India
e-mail: lakshmipriya10@jnu.ac.in

S. Jayalakshmy
Electronics and Communication Engineering, IFET College of Engineering, Villupuram, India
e-mail: jayalakshmyece@ifet.ac.in

S. Gangopadhyay et al. (eds.), *Pandemic Diaries*,
https://doi.org/10.1007/978-981-96-5415-4_12

12.1 Introduction

COVID-19 outbreak has resulted in global pandemic creating a severe threat to human community through fast spreading of disease and death rate spikes. World Health Organization (WHO) releases the epidemiological update on COVID-19 every week and the recent report in June 2021 states that the current number of novel coronavirus illness continues to be the same comparatively and the figure of additional deaths resume to subside globally, resulting in 2.6 million recent cases and 54,000 new deaths [1]. On a global scale, the overall number of infected cases transcend to 183 million and the mortality rate is nearly 4 million. Among this group, even though majority of the infected persons with COVID-19 undergo slight respiratory distress, few cases evolved a fatal acute inflammation in lungs with Pneumonia. Some of the common medical traits of this infectious coronavirus illness is bronchitis provoking high temperature, breathlessness, sneezes, and lung failure with air sacs of lungs filled with fluid [2, 2, 3]. In order to curb the contagious Wuhan coronavirus, expeditious medical assistance and efficient examination are the drastic need.

The pandemic spanning over two waves and the expected third wave intimidate the healthcare professionals and require a paradigm shift in diagnostic and therapeutic procedures to offer enhanced healthcare to a larger patient population simultaneously. Technological advancements offer a high-end handshake to the clinicians by the virtue of machine intelligence enabled decision support model to facilitate in their diagnosis routine. With the onset of COVID-19 outbreak, huge research is being conducted to devise fast and accurate diagnosis mechanisms from the chest screening modalities. Deep learning, which is being employed in multitude of medical imaging diagnoses, is employed in COVID-19 diagnosis to assist radiologists and physicians in their clinical examinations. Deployment of deep learning models in clinical practice has resulted in improved diagnostic accuracy within a short span of time at a reduced cost. The technology is also expediting towards non-invasive method of diagnosis through imaging modalities thereby avoiding the need for painful needle biopsies in majority of diseases. Despite these benefits, the technology lacks explanation on why a particular decision is made by the network. Eventually, the lack of interpretation doubts the credibility of the deep learning systems in medical diagnosis, where the decision made by the system plays a crucial role in patient management. To improve the trustworthiness of the system, the black-box model has to be made transparent to the humans, so that the network behaviour can be understood and reasoning behind the decision can be explained. The resulting model can be claimed as explainable deep learning systems and will be trusted by the physicians in their diagnostic routine.

The transparency of the network can be improved by providing visual explanations behind the decision through explainable AI (XAI) [4] approaches. These approaches basically aim at highlighting a region in the input image of the convolution neural network (CNN) to provide insight on which region of the image has contributed for the network's decision. This visual explanation is provided by identifying the set of neurons activated heavily during forward pass of the prediction. The discrimination of highly activated neurons from others is done by observing the weights of the input of a CNN in the last layer of convolution and then producing a coarse localization map which when overlaid on the input image provides visual reasoning behind the network's decision.

In this chapter, visual explanation of prediction made by the CNN architecture in classifying chest X-rays (CXR) corresponding to COVID-19, viral pneumonia and lung opacity. Experimentation on CXR classification is carried out using AlexNet, SqueezeNet, and ShuffleNet architectures and the network's decision interpretation is performed using gradient-weighted class activation mapping (Grad-CAM), occlusion sensitivity mapping and local interpretable model-agnostic explanations (LIME).

Remainder of this chapter is organized as follows: Related works on COVID-19 diagnosis using various deep neural nets and network visualization schemes were presented in Sect. 12.2. The structure of pre-trained CNNs used in this chapter is elaborated in Sect. 12.3 and the technical details on the network visualization algorithms are presented in Sect. 12.4. Section 12.5 presents the proposed XAI framework for COVID-19 diagnosis. In Sect. 12.6, the outcomes of the proposed methodology along with the inferences are outlined. Finally, the findings are concluded in Sect. 12.7.

12.2 Survey of Literature

The widely used screening method for COVID-19 illness is real-time Reverse Transcription Polymerase chain reaction (RT-PCR) test [5]. In spite of the fact that it is most common, this procedure is hand-operated, expensive, complex, and extended process [6]. In addition, the significant lack in the supply of this test kits, results in slowing down the preventive action. Other screening methods for COVID-19 pneumonia include examination of physical signs, medical history, and chest radiography. Among this, X-Ray imaging is the significant method of diagnosing as it is easily accessible. The research findings in the literature [7–11] report that a big part of infected cases have identical characteristics on X-ray images of the lower part of lungs. Since the x-ray picture of other viral infections resembles similar, radiotherapists find it hard to figure out and differentiate coronavirus infection.

In recent years, the emphasis on machine intelligence, deep learning (DL) techniques and algorithms have progressively acquired considerable significance in dealing various challenges in all real-life challenges including healthcare applications [12–15]. In specific, CNN-based DL methods have demonstrated superior performance. Ever since, this approach has been broadly accepted by scientific community for learning and training different models [16]. Exploitation of several transfer learning architectures has substantially improved the performance of deep neural nets for relatively limited number of images. Transfer learning-based ensemble model was developed for identifying pneumonia infections in CXR with an accuracy of 96.4% on Guangzhou Women and Children's Medical Center dataset [17]. In 2019, Gómez et al. [18] utilized CNN to improve the image characteristics in shaded images captured from video endoscopy for detecting larynx disorders. In 2018, the authors experimented deep CNN VGG-16 pre-trained model with Japanese society of radiological technology database for detecting bacterial and viral pneumonia on CXR and provided a better sensitivity and specificity [19].

Classification of eight types of common thoracic disorders from CXR images using CNN was carried out on Chest X-ray 8 dataset [20]. Shin et al [21] exploited three deep CNN namely CifarNet, AlexNet, and GoogLeNet in the identification of thoraco-abdominal lymph node and classification of interstitial lung diseases and yielded good performance. In [22], the authors proposed CheXNeXt a deep insightful 121-layer CNN to identify fourteen disorders which include pulmonary masses, pneumonia, and nodules in CXR and produced 88.85% accuracy. Prominent pre-trained architectures viz. AlexNet and GoogLeNet were exploited in classifying healthy and pulmonary TB [23].

Few works on network visualization are available in literature for COVID-19 datasets. Visualization of the network by the virtue of Grad-CAM was implemented on chest X-ray CXR dataset images with three classes: normal, pneumonia, and COVID-19 [24, 25]. Similarly, the Grad-CAM visualization was also implemented on lung CT images acquired from COVID-19 patients [26]. A first XAI-based open source software was developed using DenseNet201 and Grad-CAM visualization to assist radiologists across the globe in COVID-19 diagnosis from pulmonary CTs with average lesion classification accuracy of 84% [27]. Interpretation of COVID-19 images during diagnosis was carried using gradient saliency maps built from the classification scores of the deep learning-based classifier [28]. Interpretation using occlusion sensitivity maps was carried out on CXR and lung CT images for the diagnosis of corona pulmonary inflammations [29].

Most of the existing literatures on CNN-based image classification have focussed on improving the classification performance and have not paid attention on understanding the network's decision. In this work, it is proposed to build an intelligent trustworthy deep learning model for COVID-19 diagnosis from CXR by integrating an XAI approach with pre-trained CNN for prediction. A hybrid dataset of COVID-19 CXR images taken from Kaggle is used in this work to identify the clinical imaging features corresponding to healthy, lung opacity, COVID-19, and viral pneumonia infection. This work is expected to provide trustworthy deep learning model that provides transparent explanations for the diagnostic decision during the course of COVID-19 treatments. This chapter attempts to explore the prediction performance of three different CNN architectures viz. AlexNet, SqueezeNet, and ShuffleNet on CXR images. Subsequently, the network visualization will be carried out using three approaches viz. gradient-weighted class activation mapping (Grad-CAM), occlusion sensitivity mapping, and local interpretable model-agnostic explanations (LIME) and their performances will be compared.

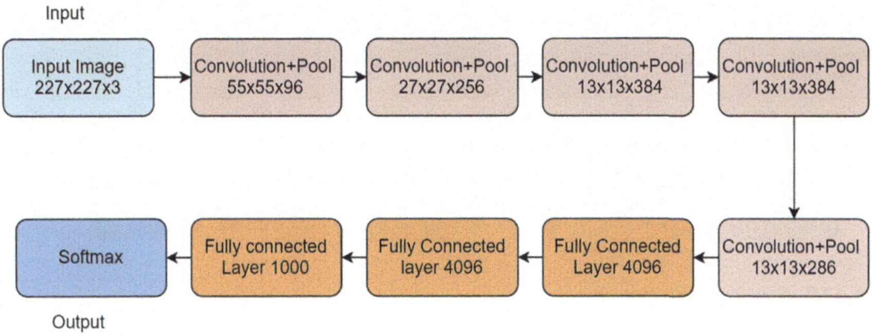

Fig. 12.1 AlexNet architecture

12.3 Pre-Trained Convolution Neural Network

12.3.1 AlexNet

The first convolution network which had significant influence in the field of machine learning is AlexNet [16]. This is a eight-layered pre-trained comprising five convolution layers, three maxpool layers to extract the features followed by two fully connected layers with softmax activation to perform classification. Each of the five convolutional layers is composed of ReLu activation functions and convolutional filters. AlexNet accepts the images as input with a size of 227×227 with stochastic gradient momentum as learning algorithm. Figure 12.1 shows the architecture of AlexNet model.

12.3.2 SqueezeNet

SqueezeNet is an 18-layer deep CNN which utilizes some of the formulating strategies to decrease certain parameters by employing fire modules [30]. A fire module is composed of a 1×1 squeeze layer of convolution providing inputs to an expand layer, a combination of convolutional kernels of order 1 and 3. The kernels in every fire module is progressively increased all the way through the network. Figure 12.2 shows the architecture of SqueezeNet model.

12.3.3 ShufleNet

ShuffleNet is a pre-trained model which employs a mixing operation (shuffle) in conjunction with depth-wise convolutions [31]. Residual blocks from ResNet model

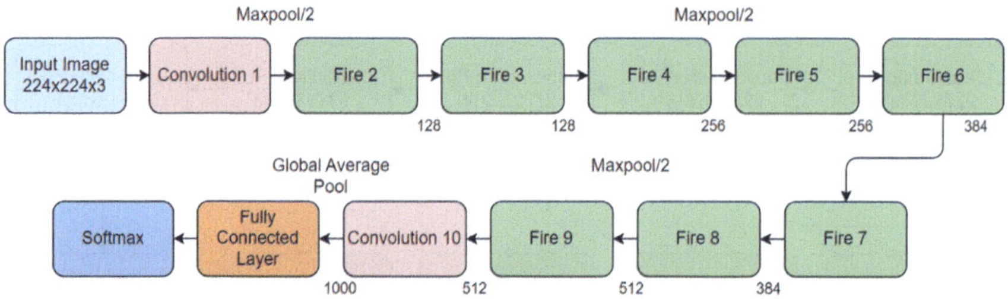

Fig. 12.2 SqueezeNet architecture

are the beginning point of ShuffleNet. This is altered with a point-wise group convolution and mixing operation. Here, the primary and secondary 1 × 1 convolutions are interchanged by group convolutions. Normally, the mixing operation is used after the first 1 × 1 convolution. Figure 12.3 shows the architecture of one ShuffleNet base unit. 17 such units are connected in cascade in the architecture of ShuffleNet. The architecture of ShuffleNet is presented in Fig. 12.4

12.4 Explainable Artificial Intelligence

The fact that the AI technology does not provide explanation on why a certain decision is made makes the technology, a black-box model in spite of its profound performance and fame. To provide explanations to the decision made by the AI system, XAI technology is emanating as a revolutionary technology to demystify the existing AI architectures, thereby making the so-called black-box model, highly transparent. The technology aims at revealing the 'Why' behind every deep learning decision by means of human understandable visual explanations.

A heterogeneous set of XAI algorithms are available in literature, namely gradient-based saliency maps [32], Bayesian case model [33], layer-wise relevance backpropagation [34], global attribution mapping [35], class activation mapping (CAM) [36], LIME [37], Grad-CAM [38], occlusion sensitivity [39], and so on. Among these algorithms, this chapter attempts to explore the functionality of three popular XAI techniques, namely Grad-CAM occlusion sensitivity mapping and LIME on CXR images using AlexNet, SqueezeNet, and ShuffleNet. These three XAI techniques are popular as they can be applied to a diverse group of CNN-based networks like CNNs comprising fully connected layers for image classification tasks, CNNs for image captioning and visual question answering and reinforcement learning application without incorporating any architectural change in the network. These visualization models render useful intuitions behind the decision of a CNN in image classification and object detection tasks. In case of incorrect prediction, the explanation provided by the visualization model assists in fine tuning the network into a more generalized model avoiding the unnecessary bias with the input dataset, thereby enhancing the network performance.

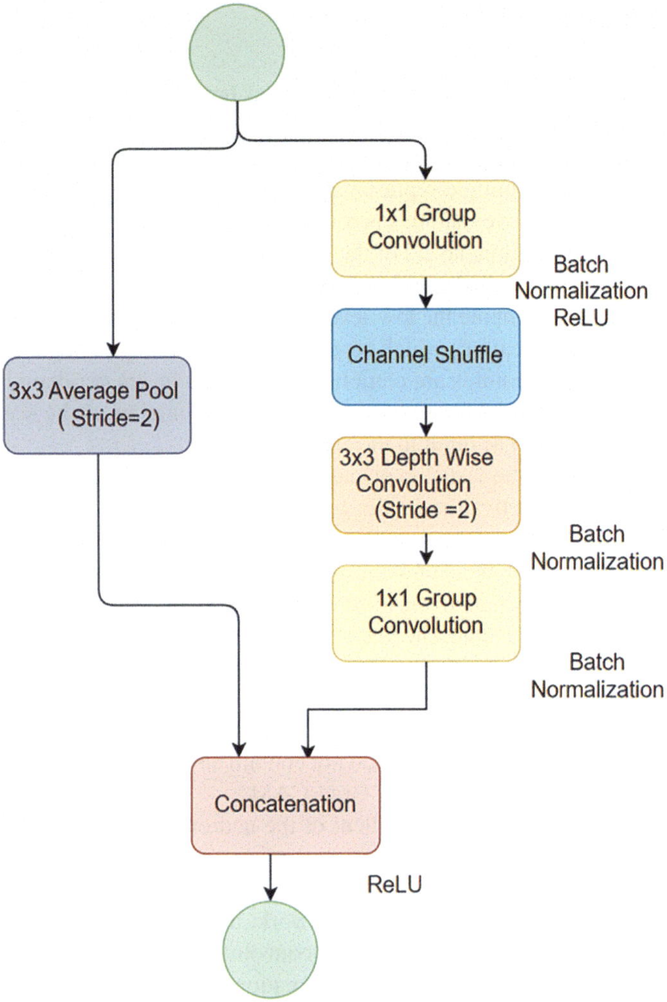

Fig. 12.3 Architecture of ShuffleNet base unit

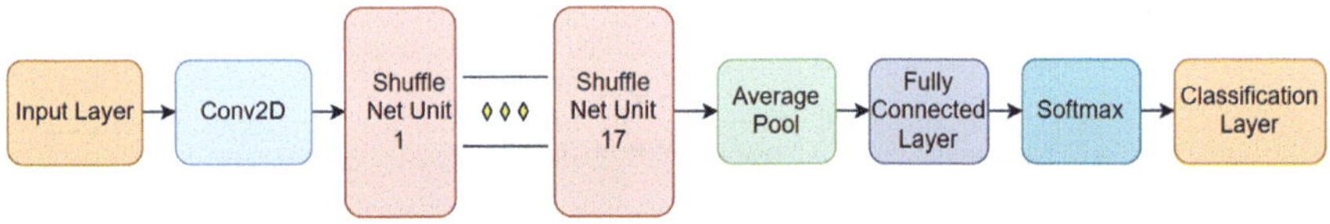

Fig. 12.4 ShuffleNet architecture

12.4.1 Grad-CAM

CNN architectures have a series of convolutional layers in which each layers are mapped with non-linear activation function backed with pooling layer that tends to capture the informative patterns from the target which is considered as input. The initial layers learn the multiple representation of the target and are then passed on to the next convolution layer and this process continues. As a result, a more abstract representation from multiple layers of the CNN is learnt. The representations learnt at the initial layers are called low-level features comprising localized edges. The middle layers tend to combine the low-level features extracted in the previous layers and produce mid-level features which correspond to a part in the image. In a similar manner, the mid-level features are combined at the final layers to obtain high-level features. These high-level features are the brief representation of the scene of the image and are finally fed as input to the classifier.

CAM [36] provides the visual explanation of the decision by identifying the set of neurons with high activation in the forward pass during the course of image classification by a CNN architecture. The activation map thus obtained will be two dimensional (2D) with a dimension same as that of the input image. The high values in the activation map correspond to the most discriminative portion of image which has led CNN in making a decision. The discriminative portion of the image is identified by overlaying the activation map on the image under consideration. Since the final convolution layer contains better spatial resolution of the scene, the activation map is derived from output of the final convolution layer. The drawback of CAM is that it can be applied only to networks that do not contain any fully connected layer. This shortcoming is overcome by means of Grad-CAM.

Grad-CAM [38] considers the gradient of the neuron weights entering a convolution layer to quantify the contribution of every neuron towards the final decision. Though Grad-CAM can assess the gradients weights associated with the neurons at the input of any convolution layer in the network, it is predominantly employed to provide explanation on decision from final convolution layer. This is because, the final convolution provides more abstract information about the input image with high degree of resolution.

The class-specific neuron weight magnitude β_k^c for the class 'c' is given in Eq. (12.1):

$$\beta_k^c = \frac{1}{Z}\sum_i \sum_j \frac{\partial y^c}{\partial A_{ij}^k}, \tag{12.1}$$

where y^c denotes the probability score corresponding to class 'c' and A^k is the kth feature map activation of the convolution layer. $\frac{\partial y^c}{\partial A_{ij}^k}$ represents the gradients computed via backpropagation along the width and height directions of the convolution kernel represented respectively by the indices i and j. $\frac{1}{Z}\sum\sum()$ denotes the global average pooling operation performed across the feature maps of one convolution layer. The

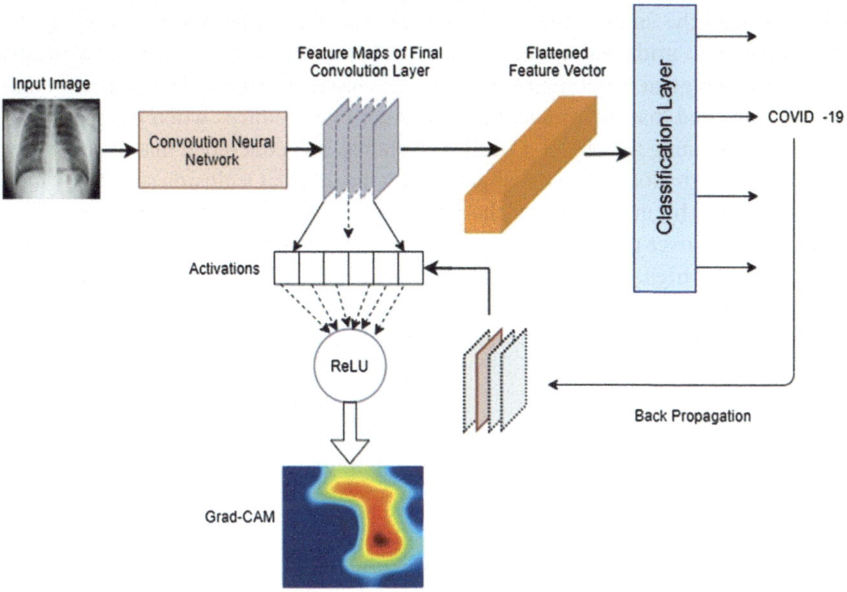

Fig. 12.5 Diagrammatic representation of Grad-CAM framework for understanding the network prediction

weighted combination of gradient activation maps (L_{GC}^{c}) in Grad-CAM approach activated by a non-linear activation function viz. rectified linear unit (ReLU) is given as

$$L_{GC}^{c} = ReLU\left(\sum_{k} \beta_{k}^{c} A^{k}\right) \tag{12.2}$$

The resultant is a heat map matrix of dimension same as that of the feature representations of the corresponding convolution layer. The Grad-CAM approach is now able to identify the highly activated neurons towards the decision in favour of a class 'c'. Still it lacks localization of the relevant regions in the input image as the dimensions of the input image and feature activation map are not the same. In order to achieve this, feature map is then back propagated to the rectified convolutional feature maps of interest followed by upsampling of L_{GC}^{c} using bilinear interpolation. This is clearly depicted in Fig. 12.5.

12.4.2 Occlusion Sensitivity Maps

Occlusion sensitivity is yet another visualization technique which tends to identify whether the network decision is based on identifying the location of the object or looking around the surrounding of the object. The technique basically divides the entire image into grids and systematically occludes one region at a time by means of a grey rectangular patch and observes the classifier output. This experimentation demonstrated that the probability of correct class reduces when the portion of the image containing the object is occluded. Hence, it is concluded that the classifier decision is made by localizing the object within the image. Occlusion sensitivity map was constructed by the observing the probability scores computed with occluding masks spanning over the entire image and the position of the occluding mask corresponding to the highest probability score during classification is identified as the occlusion sensitivity map. The occlusion sensitivity map thus obtained when overlaid on the input image recognizes the region within the image contributing towards the decision.

12.4.3 LIME

LIME model basically attempts to spot-out an interpretable model by identifying a patch of pixels called super-pixel that contributes maximum towards a class output. The super-pixel is identified with the help of a binary vector on a local patch level basis and hence the term local. As a first step in LIME, 'n' pixels are randomly picked to form a super-pixel and a prediction is made using the deep learning model used for classification. A similarity score is then calculated between the predictions made with super-pixel and the actual observations. The procedure is repeated for different permutation combination of the selected super-pixels. Resultant is the set of class prediction scores for original image and the fake image created using super-pixels. The prediction class scores for the fake data depends on the size of super-pixels chosen from the original image. Highly descriptive features which contributed towards enhanced prediction on the data created by the super-pixels are identified and are used to derive the explanations for the local behaviour of the deep learning model used.

12.5 XAI Framework for COVID-19 Diagnosis

CXR images play a vital role during preliminary investigation in the diagnosis of COVID-19 infections. CXR images are used as an alternative screening modality in the diagnosis of COVID-19 against the large demand for test kits during the pandemic outbreak. The framework proposed in this chapter aims at providing visual explanation for the decision made by the DL network towards COVID-19 diagnosis from the CXR images. The system model for the proposed framework is presented in Fig. 12.6.

In order to provide visual explanations, a classifier model has to be built for diagnosis from CXR images. In this framework, the classifier model is built using

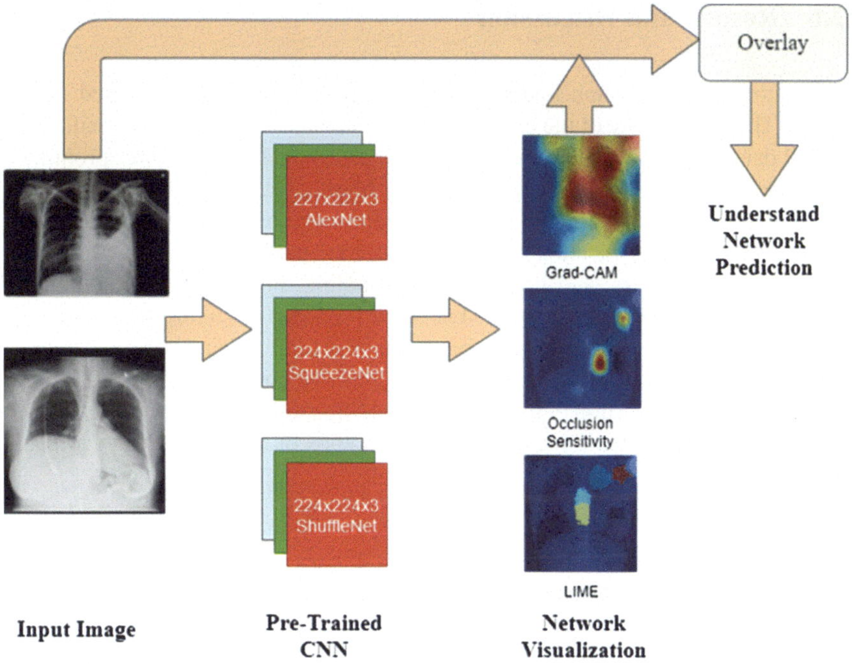

Fig. 12.6. XAI Framework for COVID-19 Diagnosis

pre-trained CNNs trained with ImageNet database are used. The performance of the proposed framework has been evaluated and compared with AlexNet, SqueezeNet, and ShuffleNet as classifier. These three networks are chosen as they have manifested themselves to be effective for image classification with minimal learnable parameters. The images used in this study are grayscale with dimensions 299 × 299. Since the pre-trained networks accept three-dimensional (3D) images as input, the training and testing images used in this framework are converted to 3D images by concatenating the 2D images in three planes. Also, the images are resized to 227 × 227 × 3 and 224 × 224 × 3 as required for AlexNet, SqueezeNet, and ShuffleNet architectures. The output of the classifier model built separately using the three pre-trained networks is compared in terms of classification accuracy, training time, and number of learnable parameters. Network visualization for every test image following the classification by the trained network is done using Grad-CAM and occlusion sensitivity approaches and the outcomes are assessed subjectively. The output of the Grad-CAM and occlusion sensitivity will be in the form of class-specific heatmap visualization which can be used to demystify the CNN network to understand where exactly the network looks in the image to make a decision.

12.6 Results and Discussion

The outcomes and research findings of the proposed model are presented in this section. The first module of the proposed system model is training and classification of CXR dataset using CNN. In this work, this is accomplished using three prominent pre-trained CNNs: AlexNet, SqueezeNet, and ShuffleNet. The performance of the classifiers using the three networks is compared in terms of, training progress graph, confusion matrix, classification accuracy, class-wise performance metrics, number of learnable parameters, and time taken for training. Following the training and classification, demystifying the trained network is carried out in the second module using Grad-CAM and occlusion sensitivity map. Network visualization implemented in the second module of this work provides subjective output in the form of image which highlights the portion of the image contributing towards the decision. The two modules are implemented using deep learning toolbox of MATLAB 2021a software on NVIDIA RTX3070 GPU.

12.6.1 COVID-19 Chest X-Ray Dataset

The images used in this review are taken from Kaggle's COVID-19 CXR database. The dataset comprises four classes of images, namely normal, COVID-19, lung opacity, and viral pneumonia. This dataset is a hybrid dataset wherein the CXR images are collected from nine different CXR datasets: Radiological Society of North America (RSNA) [40], BIMCV COVID-19 CXR dataset [41], Germany medical school CXR dataset [42], SIRM [43], Eurorad [44], GitHub COVID-19 CXR dataset [45, 46], and Kaggle Pneumonia CXR dataset [47].

The dataset contains a total of 21,164 images comprising 10,192, 3615, 6012, and 1345, respectively, for normal, COVID-19, lung opacity, and viral pneumonia. Among the 21,164 images, in this work, 4000 images (1000 per class) are considered for training and 1200 images (300 for every class) are considered for testing. All the images are available as portable network graphics (PNG) format. The visual characteristics in the CXR images for COVID-19 include retino-nodular opacities and multifocal bilateral ground class opacity [48]. Lung inflammation in the form of infiltration and increased attenuation will be visible in CXR for other pulmonary disorders. Sample CXR images belonging to four classes in the dataset are presented in Fig. 12.7.

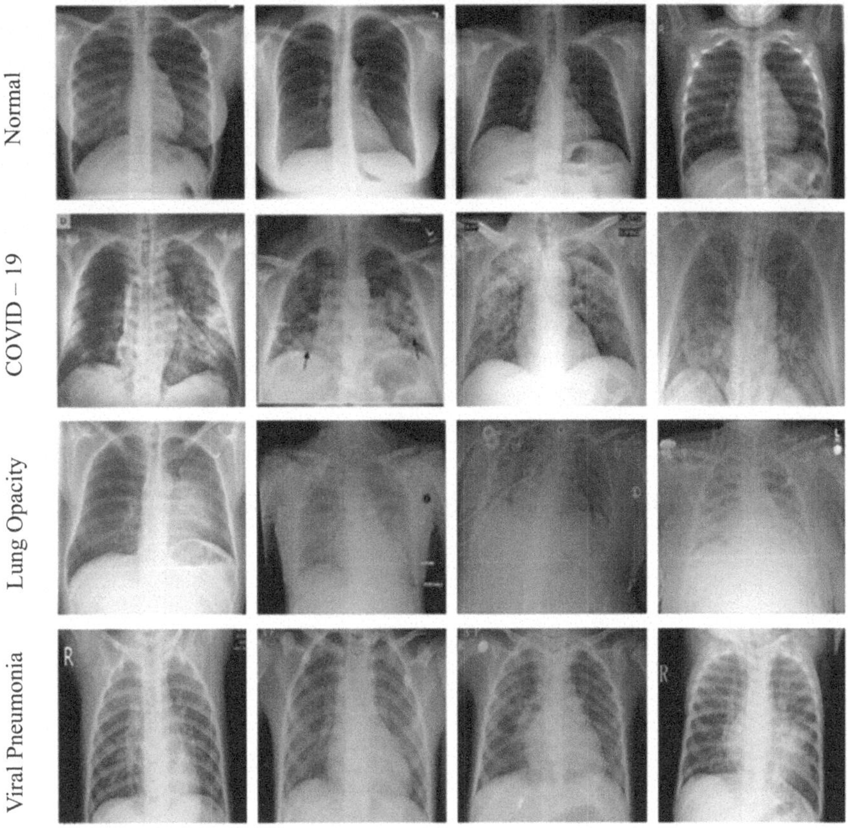

Fig. 12.7 Sample CXR dataset images

12.6.2 Assessment of Classification Using Pre-Trained CNNs

In order to make a fair comparison, hyper-parameters, the parameters defining the network model during training are maintained the same for all three networks considered in this work. The hyper-parameters defining the learning environment of the network are tabulated in Table 12.1.

Tracking the network performance during training and validation is essential for post-hoc analysis in checking how quickly the network classification accuracy is improving. This can be done by plotting the training progress to get information on accuracy and loss produced by the network during training and validation. Additionally, information on overfitting if any, can be understood from the training progress plot. The training progress plots obtained during training of AlexNet, SqueezeNet, and ShuffleNet are presented in Figs. 12.8, 12.9 and 12.10. It is seen that of the three networks, the accuracy and loss saturate earlier in ShuffleNet.

Table 12.1 Hyper-parameter settings for network training using All Three Architectures

Hyper-parameter	Value
No. of epochs	20
Learning rate	0.0001
Weight updation algorithm	Adaptive moments (Adam)
Mini batch size	100
Iterations per epoch	28
Total iterations	560
Learning rate decay factor	0.9
Squared gradient decay factor	0.999
Regularization technique	L2Norm
Batch normalization statistics	Population

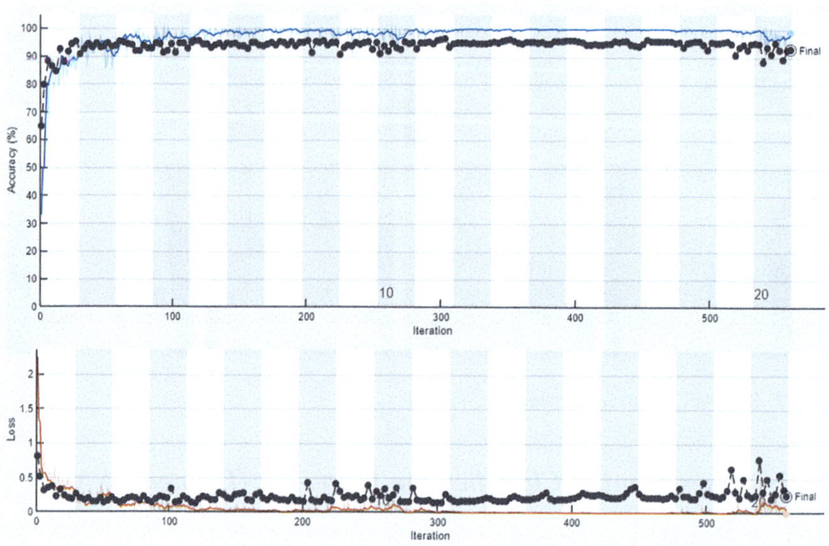

Fig. 12.8 Training progress during the classification using AlexNet architecture

In addition to training progress, the network's performance can be assessed more quantitatively through a confusion matrix. Confusion matrix provides representation about performance of a network in terms of correct and incorrect predictions. Furthermore, it summarizes the network's prediction in terms of classification accuracy (overall and class wise), precision, recall, and F1-score. Classification accuracy is a measure of statistical bias on correct prediction rate. Precision is a measure which gives insight on how often the classifier is correct while making a decision in favour of a class. Recall provides information on how many of the positive classes were able to predict correctly by the classifier under consideration. F1-score is the

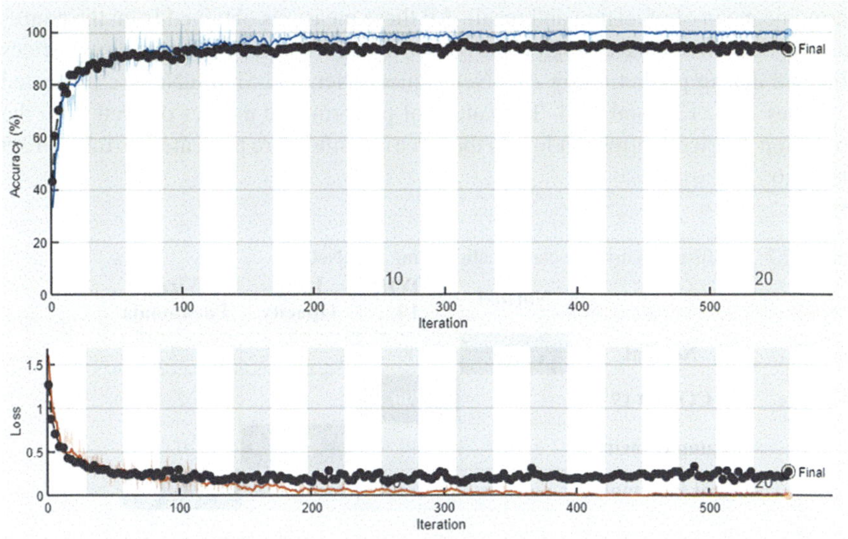

Fig. 12.9 Training progress during the classification using SqueezeNet architecture

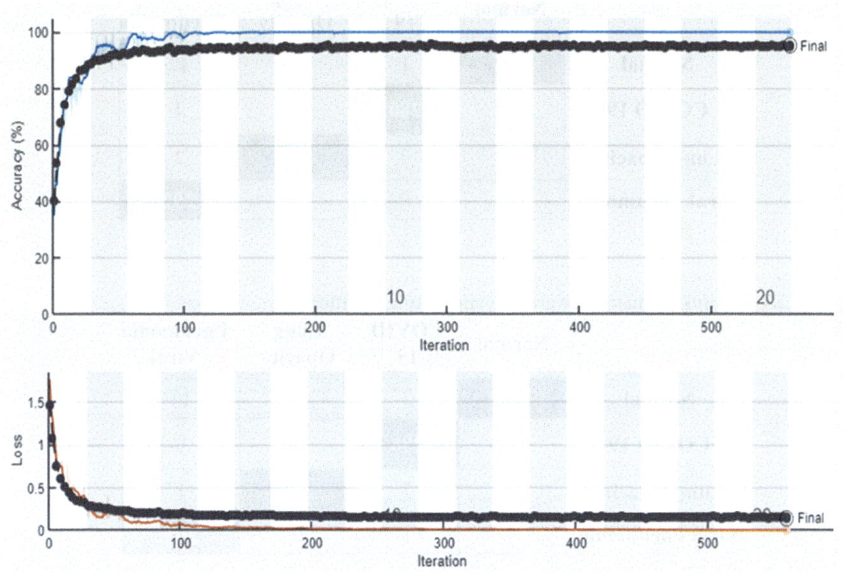

Fig. 12.10 Training progress during the classification using ShuffleNet architecture

harmonic mean of precision and recall. All these measures obtained from the confusion matrices should have a high value for better performance. The confusion matrices obtained during prediction by AlexNet, SqueezeNet, and ShuffleNet are presented in Tables 12.2, 12.3 and 12.4. The values of performance metrics derived from the confusion matrices corresponding to the three classifiers are presented in Tables 12.5 and 12.6.

Table 12.2 Confusion matrix for classification using AlexNet

	Normal	COVID 19	Lung Opacity	Viral Pneumonia
Normal	275	8	0	17
COVID 19	0	290	8	2
Lung Opacity	10	19	271	0
Viral Pneumonia	10	11	0	279

Table 12.3 Confusion matrix for classification using SqueezeNet

	Normal	COVID 19	Lung Opacity	Viral Pneumonia
Normal	286	1	1	12
COVID 19	12	260	27	1
Lung Opacity	14	0	285	1
Viral Pneumonia	8	0	0	292

Table 12.4 Confusion matrix for classification using ShuffleNet

	Normal	COVID 19	Lung Opacity	Pneumonia Viral
Normal	278	2	6	14
COVID 19	0	286	14	0
Lung Opacity	2	7	290	1
Viral Pneumonia	8	1	0	291

Table 12.5 Comparison of classification accuracies for the three architectures

Classifier	Classification accuracy (%)
AlexNet	92.92
SqueezeNet	93.58
ShuffleNet	**95.42**

Table 12.6 Comparison of class-wise metrics for the three architectures

Metric	Network	Normal	COVID-19	Lung Opacity	Viral Pneumonia
Classwise accuracy	AlexNet	96.25%	96%	96.92%	96.67%
	SqueezeNet	96%	96.58%	96.42%	***98.17%***
	ShuffleNet	***97.33%***	***98%***	***97.5%***	98%
Precision	AlexNet	0.93	0.88	***0.97***	0.94
	SqueezeNet	0.89	***1***	0.91	***0.95***
	ShuffleNet	***0.97***	0.97	0.94	***0.95***
Recall	AlexNet	0.92	*0.97*	0.90	0.93
	SqueezeNet	***0.95***	0.87	0.95	***0.97***
	ShuffleNet	0.93	0.95	***0.97***	***0.97***
F1 score	AlexNet	0.92	0.92	0.94	0.93
	SqueezeNet	0.92	0.93	0.93	***0.96***
	ShuffleNet	***0.95***	***0.96***	***0.95***	***0.96***

It is evident from the classification metrics presented in Tables 12.5 and 12.6 that ShuffleNet records better performance compared to AlexNet and SqueezeNet in terms of all metrics considered for assessment. The improvement in classification performance of ShuffleNet is due to its efficient architecture comprising channel shuffle operation. The network by the virtue of increased feature map channels tends to encode from information from the input images. The channel shuffle operation enables the flow of inter-group information across multiple group convolution layers thereby enhancing the classification performance. It is proved in [31] that the network's classification performance increases with increase in group. Though ShuffleNet is specially designed for mobile devices, attracted by its performance, the architecture is considered in this chapter for assessing the prediction performance of CXR images. As far as medical diagnosis is concerned, diagnostic accuracy is more important and hence to achieve better accuracy, the number of groups is considerably increased in this work. This has increased the computational as well as time complexity of the network during training. Since training will be done only once, the overhead on increased computation and time complexities are acceptable in order to achieve high prediction accuracy (Table 12.7).

Table 12.7 Comparison of time and computational complexities for the three architectures

Classifier	Time Taken for training	No. of learnables
AlexNet	211 min 32 s	5,68,84,612
SqueezeNet	303 min 37 s	7,40,932
ShuffleNet	470 min 16 s	8,63,892

12.6.3 Assessment of Network Visualization

The outcome of second module of this work is presented in this section. Network visualization essentially provides visual explanation on the factor that influenced the network's decision. The network's visualization and interpretability are explored using Grad-CAM and occlusion sensitivity maps in this work. The outcome of this module is presented in Figs. 12.11, 12.12 and 12.13. Two sample images from the test dataset are chosen for every class and the prediction mapping is presented here.

Figures 12.11, 12.12 and 12.13 show the outcome of network visualization module by overlapping the activation and occlusion sensitivity maps onto the input image. The red-coloured portion of the image corresponds to the strongest activation followed respectively by yellow and cyan colours in the order of decreasing importance. The dark-blue portion of the map resembles the neutral information and does not contribute towards classification decision. Observing the heatmaps produced by Grad-CAM, occlusion sensitivity maps, and LIME approaches, there is a considerable coincidence in the decision region of produced by three approaches for most cases considered in this work. The visualization provides the answer for why the deep neural network makes a particular classification decision. The insights provided by Grad-CAM, occlusion sensitivity maps, and LIME help to improve the transparency of the deep learning model which was so far considered to be a black-box model by ensuring that whether the network focuses on the correct part within the input image. Additionally, they help to diagnose network failure in case of incorrect prediction and correct unnecessary biases learnt by the deep neural network. The shortcoming can be rectified by retraining the network with more data to unfold a deeper range of the misclassified class. However, the explanations provided by these network visualization approaches are subjective and are human understandable, and hence the performance cannot be concretely quantified. The visual interpretation by experts varies from person to person. The network visualization approaches in general culminate the proficiency of deep learning technology in knowledge transfer between human and machine interpretation to facilitate healthcare practitioners in their diagnostic procedure.

12.7 Conclusion

In this chapter, an elite solution for the challenges faced by the clinicians in COVID-19 diagnosis during the pandemic has been proposed. An efficient computer-aided diagnosis (CAD) model for the prediction of COVID-19 from imaging-based screening methodology has been formulated. The performance of three popular CNNs is compared in this study and demonstrated that ShuffleNet architecture when deployed for COVID-19 diagnosis from CXR images will strengthen the screening

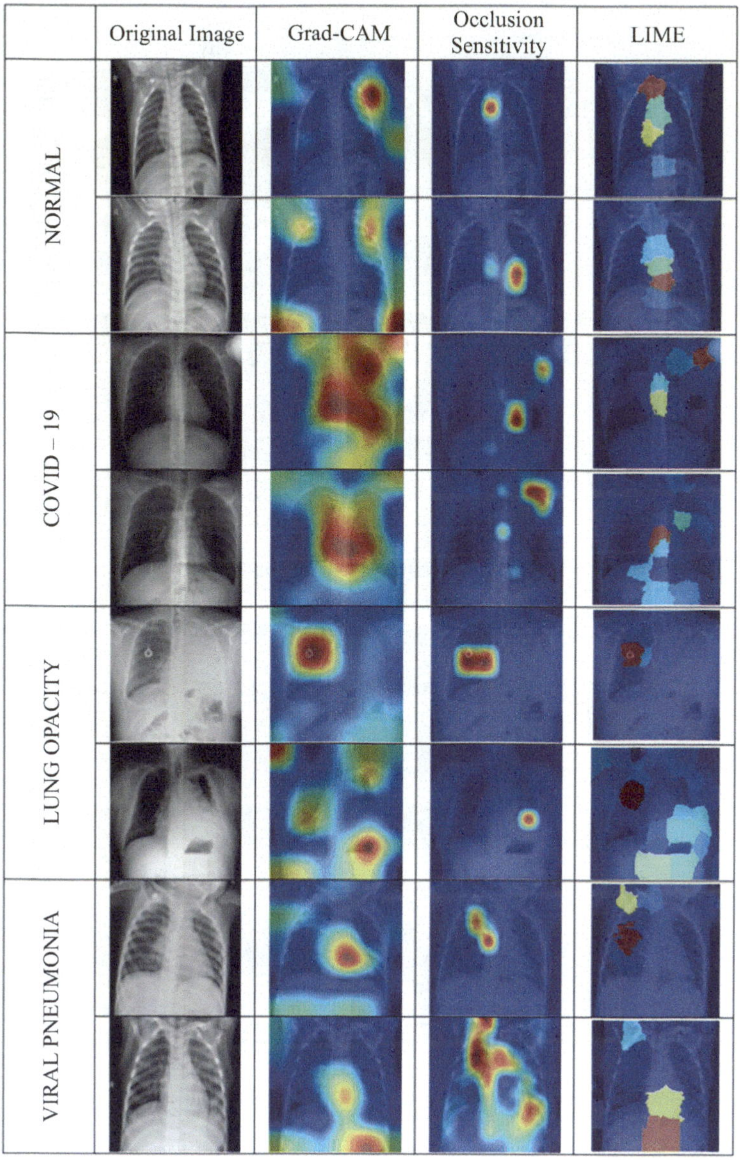

Fig. 12.11 Network visualization provided by AlexNet using Grad-CAM, occlusion sensitivity, and LIME

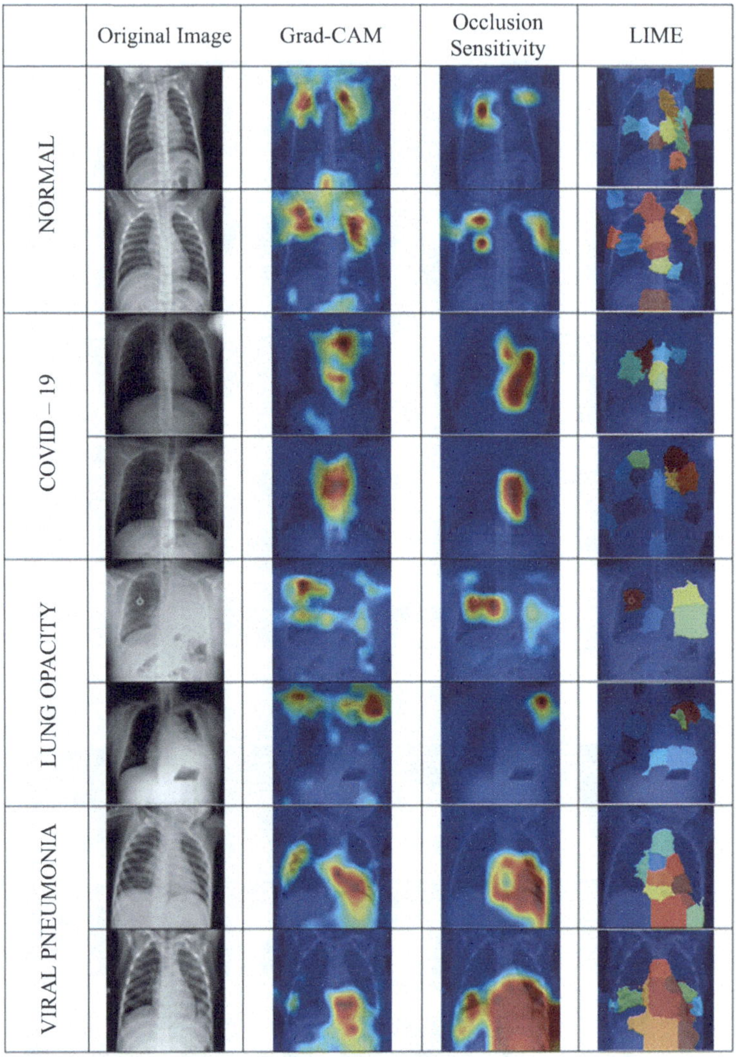

Fig. 12.12 Network visualization provided by SqueezeNet using Grad-CAM, occlusion sensitivity, and LIME

process with enhanced prediction accuracy. The network visualization strategy implemented on top of the CAD system will reveal the 'why' behind the network's prediction. Although there is no concrete mechanism to quantify the outcome of visual interpretation proposed in this study, a clinical correlation with the involvement of expert clinician can be performed as a separate study to assess the performance of the interpretation techniques on every single image and the method performing closer to real-time diagnosis by the clinician can be identified. The visual interpretation

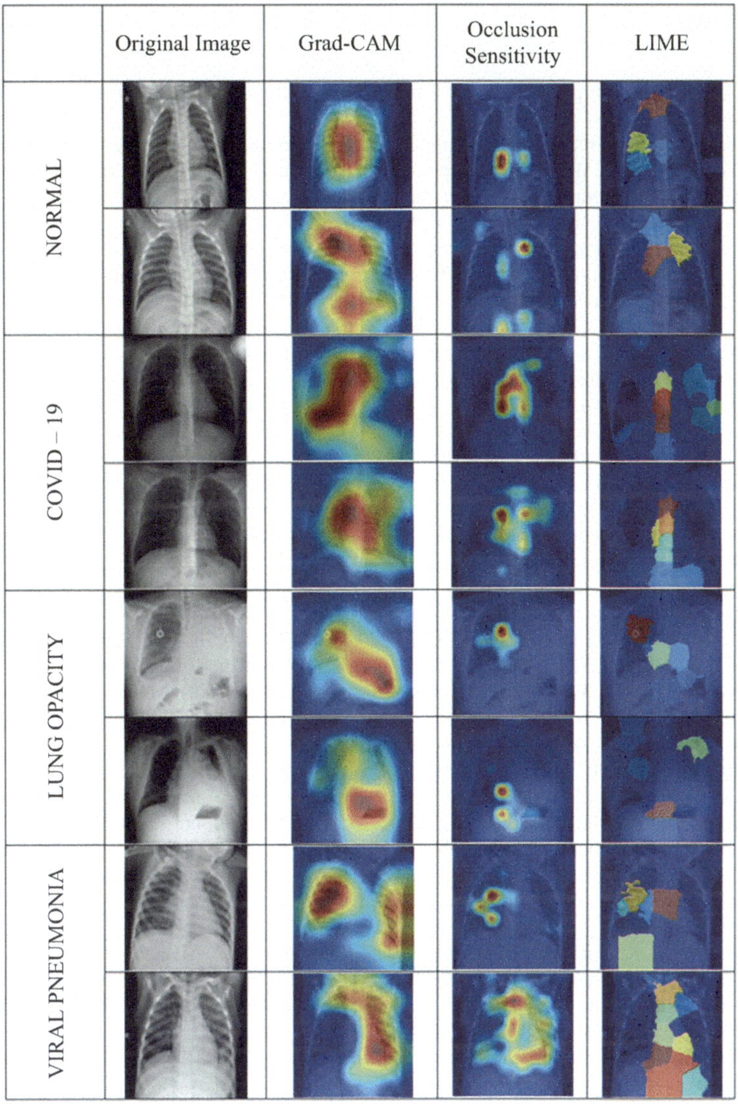

Fig. 12.13 Network visualization provided by ShuffleNet using Grad-CAM, occlusion sensitivity, and LIME

offered by the proposed XAI framework for COVID-19 diagnosis from CXR images will encourage physicians to trust the CAD system thereby building a trustworthy diagnostic system.

References

1. https://www.who.int/publications/m/item/weekly-epidemiological-update-on-covid-19---29-june-2021.
2. Wang, D., Hu, B., Hu, C., Zhu, F., Liu, X., Zhang, J., Wang, B., Xiang, H., Cheng, Z., Xiong, Y., Zhao, Y., Li, Y., Wang, X., & Peng, Z. (2020). Clinical characteristics of 138 hospitalized patients with 2019 novel coronavirus–infected pneumonia in Wuhan, China. *JAMA, 323*(11), 1061. https://doi.org/10.1001/jama.2020.1585
3. Spiess, B. (2020). Faculty opinions recommendation of clinical features of patients infected with 2019 novel coronavirus in Wuhan, China. *Faculty Opinions—Post-Publication Peer Review of the Biomedical Literature.* https://doi.org/10.3410/f.737258313.793572876
4. Barredo Arrieta, A., Díaz-Rodríguez, N., Del Ser, J., Bennetot, A., Tabik, S., Barbado, A., Garcia, S., Gil-Lopez, S., Molina, D., Benjamins, R., Chatila, R., & Herrera, F. (2020). Explainable Artificial Intelligence (XAI): Concepts, taxonomies, opportunities and challenges toward responsible AI. *Information Fusion, 58*, 82–115. https://doi.org/10.1016/j.inffus.2019.12.012
5. Wang, W., Xu, Y., Gao, R., Lu, R., Han, K., Wu, G., & Tan, W. (2020). Detection of sars-cov-2 in different types of clinical specimens. *JAMA.* https://doi.org/10.1001/jama.2020.3786
6. Yang, T., Wang, Y.-C., Shen, C.-F., & Cheng, C.-M. (2020). Point-of-Care rna-based diagnostic device for COVID-19. *Diagnostics, 10*(3), 165. https://doi.org/10.3390/diagnostics10030165
7. Corman, V. M., Landt, O., Kaiser, M., Molenkamp, R., Meijer, A., Chu, D. K., Bleicker, T., Brünink, S., Schneider, J., Schmidt, M. L., Mulders, D. G., Haagmans, B. L., van der Veer, B., van den Brink, S., Wijsman, L., Goderski, G., Romette, J.-L., Ellis, J., Zambon, M., … Drosten, C. (2020). Detection of 2019 novel coronavirus (2019-nCoV) by real-time RT-PCR. *Eurosurveillance, 25*(3). https://doi.org/10.2807/1560-7917.es.2020.25.3.2000045
8. Chu, D. K. W., Pan, Y., Cheng, S. M. S., Hui, K. P. Y., Krishnan, P., Liu, Y., Ng, D. Y. M., Wan, C. K. C., Yang, P., Wang, Q., Peiris, M., & Poon, L. L. M. (2020). Molecular diagnosis of a novel coronavirus (2019-ncov) causing an outbreak of pneumonia. *Clinical Chemistry, 66*(4), 549–555. https://doi.org/10.1093/clinchem/hvaa029
9. Zhang, N., Wang, L., Deng, X., Liang, R., Su, M., He, C., Hu, L., Su, Y., Ren, J., Yu, F., Du, L., & Jiang, S. (2020). Recent advances in the detection of respiratory virus infection in humans. *Journal of Medical Virology, 92*(4), 408–417. https://doi.org/10.1002/jmv.25674
10. Chung, M., Bernheim, A., Mei, X., Zhang, N., Huang, M., Zeng, X., Cui, J., Xu, W., Yang, Y., Fayad, Z. A., Jacobi, A., Li, K., Li, S., & Shan, H. (2020). CT imaging features of 2019 novel coronavirus (2019-ncov). *Radiology, 295*(1), 202–207. https://doi.org/10.1148/radiol.2020200230
11. Hosseiny, M., Kooraki, S., Gholamrezanezhad, A., Reddy, S., & Myers, L. (2020). Radiology perspective of coronavirus disease 2019 (COVID-19): Lessons from severe acute respiratory syndrome and middle east respiratory syndrome. *American Journal of Roentgenology, 214*(5), 1078–1082. https://doi.org/10.2214/ajr.20.22969
12. Miotto, R., Wang, F., Wang, S., Jiang, X., & Dudley, J. T. (2017). Deep learning for healthcare: Review, opportunities and challenges. *Briefings in Bioinformatics, 19*(6), 1236–1246. https://doi.org/10.1093/bib/bbx044
13. Yang, S., Zhu, F., Ling, X., Liu, Q., & Zhao, P. (2021). Intelligent health care: Applications of deep learning in computational medicine. *Frontiers in Genetics, 12.* https://doi.org/10.3389/fgene.2021.607471
14. M. S. L. N. R. (2017). Effective method of image retrieval using BTC with gabor wavelet matrix. *International Journal on Recent and Innovation Trends in Computing and Communication, 5*(5), 453–458. https://doi.org/10.17762/ijritcc.v5i5.544
15. Sivakumar, S., Rajalakshmi, R., Prakash, K. B., Kanna, B. R., & Karthikeyan, C. (2020). Virtual vision architecture for VIP in ubiquitous computing. *Technological Trends in Improved Mobility of the Visually Impaired*, 145–179.
16. Krizhevsky, A., Sutskever, I., & Hinton, G. E. (2017). ImageNet classification with deep convolutional neural networks. *Communications of the ACM, 60*(6), 84–90. https://doi.org/10.1145/3065386

17. Chouhan, V., Singh, S. K., Khamparia, A., Gupta, D., Tiwari, P., Moreira, C., Damaševičius, R., & de Albuquerque, V. H. C. (2020). A novel transfer learning based approach for pneumonia detection in chest x-ray images. *Applied Sciences, 10*(2), 559. https://doi.org/10.3390/app10020559
18. Gómez, P., Semmler, M., Schützenberger, A., Bohr, C., & Döllinger, M. (2019). Low-light image enhancement of high-speed endoscopic videos using a convolutional neural network. *Medical & Biological Engineering & Computing, 57*(7), 1451–1463. https://doi.org/10.1007/s11517-019-01965-4
19. Gu, X., Pan, L., Liang, H., & Yang, R. (2018). Classification of bacterial and viral childhood pneumonia using deep learning in chest radiography. *Proceedings of the 3rd International Conference on Multimedia and Image Processing*, 88–93. https://doi.org/10.1145/3195588.3195597
20. Wang, X., Peng, Y., Lu, L., Lu, Z., Bagheri, M., & Summers, R. M. (2017). ChestX-Ray8: Hospital-Scale chest x-ray database and benchmarks on weakly-supervised classification and localization of common thorax diseases. *IEEE Conference on Computer Vision and Pattern Recognition (CVPR), 2017*, 3462–3471. https://doi.org/10.1109/cvpr.2017.369
21. Shin, H.-C., Roth, H. R., Gao, M., Lu, L., Xu, Z., Nogues, I., Yao, J., Mollura, D., & Summers, R. M. (2016). Deep convolutional neural networks for computer-aided detection: CNN architectures, dataset characteristics and transfer learning. *IEEE Transactions on Medical Imaging, 35*(5), 1285–1298. https://doi.org/10.1109/tmi.2016.2528162
22. Rajpurkar, P., Irvin, J., Ball, R. L., Zhu, K., Yang, B., Mehta, H., Duan, T., Ding, D., Bagul, A., Langlotz, C. P., Patel, B. N., Yeom, K. W., Shpanskaya, K., Blankenberg, F. G., Seekins, J., Amrhein, T. J., Mong, D. A., Halabi, S. S., Zucker, E. J., ... Lungren, M. P. (2018). Deep learning for chest radiograph diagnosis: A retrospective comparison of the CheXNeXt algorithm to practicing radiologists. *PLOS Medicine, 15*(11), e1002686. https://doi.org/10.1371/journal.pmed.1002686
23. Lakhani, P., & Sundaram, B. (2017). Deep learning at chest radiography: Automated classification of pulmonary tuberculosis by using convolutional neural networks. *Radiology, 284*(2), 574–582. https://doi.org/10.1148/radiol.2017162326
24. Zebin, T., & Rezvy, S. (2020). COVID-19 detection and disease progression visualization: Deep learning on chest X-rays for classification and coarse localization. *Applied Intelligence, 51*(2), 1010–1021. https://doi.org/10.1007/s10489-020-01867-1
25. Bhatt, A., Ganatra, A., & Kotecha, K. (2021). COVID-19 pulmonary consolidations detection in chest X-ray using progressive resizing and transfer learning techniques. *Heliyon, 7*(6), e07211. https://doi.org/10.1016/j.heliyon.2021.e07211
26. Harmon, S. A., Sanford, T. H., Xu, S., Turkbey, E. B., Roth, H., Xu, Z., Yang, D., Myronenko, A., Anderson, V., Amalou, A., Blain, M., Kassin, M., Long, D., Varble, N., Walker, S. M., Bagci, U., Ierardi, A. M., Stellato, E., Plensich, G. G., ... Turkbey, B. (2020). Artificial intelligence for the detection of COVID-19 pneumonia on chest CT using multinational datasets. *Nature Communications, 11*(1). https://doi.org/10.1038/s41467-020-17971-2
27. Pennisi, M., Kavasidis, I., Spampinato, C., Schinina, V., Palazzo, S., Salanitri, F. P., Bellitto, G., Rundo, F., Aldinucci, M., Cristofaro, M., Campioni, P., Pianura, E., Di Stefano, F., Petrone, A., Albarello, F., Ippolito, G., Cuzzocrea, S., & Conoci, S. (2021). An explainable AI system for automated COVID-19 assessment and lesion categorization from CT-scans. *Artificial Intelligence in Medicine, 118*, 102114. https://doi.org/10.1016/j.artmed.2021.102114
28. DeGrave, A. J., Janizek, J. D., & Lee, S.-I. (2021). AI for radiographic COVID-19 detection selects shortcuts over signal. *Nature Machine Intelligence, 3*(7), 610–619. https://doi.org/10.1038/s42256-021-00338-7
29. Ahmad, N., Aminu, M., & Noor, M. H. M. (2021). *Covid-19 detection via deep neural network and occlusion sensitivity maps*. Institute of Electrical and Electronics Engineers (IEEE). https://doi.org/10.36227/techrxiv.14100890.v1
30. Iandola, F. N., Han, S., Moskewicz, M. W., Ashraf, K., Dally, W. J., & Keutzer, K. (2016, February 24). *SqueezeNet: AlexNet-level accuracy with 50x fewer parameters and <0.5MB model size*. arXiv.Org. https://arxiv.org/abs/1602.07360

31. Zhang, X., Zhou, X., Lin, M., & Sun, J. (2018, June). ShuffleNet: An extremely efficient convolutional neural network for mobile devices. *2018 IEEE/CVF conference on computer vision and pattern recognition.* https://doi.org/10.1109/cvpr.2018.00716
32. Simonyan, K., Vedaldi, A., & Zisserman, A. (2013, December 20). *Deep inside convolutional networks: Visualising image classification models and saliency maps.* arXiv.Org. https://arxiv.org/abs/1312.6034
33. Kim, B., Rudin, C., & Shah, J. (2015, March 3). *The bayesian case model: A generative approach for case-based reasoning and prototype classification.* arXiv.Org. https://arxiv.org/abs/1503.01161
34. Bach, S., Binder, A., Montavon, G., Klauschen, F., Müller, K.-R., & Samek, W. (2015). On pixel-wise explanations for non-linear classifier decisions by layer-wise relevance propagation. *PLoS ONE, 10*(7), e0130140. https://doi.org/10.1371/journal.pone.0130140
35. Caruana, R., Lou, Y., Gehrke, J., Koch, P., Sturm, M., & Elhadad, N. (2015, August 10). Intelligible models for healthcare. *Proceedings of the 21th ACM SIGKDD International Conference on Knowledge Discovery and Data Mining.* https://doi.org/10.1145/2783258.2788613
36. Zhou, B., Khosla, A., Lapedriza, A., Oliva, A., & Torralba, A. (2016, June). Learning deep features for discriminative localization. *2016 IEEE Conference on Computer Vision and Pattern Recognition (CVPR).* https://doi.org/10.1109/cvpr.2016.319
37. Ribeiro, M., Singh, S., & Guestrin, C. (2016). "Why should I trust you?": Explaining the predictions of any classifier. *Proceedings of the 2016 conference of the North American chapter of the association for computational linguistics: Demonstrations.* https://doi.org/10.18653/v1/n16-3020
38. Selvaraju, R. R., Cogswell, M., Das, A., Vedantam, R., Parikh, D., & Batra, D. (2017). Grad-CAM: Visual explanations from deep networks via gradient-based localization. *IEEE International Conference on Computer Vision (ICCV), 2017*, 618–626. https://doi.org/10.1109/iccv.2017.74
39. Habibi Aghdam, H., & Jahani Heravi, E. (2017). Visualizing neural networks. In *Guide to Convolutional Neural Networks* (pp. 247–258). Springer International Publishing. https://doi.org/10.1007/978-3-319-57550-6_7
40. Rahman, T. (n.d.). *COVID-19 radiography database.* Kaggle. Retrieved December 6, 2023, from https://www.kaggle.com/tawsifurrahman/covid19-radiography-database
41. *RSNA pneumonia detection challenge.* (n.d.). Kaggle. Retrieved February 16, 2024, from https://www.kaggle.com/c/rsna-pneumonia-detection-challenge/data
42. *BIMCV-COVID19 – BIMCV.* (n.d.). Retrieved December 6, 2024, from https://bimcv.cipf.es/bimcv-projects/bimcv-covid19/#1590858128006-9e640421-6711
43. ml-workgroup. (n.d.). *GitHub - Ml-workgroup/covid-19-image-repository: Anonymized dataset of COVID-19 cases with a focus on radiological imaging. This includes images (x-ray / ct) with extensive metadata, such as admission-, ICU-, laboratory-, and patient master-data.* GitHub. Retrieved December 6, 2024, from https://github.com/ml-workgroup/covid-19-image-repository
44. *SIRM - Società Italiana di Radiologia Medica e Interventistica.* (n.d.). SIRM. Retrieved December 6, 2024, from https://sirm.org/category/senza-categoria/covid-19/
45. *Eurorad.org.* (n.d.). Eurorad - Brought to You by the ESR. Retrieved December 6, 2024, from https://eurorad.org
46. ieee8023. (n.d.). *GitHub - Ieee8023/covid-chestxray-dataset: We are building an open database of COVID-19 cases with chest X-ray or CT images.* GitHub. Retrieved December 6, 2024, from https://github.com/ieee8023/covid-chestxray-dataset
47. Mooney, P. (n.d.). *Chest x-ray images (pneumonia).* Kaggle. Retrieved December 6, 2024, from https://www.kaggle.com/datasets/paultimothymooney/chest-xray-pneumonia
48. Chandra, T. B., Verma, K., Singh, B. K., Jain, D., & Netam, S. S. (2021). Coronavirus disease (COVID-19) detection in Chest X-Ray images using majority voting based classifier ensemble. *Expert Systems with Applications, 165*, 113909. https://doi.org/10.1016/j.eswa.2020.113909

B. Lakshmipriya received her Ph.D from Pondicherry University, Puducherry, India in 2020 and subsequently pursued her Post Doctoral Fellowship at Jawaharlal Institute of Postgraduate Medical Education and Research (JIPMER), Puducherry sponsored by Science and Engineering Board of India during 2021–2023. She is currently working as Associate Professor in School of Engineering, Jawaharlal Nehru University, New Delhi, India. Her research interests include biomedical image and signal processing, machine learning and deep learning. She is life member of ISTE and senior member of IEEE.

S. Jayalakshmy received her Ph.D from Pondicherry University, Puducherry, India, in the year 2023. She has been in teaching for more than 16 years. Currently, she is working as an Associate professor in the Electronics and Communication Engineering (ECE) Department at the IFET College of Engineering, Villupuram. Her areas of interest include Biomedical signal processing, Machine learning, and Deep learning. She is a life member of the Indian Society for Technical Education (ISTE) and a senior member of IEEE.

Chapter 13
Digital Transformation Required by the Brick and Mortar Retailers to be the Customers' Choice in the Post-COVID-19 Era

Sanjay Das, Soumen Santra, and Anirban Sarkar

Abstract The pandemic transformed consumer behaviour, emphasising safety and contactless shopping. To remain relevant, Brick and Mortar (B&M) retailers must innovate by integrating physical and digital experiences. This paper proposes the use of artificial intelligence (AI) to transform B&M retailers, enhancing their efficiency and appeal to modern consumers. The goal is to create a unique walk-in shopping experience that combines physical and digital elements. The study suggests several AI-based implementations. A machine learning-(ML)based K-means unsupervised learning algorithm to analyse customer data, including location and purchase patterns. The K-Nearest Neighbours (KNN) clustering technique guides prospects to nearby stores. An AI-based Voice Assistant (AIVA) at B&M stores to interact with customers, provide product information, and assist sales personnel in product demonstrations. A Satisfaction Assessment System (SAS) using AI to detect customer emotions through facial images upon leaving the store. Digital marketing communication (DMarCom) strategies based on SAS feedback to inform customers about products and services and address customer satisfaction levels. Implementing AI-based K-means clustering will help identify target customer segments. DMarCom efforts will create customer awareness through various digital channels. AIVA will enhance the purchasing experience by explaining product features, while SAS will assess customer satisfaction post-purchase. The data from SAS will be used to improve marketing strategies and customer interactions. By leveraging AI technologies such as clustering algorithms, voice assistants, emotion detection systems, and

S. Das
Department of Management and Marketing, West Bengal State University, Barasat, West Bengal, India

S. Santra
Department of Master of Computer Application, Techno International New Town, Kolkata, West Bengal, India

A. Sarkar (✉)
Department of Management and Marketing, West Bengal State University, Barasat, West Bengal, India
e-mail: anirban_sarkar77@wbsu.ac.in

S. Gangopadhyay et al. (eds.), *Pandemic Diaries*,
https://doi.org/10.1007/978-981-96-5415-4_13

digital marketing communication, B&M retailers can enhance their efficiency, attract more customers, and ensure long-lasting engagement in a competitive market.

Keywords Virtual assistance · Clustering learning · Bot algorithms · Industry market 5.0 · Human-centric learning · Voice assistant · Emotion detection system · Digital transformation

13.1 Introduction

The COVID-19 pandemic significantly impacted many countries, including India. Businesses of all types were affected due to the restrictions imposed by the authorities, precautions advised by the medical organisations, and anxiety about contamination that developed among the people. To survive in that constrained situation, exploring alternative ways was undertaken at all levels to conduct the necessary activities. With the implementation of lockdowns and social distancing procedures, remote work became the standard mode for many sectors, including education, offices, and markets. Physical stores faced the challenge of survival because of loss of sales owing to reduced footfall accompanied by competition from e-commerce as the consumers turned to online shopping, finding it safer in all respects. Customers preferred to buy online to remain safe at home due to fear of the pandemic [7]. Thus, consumer behaviour has been transformed, safety has been prioritised, and contactless online shopping has been practised. Therefore, online buying frequency decreased during COVID-19 [6]. Consequently, the Brick & Mortar (B&M) stores started facing more severe competition from e-commerce. E-commerce operators have already conditioned their customers by offering a comprehensive product range at a competitive price, with the necessary information and convenience of 24×7 shopping. The impact of COVID-19 and offerings of e-commerce together placed a challenge of footfall, revenue, profitability and finally, survival in front of traditional B&M stores. B&M stores must re-establish relevance to today's prospects to survive and thrive in the competition. To support the B&M retailers who have been playing a significant role in society by providing service and generating employment, this paper has proposed a **digital transformation strategy** for these traditional retailers, implementing artificial intelligence (AI)-based tools to blend physical and digital experiences for ensuring an enjoyable and attractive shopping experience to their customers to achieve long-lasting engagement. While digital shift refers to the transition of business processes from traditional methods to digital ones adopting new technologies, digital transformation is a comprehensive process of redesigning the entire operational process and strategies leveraging digital technologies to improve profitability and enhance customer satisfaction through improved competitiveness. It is about a change shaped by developing a new digital business model employing digital technologies, which creates more value for the firm [18].

Objective: To propose a set of artificial intelligence (AI) systems and procedures that can be implemented at retail outlets to transform the B&M store into a preferred destination for today's prospects.

13.2 Review of Literature

13.2.1 With the spread of online, though physical B&M stores have undergone erosion of business, many have shown that combining stationary retail formats with an integrated online channel can be effective [16]. In the prevailing changed market scenario, realising that it would not be feasible for them to compete efficiently on the price or assortment front with online retailers, the retailers focus on customers' experience and engagement achieved because of the same. The authors also mentioned that B&M stores may succeed as a vital interaction point when customers use online and offline in a multichannel decision-making process.

13.2.2 Researchers have emphasised the potential of virtual reality (VR) for adopting this technology in marketing and retail sales. According to them, VR is believed to improve the customers' shopping experience at traditional B&M retail stores. Multiple interactive VR devices make it happen, so the viewers feel like they are navigating in a three-dimensional imaginary environment. To further enhance customers' experience, the authors have suggested deploying sound systems and haptic devices [21].

13.2.3 Researchers have stressed the importance of personalisation at physical stores in three stages: identification, implementation, and interaction. Identification is the starting point; at this stage, customers' preferences will be identified to offer a tailored marketing mix. The authors have recommended the implementation of technology-assisted human interaction with the active involvement of knowledgeable sales personnel, wherein AI can contribute to technology-based personalisation. During the process, the traits of the customers concerning their willingness to participate in personalisation must also be kept in mind [17].

13.2.4 Researchers have clarified that, as COVID-19 has changed multiple market parameters, retailers need to relook at the changed market needs and consider offering a system merging online and offline supported by the combination of cloud platforms and 4th-generation artificial intelligence (AI). According to the authors, this, along with technologies like big data analysis and machine learning, would help collect consumer behaviour data to offer their customers a unique shopping experience [10].

13.2.5 In the case of artificial intelligence (AI)-enabled voice assistant, empathy must be recognised as a core design element because families would respond better to the voice assistant when AI empathy is present. As these empathetic voice assistants are expected to assist prospects in decision-making, the stakeholders of the retail stores implementing them may hope for higher selling efficiency. The authors have further suggested that AI agents should be enabled to understand users' perspectives when conducting marketing interactions [12].

13.2.6 Researchers have investigated business model innovation (BMI) supported by digital technology triggered by the challenges posed by COVID-19 to retailers operating in the food sectors. Based on digital technology-related activities initiated by Walmart and Carrefour, the authors proposed a BMI model to respond to the crisis and tackle long-term business opportunities. This study shows how innovative digital techniques can be adapted to reconfigure existing businesses in three steps: optimisation, transformation, and extension [11].

13.2.7 The researcher has presented that a "follow the digital" methodology is expected to help us adapt to volatile processes and unpredictable routines involving geographical networks framed by human and non-human agents. As per the author, digital technology has emerged as a critical commodity in contemporary life, impacting entities and their activities in social spaces and changing various things. As a result, digital technologies are constantly getting incorporated into identities and activities in society via interfaces like platforms, software, and hardware [9].

13.2.8 Researchers mentioned that, while COVID-19 had been profoundly impacting people in every activity of life, including working and eating, they conducted qualitative interviews of sixteen individuals in Germany to investigate the influence of the pandemic on food behaviour and understand the role of technology concerning food. It was found that COVID-19 brought changes in eating, too. This was done to understand the changes caused by the pandemic and to tackle human behaviour and normalcy in future pandemics [20].

13.2.9 Researchers revealed Amazon's endeavour in using Soft Innovation Resources (SIRs) for developing Advanced Digital Fashions (ADFs) with AI-oriented R&D. The authors mentioned Amazon's perfect timing of securing digital solutions in developing multiple ADFs, which was possible due to their advanced AI-based institutional system. Such parallel developments have been recommended to resolve the dilemma between the decline in productivity and the expansion of R&D (research and development) [19].

13.2.10 Researchers have explained "Market Shaping", mentioning that market shaping involves purposeful activities and operations competitors undertake to change the market. According to the authors, such market shaping can make the market decline, grow, or cause market resilience. The authors have clarified that it is essential for the competitors to understand the shaping of the market well and collaborate as required with other market players if required at critical times or at times of disturbances [2].

13.3 Methods

The digital transformation strategy proposed in this paper for the B&M stores encompasses artificial intelligence (AI)-based multiple systems along with digital marketing communication processes to attract the prospects to their shop for a novel and interactive shopping experience, which would be appraised by a Satisfaction Assessment System (SAS) for further improvement.

A judicious man–machine blend has been conceived as the theme of the digital transformation strategy for AI-enabled stores. The data about the customers captured and arranged by the store's personnel will be processed by an AI-based system. Interaction between artificial intelligence-based voice assistants (AIVA) will happen with the sales personnel and customers. Capturing human emotion would be done by an AI-based system. In contrast, the stakeholders of B&M stores would act on the result using DMarCom (Digital Marketing Commutation) for continuous improvement. Activities involved in the four processes of the entire system, along with the objective of the respective process, have been mentioned in Table 13.1 and shown in Fig. 13.1

13.3.1 K-Means Clustering

Potential prospects are identified and clustered using AI-powered K-means clustering to address them with appropriate marketing communication to inform them about the products and services available at the B&M store in their neighbourhood so that they are motivated to visit the store.

Details of the Process

Clustering is an unsupervised machine learning (ML) algorithm that organises and classifies different data points or clusters based on similarities or patterns. K-means clustering is used to group data points into clusters based on their features, where each data point belongs to the cluster with the nearest mean. Unsupervised ML teaches the computer to use unlabelled, unclassified data sets and enables its operation on raw data sets without supervision. The goal of clustering is to divide the population or set of data points into several groups so that data points belonging to a particular group are different from those of other data points present in another group, and both are comparable.

13.3.1.1 Algorithm for K-Means Clustering

Starting phase: The kth number of clusters should be selected. The initial centroids (the points that symbolise the centre of each cluster) should be chosen randomly for the kth cluster.

Table 13.1 Processes and the activities

Step	Process	Objective	Activity
13.3.1	K-means Clustering	Identify target segment customers to motivate them to visit the B&M store	Applying AI-powered K-means clustering, potential prospects are identified as target segments and clustered for addressing with appropriate marketing communication to update them about the products and services of the B&M store in their neighbourhood
13.3.2	AIVA	Delight customers by explaining product features in a human voice and responding to their queries	Artificial intelligence-based voice assistant (AIVA) at the B&M store would a) respond to the customers with the latest product information to create the desire to purchase and b) train the sales personnel on details of the products and the services
13.3.3	SAS (Satisfaction Assessment System)	Assessment of customers' satisfaction level based on the shopping experience	AI-based emotion detection system (EDS) would identify customers' emotional condition from the real-time image of the customer's face using DeepFace, a deep learning facial recognition system built on deep convolutional neural network (CNN) architecture. Customers' satisfaction levels will be assessed based on their emotional condition
13.3.4	DMarCom (Digital Marketing Commutation)	Customer engagement through digital marketing: Communicate with the customers before and after purchasing	Communicate with the customers through website, e-mail, phone, and social media as required: a. About products and promotional offers b. Services of AIVA c. According to the SAS output, the number of delighted customers should be maximised

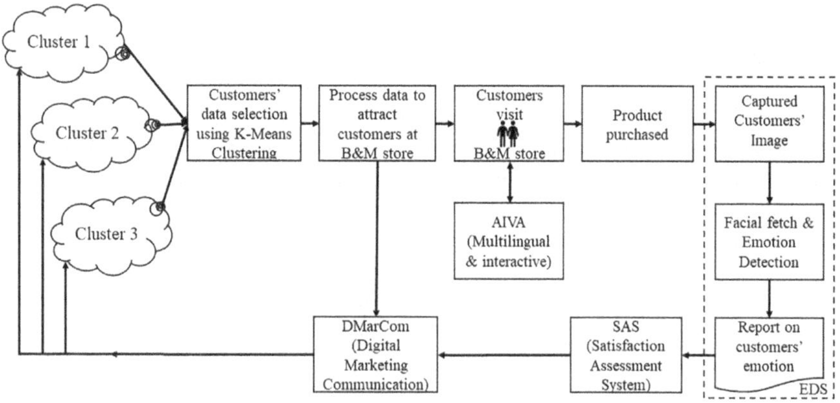

Fig. 13.1 Full system with AI-based processes and DMarCom

Centroid selection: Each data point should be assigned to the closest centroid. Thus, kth clusters are produced.

Update: The cluster centroids can be recalculated by finding the meaning of all the data points assigned to each centroid.

Iteration: The assignment and update processes will be repeated, and the centroid positions will be adjusted until they no longer undergo significant changes, or the allotted number of iterations is reached.

Finishing phase: The algorithm causes the clusters to stabilise when they converge, and no further changes occur in the assignments.

The Euclidean distance is used to measure similarity, and the algorithm categorises items into kth groups or clusters based on this measurement:

13.3.1.2 Implementation of K-Means Clustering in Python

```
Import numpy as np.
Import matplotlib. pyplot as plt
From sklearn. Datasets import make_ blobs
X,y = make_ blobs(n_ samples = 500, n_ features = 2,centers =
3, random_ state = 23)
fig = plt. Figure (0)
plt. grid(True)
plt. scatter(x[:, 0], x[:, 1])
plt. show()
```

With the help of the K-means unsupervised learning algorithm as above, the data on customers' details with their location will be analysed to create a graphical representation of their relative location around the B&M store.

13.3.2 Artificial Intelligence-Based Voice Assistant (AIVA) Would Receive the Requirements from Customers Directly as They Speak on a Microphone

AI-based automatic speech recognition (ASR) would convert speech to text and process the requirements to provide the output as shown in the Fig. 13.2. AIVA would be updated 24 × 7 through the cloud platform with the latest product information to be enabled to suggest the product that matches customers' requirements and explain its features, highlighting the specifications to create further interest driving toward conversion. The salesperson will give a live demonstration of the features narrated by AIVA, offering a close "Touch and feel" of the product to maximise the desire to purchase.

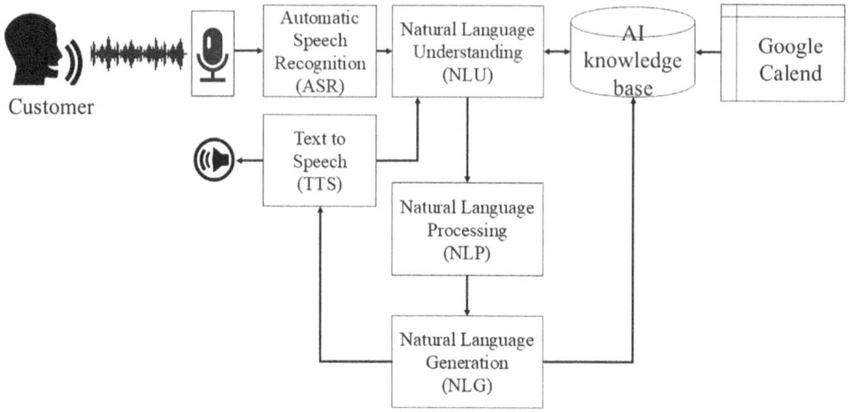

Fig. 13.2 Artificial intelligence-based voice assistant (AIVA)

The details of the AIVA methodology as follows:

13.3.2.1 Gathering and preparing data: To train models on various accents, tones, and speech patterns, then collect multiple voice samples. It can generate training datasets and translate customers' voice data into text. The data labels are normalised per the requirements and variety of customers. It follows entity extraction or intent recognition from the data.

13.3.2.2 Speech Recognition: It uses the technology of transformer-based architectures or Deep Neural Networks (DNN) to translate spoken language into text which can handle background noise, standardise audio signals, and accommodate various speaking enunciations.

13.3.2.3 Natural Language Understanding: Using the text as input from input device would ascertain the user's intention through intent recognition such as names, dates, and locations are to be recognised and taken from the text. To give logical answers, the context must be consistent across several exchanges.

13.3.2.4 Dialogue Management Policy: A subsequent course of action will depend on user input and context by creating rules or utilising machine learning. The conversation's status must be monitored to give pertinent answers and reminders.

13.3.2.5 Natural Language Generation: The system, i.e., NLG, generates natural language responses that resemble a human. The answers must be adapted according to the user's preferences and past exchanges.

13.3.2.6 Text-to-Speech (TTS): The system would interface with an American English text-to-speech engine. It is intended to develop engines that can convert across languages in future to increase the accessibility of text-to-speech systems, which can be further improved to make it a web-based real-time synthesis system models such as WaveNet or Tacotron must be used to translate written responses into spoken language. The voice characteristics (such as pitch and tone) must be modified to fit various uses [14].

13.3.2.7 Coordination and Implementation: API Development: APIs to be provided to allow voice help to be integrated into other platforms and applications. The user interfaces that enable seamless voice interactions will be created. It must be ensured that the system is accurate, natural, and meets user needs by continuously testing and improving it.

13.3.2.8 Privacy and Ethics-Data Security: Ensure user information is safeguarded, and privacy issues are handled.

In the Fig. 13.3 the study has shown the mechanism of the model for AIVA. The operational procedure as below has been recommended-

13.3.2.9 Importing necessary Python Libraries: It has been used for speech recognition, text-to-speech conversion, playing songs on YouTube, to get the current time and fetching information from Wikipedia.

13.3.2.10 Initializing Speech Recognizer and Text-to-Speech Engine: It initializes a speech recognizer and a text-to-speech engine ('engine') using 'pyttsx3'. It also sets the voice for the assistant.

13.3.2.11 Defining Functions: A function that speaks the provided text using the text-to-speech engine to capture the voice input from the user using a microphone and returns the recognized command. The primary function processes the user's voice command and performs actions based on the recognized commands.

13.3.2.12 Voice Input Processing: It would be capturing voice input using the microphone. It uses to listen and recognize the voice using Google's speech recognition. The function would process the command and perform actions based on specific keywords in the command. If the command contains 'play,' it assumes the user wants to play a song on YouTube. If the command contains 'time,' it fetches the current time using 'datetime.' If the command contains 'what is,' it fetches information about the specified topic using Wikipedia.

13.3.2.13 Main Execution: The main block initializes the assistant by greeting the user and continuously listens for the user's commands using a 'while' loop which processes each command and takes appropriate actions.

13.3.3 Satisfaction Assessment System (SAS)

Assessment of customers' satisfaction level is essential in any business to ensure continuous improvement. B&M stores are being recommended to improve-related and service-related attributes, which is expected to impact customer satisfaction positively [5]. Enhanced customer satisfaction can lead to customer retention through repeat purchases [1]. A decent in-shop experience, even without conversion, sows the seed of desire for future purchase. Hence, it would be prudent to assess the level of satisfaction and utilise the feedback information for further improvement.

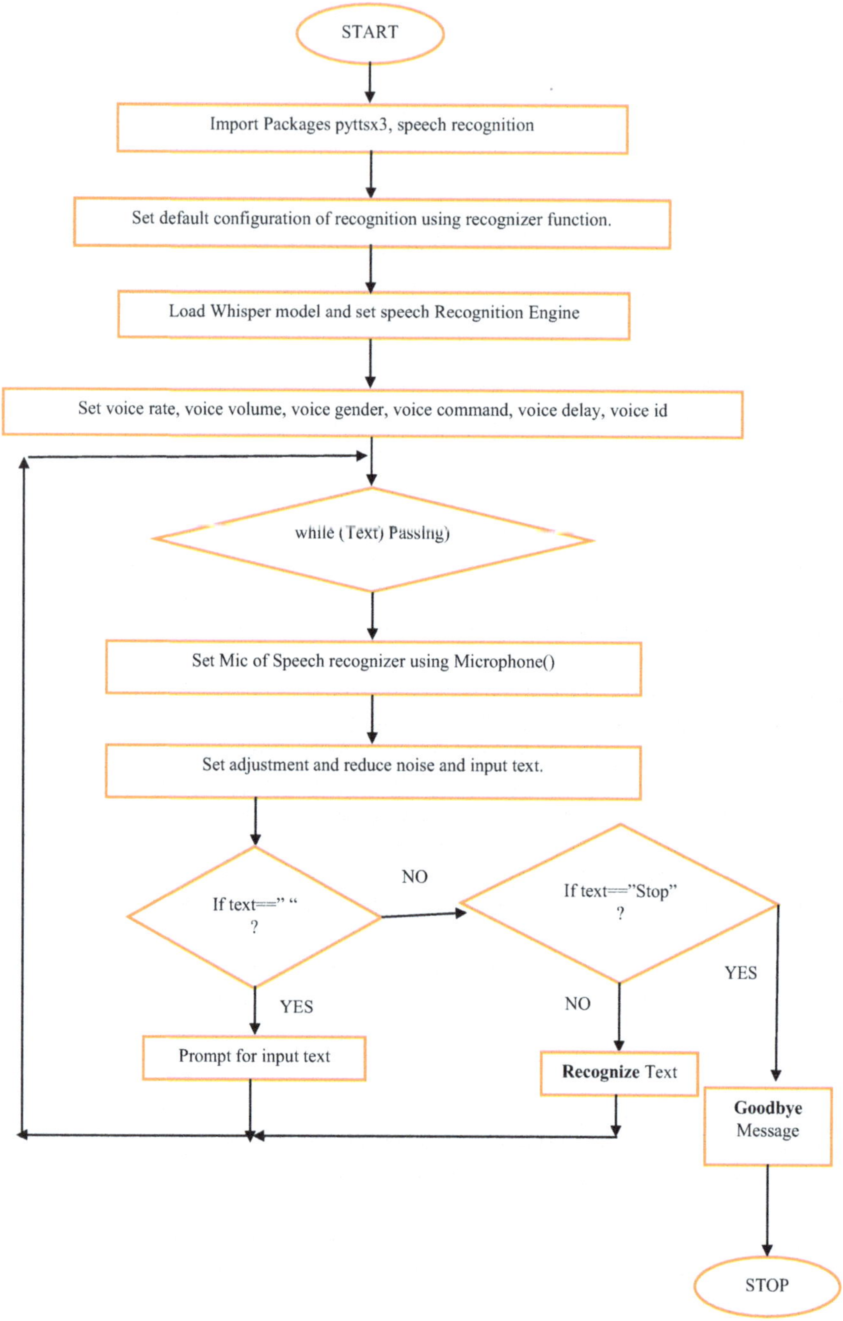

Fig. 13.3 Flow chart of the model for AIVA

Traditional practices of assessment of customers' satisfaction involved interaction over the phone and e-mail. But in the case of communication made by human beings, only 7% is conveyed by uttered words, 38% by voice intonation, and a significant 55% gets communicated by facial expression [13]. That is why an AI-based emotion detection system (EDS) has been proposed for the B&M retailers in this paper, wherein customers' real-time emotional condition while leaving the store would be ascertained by the system, which will indicate the level of satisfaction attained by the customer through the in-shop experience. The effect of satisfaction on a human being is expected to reflect on the face. That is why the facial image of the customer would be captured to identify the emotional condition and, thereby, the customer's satisfaction level. Taking the image of the face as an input, emotion would be detected by deploying machine learning with a Python programming model using Google Colaboratory. The system would consider the parameters below to ascertain the emotional states-

a. Facial expressions.
b. Eyebrow shape detection.
c. Shape of eye blob detection.
d. Shape of mouth blob detection.
e. Forehead's folds detection.
f. Dispersion of different face components.

The model proposed would point at one of the following seven emotional states of the customer to indicate the emotional condition depending on the parameters above–

i. Neutral
ii. Happy
iii. Surprise
iv. Disgust
v. Fear
vi. Anger
vii. Sad

DeepFace, a facial recognition system originated by Meta, would be deployed based on Convolutional Neural Networks (CNN) architecture. For the classification of images, Convolutional Neural Networks (CNN) are among the algorithms that are mainly used [4] Click or tap here to enter text.

The steps recommended for EDS are as follows:

Step I: Import all dependencies to make the real-life application. "DeepFace" is the lightweight main framework for face recognition and facial attribute analysis. Its accuracy level is approximately 97%. The two most significant imported library files are:

The "OpenCV" library is used for image processing and finally.

"Matplotlib" is used for data visualisation and graphical plotting.

Step II: Import the "IPython.display" module, which is used to write JavaScript to get permission for the camera from the browser. An alert box will appear, asking for permission from the user to access the device camera.

Step III: Import the "google. colab. output" module, which is used for interacting with the output interface of the colab cell.
Step IV: Import "base64" to encode binary files, such as images, videos, etc., within the scripts.
Step V: The model has to ask for the user's permission to take the user's image, which will be saved for further analysis in Google Colaboratory.
Step VI: The captured image can be processed and converted into BGR or RGB using OpenCV's cv2 module.
Step VII: The image must be plotted using the matplotlib library to use in DeepFace for further facial analytical functionalities.
Step VIII: The image must be analysed using the DeepFace pre-trained model to get emotional analytics using the abovementioned seven parameters. Only the dominant or the highest-scored emotional parameter will be considered the current user's emotional state.

13.3.4 Digital Marketing Communication (DMarCom)

B&M retailers need to have a digital face for their customers to match the consumer behaviour modified by the digital shift. Therefore, as a part of the "Digital Transformation Strategy", the B&M stores need to adopt "Digital Marketing" involving the actions recommended below-

a. Website—to be launched offering virtual visits to the stores covering product and service details.
b. E-mail marketing: E-mails should be sent to the
    I. Prospective target segment customers identified through K-means clustering to create awareness about AI-based services like AIVA along with the details of products, their prices and related promotional offers.
    II. Customers found to be 'Happy' in SAS are congratulated for enrolling them in loyalty programs.
    III. For customers not found as 'Happy' in SAS, the reason for dissatisfaction is to be identified as offering a product or service through special offers.
c. Content Marketing: Contents depicting AIVA in action should be uploaded to YouTube.
d. Social Media Marketing (SMM): Synergy of Integrated Marketing Communication (IMC) should be achieved by spreading selected content through social media.
e. WhatsApp: The concise version of e-mail should be sent to customers by WhatsApp for faster response regarding promotional offers and customer feedback.

Digital transformation that the traditional B&M stores are recommended to undergo needs to be customer-oriented, for which the digital payment system using a Unified

Payments Interface, i.e., UPI, should be implemented. UPI payment has become a common feature of B&M stores in India and is expected to overtake cards and internet banking in future [8]. The availability of the QR code (or Quick Response code) in the store for payment shows the "Digital Literacy" of the store authority, indicating that they understand and acknowledge the prevailing digital awareness in society. UPI payment system, therefore, must be made available at the B&M stores along with card payment so that the customers who have already gotten tuned to online payment post-COVID-19 period are encouraged to visit.

13.4 Result and Discussion

13.4.1 Result-K-Means Clustering

Customer locations are graphically represented in the scatter plot using Euclidean distance, showing the clustering of customers by location around the centroid.

Figure 13.4 shows the cluster formation of data from the imported dataset, for which the following algorithm has been used.

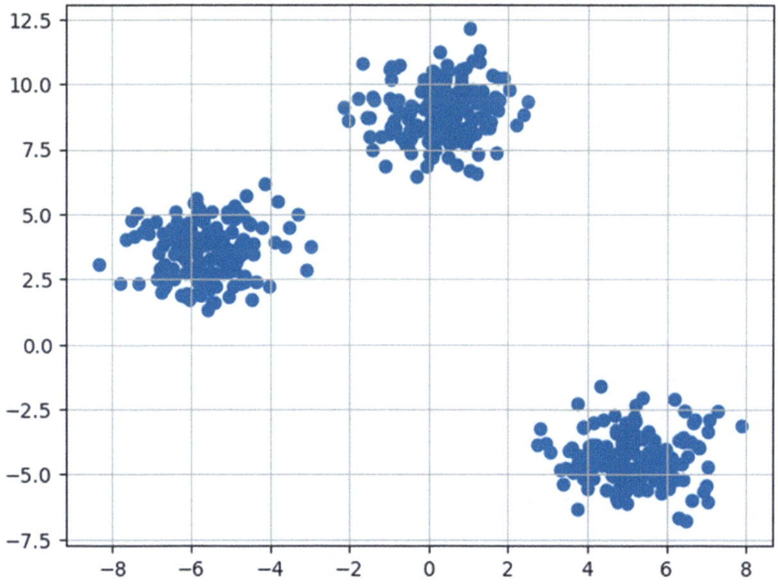

Fig. 13.4 Cluster form of customer segmentation data using K-means

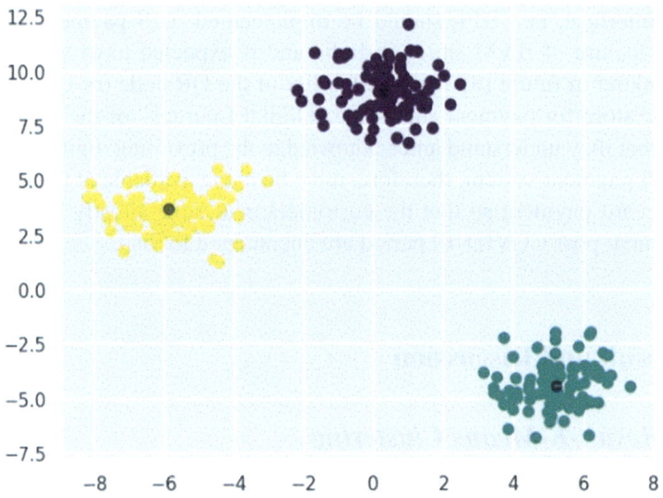

Fig. 13.5 Cluster form of customer segmentation data with centroid using K-means

```
kmeans=KMeans(n_clusters = 3)
kmeans.fit(x)
y_kmeans = kmeans.predict(x)
plt. grid(True)
plt. scatter(x[: , 0], x[: , 1],c=y_kmeans,cmap='viridis')
centers = kmeans.cluster_centers_
plt.    scatter(centers[:    ,    0],    centers[:    ,
1],c='black',alpha=0.5)
plt. show()
```

In Fig. 13.5, centroids of each cluster are shown by which the model would predict the concentration of willing **customers of the products.**

13.4.2 Result-AIVA

AIVA would provide results in the following areas:

a. **Responding to customers**: Reply in human voice with the latest updates to the voiced queries of the customers visiting the store.
b. **Support salespersons** during product demonstration: Salespersons could utilise AIVA output to endorse their explanation during in-shop demonstrations to convince their prospects of efficient conversion.

c. **Digitally empowered**: Influenced by the COVID-19-triggered "Digital shift" of consumer behaviour, many of today's customers prefer to interact with technically updated entities. B&M stores, in the presence of AIVA, would potentially be their preferred destination.
d. **Reply to the competition of e-commerce**: AIVA would be a reply on behalf of B&M retailers to the information availability at e-commerce sites.
e. **Product training of the sales personnel**: The sales personnel can be done efficiently utilising the AI-enabled exhaustive knowledge base of AIVA and its ability to provide voiced responses to the questions asked. The training activity can be scheduled during off-peak hours and weekly holidays or according to the availability of salespersons. The quality of content and delivery will remain at its best. The absenteeism and turnover of trained salespersons, which have been an operational challenge for the stores for ages, could be mainly mitigated by implementing AIVA.
f. **Enhance footfall** of prospects and customers: B&M stores could position AIVA as a point of attraction for the prospects and a USP (Unique Selling Proposition) for the store.

13.4.3 Result-Satisfaction Assessment System (SAS)

The AI-powered emotion detection system (EDS) model proposed for the B&M stores would ascertain the satisfaction level of the customers from their facial image captured on the verge of leaving the store. AI-based analysis of the captured face image would reveal the customer's feelings. By focusing on customers' real-time [4] emotions resulting from in-shop experience, the B&M store authority could project a "Human Face" on behalf of the store. Such an approach of expressing concern for the customers' feelings, which is missing in the case of e-commerce retailers, would provide a competitive edge for the B&M retailers. EDS would identify the dominant emotion in the customer's face and point at one of the seven emotional states, i.e., Neutral, Happy, Surprise, Disgust, Fear, Anger, and Sad, that the AI-based system can reveal.

Actions recommended for the B&M stores for customers with different emotional states, i.e., Neutral, Happy, Surprise, Disgust, Fear, Anger, and Sad, as detected by EDS, have been mentioned in Table 13.2. Considering the level of satisfaction or dissatisfaction that might have caused the emotion, scores ranging from 1 to 7 on the Likert scale have been assigned to each emotion type for the Satisfaction Assessment System (SAS). The Likert scale is a functional tool for studying emotions [15]. In the Likert scale used for SAS, seven would indicate the highest satisfaction corresponding to the emotional status 'Happy ', and one would indicate maximum dissatisfaction corresponding to the emotional status 'Disgust'. Customers flagged with emotion type "Neutral' would be considered neither satisfied nor dissatisfied and therefore assigned a score of 3. Mapping of balance emotional conditions, i.e., 'Surprise,' 'Anger,' 'Fear', and 'Sad' with the scores have been mentioned in Table 13.2.

Happy customers are the strength of any store, and these customers need to be consistently approached with attractive offers to ensure repeat purchases. Customers recorded with the emotions' Surprise,' 'Fear', and 'Neutral' are considered opportunities. Hence, their needs are to be explored to offer an appropriate value proposition to engage them. Cases flagged with emotion type' Anger' are to be treated as 'threat' as these customers might shift to other retailers. They are to be contacted with care to understand the reason for their anger and resolve it to their satisfaction. Similar would be the situation for customers flagged with emotion types 'Disgust' and 'Sad' whose perception is the result of weakness of the B&M store. For these cases, a detailed "Root cause analysis" involving the customers would be necessary, and a personalised service with suitable promotional offers will be provided as a retention measure.

13.4.4 Result-Digital Marketing Communication (DMarCom)

DMarCom would be working toward customer engagement in two stages, as shown in Fig. 13.1. In the first stage, the result achieved by DMarCom can be assessed by the count of footfall that could be generated by the marketing communication sent to the target segment about products details and usage of AIVA. In the second stage, DMarCom is utilised to communicate with customers with different emotions, thereby contrasting satisfaction levels to create and maintain the bonds for maximising retention. While the B&M retailers would communicate with the customers through e-mail, WhatsApp or SMS, and social media, inviting them to the stores with personalised promotional offers, they would also draw their customers' attention toward the "Digital Maturity" they attained. Digital maturity evaluates an organisation's ability toward creating value by implementing digital technology [3]. As digital consumer behaviour was found to be a result of the COVID-19 digital shift, the "Digital Maturity" shown by modernised B&M stores are likely to be appreciated by today's customers.

13.5 Conclusion

E-commerce sites offer information and convenience to draw customers. AI-powered B&M stores would provide information through AIVA in a human–machine-blended mode and the opportunity to "Touch and feel" the products. The explanation and personalised advice of experienced salespersons will further support this. The synergistic effect of all these facilities is expected to create an environment conducive

Table 13.2 Actions recommended based on emotions detected

Emotion detected	Emotion type	Score assigned	Flag	Approach	Action required
Happy	Completely satisfied	7	Strength	Congratulate	Offer enrolment in the loyalty programme and invite attendees to events, such as gift/free home delivery of high-value orders
Surprise	Mostly satisfied	6	Opportunity	Explore	Communicate over E-mail & WhatsApp to understand what was liked most and what needs to be done to make that even better
Fear	Somewhat satisfied	5	Opportunity	Explore	Communicate over E-mail, WhatsApp, or SMS, seeking suggestions to improve
Neutral	Neither satisfied nor dissatisfied	4	Opportunity	Explore	Contact over WhatsApp/SMS to understand the areas of likes and dislikes
Anger	Somewhat dissatisfied	3	Threat	Explore	Discuss in person over the phone to understand the details of experience and offer promotional benefits like discount or free WhatsApp order delivery

(continued)

Table 13.2 (continued)

Emotion detected	Emotion type	Score assigned	Flag	Approach	Action required
Sad	Mostly dissatisfied	2	Weakness	Root cause analysis	Discuss in person over the phone to understand the details of experience and offer promotional benefits like discount or free WhatsApp order delivery
Disgust	Completely dissatisfied	1	Weakness	Root cause analysis	Call to reconfirm the feeling and promise to offer escorted shopping on the next visit

enough for making purchase decisions. Thus, K-means clustering, AIVA, and SAS are expected to trigger a reverse churn that would motivate customers to move toward B&M stores, again turning away from e-commerce. Customers are to be informed about capturing their image and the reason for the same assuring them about privacy. With the implementation of AI-based systems proposed in this paper along with digital marketing communication, the Brick-and-Mortar stores are expected to effectively draw the attention of today's prospects, many of whom have undergone a change in consumer behaviour owing to the "Digital Shift" triggered by COVID-19.

References

1. Aityassine, F. L. Y. (2022). Customer satisfaction, customer delight, customer retention and customer loyalty: Borderlines and insights. *Uncertain Supply Chain Management, 10*(3). https://doi.org/10.5267/j.uscm.2022.3.005
2. Beninger, S., & Francis, J. N. P. (2021). Collective market shaping by competitors and its contribution to market resilience. *Journal of Business Research, 122*. https://doi.org/10.1016/j.jbusres.2020.09.005
3. Boston Consultancy Group. (2024). *Digital Maturity*. Digital, Technology, and Data. https://www.bcg.com/capabilities/digital-technology-data/digital-maturity
4. Chiurco, A., Frangella, J., Longo, F., Nicoletti, L., Padovano, A., Solina, V., Mirabelli, G., & Citraro, C. (2022). Real-time Detection of Worker's Emotions for Advanced Human-Robot Interaction during Collaborative Tasks in Smart Factories. *Procedia Computer Science, 200*. https://doi.org/10.1016/j.procs.2022.01.388

5. Daultani, Y., Goyal, K., & Pratap, S. (2021). An empirical investigation of the relationship between store attributes and customer satisfaction: A retail operations perspective. *Operations and Supply Chain Management, 14*(1). https://doi.org/10.31387/oscm0440289
6. Dinesh, S., & MuniRaju, Y. (2021). Scalability of e-commerce in the covid-19 era. *International Journal of Research -Granthaalayah, 9*(1). https://doi.org/10.29121/granthaalayah.v9.i1.2021.3032
7. Jílková, P., & Králová, P. (2021). Digital consumer behaviour and eCommerce trends during the COVID-19 Crisis. In *International Advances in Economic Research, 27* (1). https://doi.org/10.1007/s11294-021-09817-4
8. Kumar, A., Choudhary, Kumar, R., Mishra, S. K., Kar, S. K., & Bansal, R. (2022). The growth trajectory of UPI-based mobile payments in India: enablers and inhibitors. *Indian Journal of Finance and Banking*. https://doi.org/10.46281/ijfb.v11i1.1855
9. Liu, C. (2024). Follow the digital: Methodological thoughts on doing everyday geographies in a digital world. *Digital Geography and Society, 6*. https://doi.org/10.1016/j.diggeo.2024.100079
10. Lu, H. P., Cheng, H. L., Tzou, J. C., & Chen, C. S. (2023). Technology roadmap of AI applications in the retail industry. In *Technological forecasting and social change* (Vol. 195). https://doi.org/10.1016/j.techfore.2023.122778
11. Mancuso, I., Messeni Petruzzelli, A., & Panniello, U. (2023). Innovating agri-food business models after the Covid-19 pandemic: The impact of digital technologies on the value creation and value capture mechanisms. *Technological Forecasting and Social Change, 190*. https://doi.org/10.1016/j.techfore.2023.122404
12. Mari, A., Mandelli, A., & Algesheimer, R. (2024). Empathic voice assistants: Enhancing consumer responses in voice commerce. *Journal of Business Research, 175*. https://doi.org/10.1016/j.jbusres.2024.114566
13. Mehrabian, A. (2017). Communication without words. In *Communication Theory: Second Edition*. https://doi.org/10.4324/9781315080918-15
14. Mukherjee, P., Santra, S., Bhowmick, S., Paul, A., Chatterjee, P., & Deyasi, A. (2018). Development of GUI for text-to-speech recognition using natural language processing. *2018 2nd International Conference on Electronics, Materials Engineering and Nano-Technology, IEMENTech 2018*. https://doi.org/10.1109/IEMENTECH.2018.8465238
15. Nicoara, N. D., Varga, D., Voita-Mekeres, F., Galea-Holhos, L.-B., Andreescu, G., & Costas, L. (2023). Study of basic emotions in the general population using the likert scale. *Pharmacophore, 14*(4). https://doi.org/10.51847/tjyoah1vwm
16. Reinartz, W., Wiegand, N., & Imschloss, M. (2019). The impact of digital transformation on the retailing value chain. *International Journal of Research in Marketing, 36*(3). https://doi.org/10.1016/j.ijresmar.2018.12.002
17. Scholdra, T. P., Wichmann, J. R. K., & Reinartz, W. J. (2023). Reimagining personalization in the physical store. *Journal of Retailing, 99*(4). https://doi.org/10.1016/j.jretai.2023.11.001
18. Verhoef, P. C., Broekhuizen, T., Bart, Y., Bhattacharya, A., Qi Dong, J., Fabian, N., & Haenlein, M. (2021). Digital transformation: A multidisciplinary reflection and research agenda. *Journal of Business Research, 122*. https://doi.org/10.1016/j.jbusres.2019.09.022
19. Watanabe, C., Akhtar, W., Tou, Y., & Neittaanmäki, P. (2021). Amazon's initiative transforming a non-contact society - Digital disruption leads the way to stakeholder capitalization. *Technology in Society, 65*. https://doi.org/10.1016/j.techsoc.2021.101596
20. Weber, P., Ludwig, T., & Michel, L. K. (2023). The role of technology use in food practices during the COVID-19 pandemic. *International Journal of Gastronomy and Food Science, 31*. https://doi.org/10.1016/j.ijgfs.2023.100687
21. Xi, N., & Hamari, J. (2021). Shopping in virtual reality: A literature review and future agenda. *Journal of Business Research, 134*. https://doi.org/10.1016/j.jbus

Sanjay Das is a Research Scholar at the Department of Management & Marketing at West Bengal State University (WBSU). He also serves as a visiting faculty member at the Indian Institute of Social Welfare and Business Management (IISWBM) and as a guest teacher at the Department of Management & Marketing at West Bengal State University (WBSU). He has 33 years of corporate experience in leadership positions of reputed multinationals and major Indian organisations and more than 12 years of academic experience. His areas of interest include Digital Marketing, E-Commerce, Consumer Behaviour, and AI-based Natural Language Processing.

Soumen Santra works as an Assistant Professor in the Department of Master of Computer Application at Techno International New Town (Techno India Group) Kolkata, INDIA. He has more than 13 years of professional experience in academia and research. His work spans the field of cybersecurity using different programming languages like C, C++, Python, etc. He has published over 50 peer-reviewed research papers with 120+ Google Scholar citations. He is associated with different International and National Conferences in various aspects and has organized several technical events under the banner of the IEEE(I) Kolkata section. He is a reviewer of an International journal, TETA.

Anirban Sarkar is a Professor and Head of the Department of Management & Marketing at West Bengal State University (WBSU). He also serves as Director of the Centre for Management Studies and Director of the Sports Board. With over 20 years of academic experience, his areas of interest include Marketing, Finance, Statistics, Market Research, Consumer Behaviour, Digital Marketing, and Behavioural Finance. He has led UGC and ICSSR-sponsored research projects and recently submitted a Covid Special Project Report to ICSSR. Dr. Sarkar has presented papers at national and international conferences and published in ABDC, Scopus, SCI, Web of Science, and UGC Care listed journals. He is a Life member of several professional associations and currently holds an executive position in the Indian Economic Association (IEA).

Chapter 14
Emerging Trend of Online Purchase of Medicine in Light of COVID-19 Pandemic

Soumik Gangopadhyay, Peu Das, Shreosee Kushari, Shankha Shubhra Sarkar, Shubham Mondal, Soumojoy Sen, and Siddharth Das

Abstract COVID-19 pandemic has challenged the fundamental existence of mankind. A microscopic virus posed the greatest question to mankind about its existence and the most evolved animal of the planet earth was flabbergasted. Zero preventive option, unidentified solution, high risk of infection, unexpected mortality have compelled the society hide behind the doors due to lockdown, work from home, and quarantine as a strategic step to escape the infection. Life and livelihood faced the greatest challenge of the century. On the one hand the death tolls were rising and the other was witnessing disastrous fall in commercial value. Due to easy access, scope of online purchase of medicine enhanced to a greater scale. Further, social media has progressively being used for sharing health information and for networking

---

S. Gangopadhyay · P. Das (✉)
Department of Business Administration, Institute of Engineering and Management, University of Engineering & Management, Kolkata, India
e-mail: peu.das@iem.edu.in

S. Gangopadhyay
e-mail: gangopadhyaysoumik@gmail.com

S. Kushari · S. Das
Institute of Management, Nirma University, Ahmedabad, India
e-mail: kusharishreosee@gmail.com

S. Das
e-mail: siddharth.dasiem2023@gmail.com

S. S. Sarkar
KIIT School of Management, Bhubaneswar, India
e-mail: shubhra160shankha@gmail.com

S. Mondal
Kargo Motors, Kolkata, India
e-mail: mondal.shubham12@gmail.com

S. Sen
British Telecom, Kolkata, India
e-mail: soumojoy.semiem2023@gmail.com

S. Gangopadhyay et al. (eds.), *Pandemic Diaries*,
https://doi.org/10.1007/978-981-96-5415-4_14

among clinician and patients. Such media played an important role on the adoption of online purchase of medicine. Objective of the current study is to reveal the factors influencing the online purchase of medicine during the pandemic and to analyse the impact of social media on purchase of medicine online. Randomly selected 200 respondents were approached for their opinion with the help of a structured questionnaire contains a set of close-ended questions. The opinions were coded on a five-point scale and were statistically analysed with the help of software. 3 factors emanated namely perceived objective value, perceived subjective value and perceived consultative value of online medicine purchase. The outcome of the study can be used as a game changing strategy for online medicine retails.

Keywords COVID-19 · Pharmacy · Internet · Perceived objective value · Perceived subjective value · Perceived consultative value · Social media · Online medicine retail

14.1 Introduction

The novel silent killer of twentieth century took birth in Wuhan, China, in late 2019 and gradually started to bite the globe tenaciously. This insidious virus impacted human life from every angle from its time of birth. It troubled every individual alongside the broader society. During the first phase of 2020 World Health Organization announced it as pandemic as it had all characteristics of a virus that can be recognised as pandemic. This virus scattered through the boundless territory and became a reason for uncertainty, cases of acute illness and millions of death throughout the world. An enormous strain had been put on overall health system as it had a characteristics of transmitting through droplets of a virus infected person while sneezing, coughing, breathing and even talking also. Additionally it also flowed via contact of infected surfaces which means if a virus affected person touches a place or object it can transmit to other people's body through mouth, eyes, nose by touching it. Consequently the world witnessed the spread of COVID-19 at an alarming rate within few months from its birth. During the earlier phase healthcare centres were plagued with overflowed patients' burden in heavily affected countries like Italy, USA where most of the patients were required intense health support. As the virus commenced to flow it put the crowd in risk as it transmitted through human to human contact. Govt. announced several measures like lockdown, home quarantine, social distancing to break the infectious chain of COVID-19 as it made each crowded, and confined place risky. Those measures were essential to slow the speedy rate of virus spreading but it interrupted human life from different aspects. People were suffering physically and mentally due to anxiety, risk and insecurity. Pandemic like COVID-19's psychological impacts including mental suffering, unusual behaviour emphasise the need for better mental health support specially for the vulnerable population, healthcare stuffs to ensure social stability in long run [16]. The spread of COVID-19 virus along with various misinformation made the situation more severe [52]. This out of the blue situation combined with the norms of maintaining social distancing created a huge metamorphosis which led to the alteration in consumer's purchasing behaviour and made them to rely more on online platforms for fulfilling their every wants. This killer impacted not only the global physical health but also it has revolutionised many industries. Pharmaceutical industries are one of those industries which experienced a remarkable transformation in its traditional structure. Everything is changeable and prone to transformation. Emerging trends need new approaches. Online mode of supply chain has definitely advanced the consumer access to every product. Additionally, COVID-19 transformed the consumer psyche from physical to digital which has reversed further with the dissipation of the pandemic. Currently, it is also digital. But, it has benchmarked its approaches by practice since its evolution. But, it is more contextual rather than conceptual. The operation and market ecosystem has taken a new dimension. It was the major influencer of many changes. Even, change has evidenced changes in adoption and practices.

The concept of online purchase of medicine was in existence in pre-pandemic era also but during pandemic it got huge fame as it guaranteed that people can get required medical services from home without any chance of getting affected. Before the emergence of pandemic people were accustomed with the process of purchasing medicine by visiting the medicine shops, directly connecting with the pharmacists physically to cure illness and shopping goods in person. During the COVID-19 pandemic, contactless shopping process made e-commerce as the first choice of shopping. Studies have identified that as the pandemic unfolded, lockdown induced consumers to buy medicines and all other necessary products through online platforms. Throughout the phase abiding by the rules medicine shops remained open as a part of essential services but people avoided to visit the stores physically for shopping medicines and other health-related products out of fear of getting infected. People seek for finding alternative options made E-commerce the first and safest choice as this system enabled people to order and receive medicines from home which reduced the chance of getting infected. This caused a significant emergence of online pharmacy practices. E-commerce companies have witnessed spike of 20–30% on online orders. As a consequence e-pharmacy became popular among all other essential services. Social media's role was pivotal in influencing people to adopt this transformed way of shopping. As lockdown forced people not to come out from home, social media platforms like Facebook, twitter, WhatsApp, and many others controlled people's emotion, decision, and thought process predominantly. Social media was a good medium of keeping the population aware and updated by sharing news, updates. Pharmaceuticals companies are several merchants decided to launch promotional activities through social media platforms to promote brands and increase sales [6]. The internet has also become an ideal source for pharmacists to discover health information for healthcare providers and patients. Social media has also been used in pharmacies to advertise products and services to consumers through websites. An example of services provided through the internet includes the 'Ask-the-pharmacist' service which allows consumers to send questions about their medicine to a pharmacist through e-mail [62]. There is great potential for a pharmacy to engage in the utilisation of social media platforms to improve health outcomes, and this paper explores the influence of digital media to purchase medicines online in post-pandemic. Social marketing has influential role to help people to have sustainable lifestyle via sensible consumption [54]. It has immense potentiality and promising avenue to improve individual health [27]. Upstream causes of social problem are a prerequisite to effective social health marketing [76]. Voluntary health behaviour can be inculcated through downstream social marketing with the support of upstream social marketing by collaborating with public health researchers [28]. This is not only a digital but also a sustainable approach to reduce carbon footprint. Pandemic has enriched the crisis. Although online medicine purchase is not a by-product of pandemic but, the trend has taken a new trend before the mass. A new habit has emerged. In pandemic under such backdrop, it is essential to review and analyse the cause-effect relationship fuelling this shift. The broader objectives of this study were to reveal the factors influencing the online purchase of medicine during the pandemic, to understand the impact of the digital presence on the purchase of medicine and to analyse the impact of social media on purchasing medicine.

14.2 Survey of Literature

14.2.1 Changing Value of Health Need

Subjective knowledge and a positive attitude towards lifestyle intervention have profound association [34]. Public are interested to receive information from authentic sources [58]. With the advent of internet, there is a comparative escalation in information seeking behaviour amongst current patients regarding facility, process, treatment, staff care, doctors, diseases and the progress of health crisis from clinical establishments in an emergency [56]. With the easy access of internet-based knowledge, patients are increasingly mining diseases and health-related information [57]. Healthcare sectors are adopting advanced internet-based technologies rapidly in fieldwork to get the benefit of less cost and higher productive efficiency with a confirmation of maintaining data privacy and the standard of health service [45]. Healthcare industries initiate the use of internet as a provider of health information as well as they introduce it as a bridge that connects both the ends, healthcare consumer and supplier which influence the growth of e-health positively, hence it ensures the delivery of good quality service at low cost [46]. Medical professionals, govt. officials and decision-makers are taking initiatives to introduce technologies in health sector in a broader sense by introducing internet of things technologies for providing reliable and effective health services and COVID-19 pandemic fuelled this practice [37]. Internet plays as a game changer for transforming the framework of medical sector by giving birth of e-health as it motivates doctors to use more advanced technology like digital medical records to enhance their ability, improves patient doctor relationship, delivers more health information to the patients [15]. Progress in e-health is assuring strong relationship between physician and patients by improving interaction between them and up surging efficiency in medical decision-making process [12]. American health sector is facing a significant transformation through the implementation of the concept of e-health which ensures efficient electronic transaction of health data or information among organisations and serves the cost-effective healthcare to US population [18].

Pew Internet report June 2012 has revealed that majority of information seekers (66%) search-specific diseases or medical problems as compared to 56% users who search for doctors or other service delivery related to health professionals and only 36% enquire for hospitals or other medical facilities. Further, infrequent online mode of costly and complicated product purchase is a function of easiness, enjoyment, trust and confidence [72]. On the contrary, lack of security, lack of complete product-related information, unattractive visual displays, absence of physical examination, layout and poor scope of testing of products de-motivate people for online purchase [2]. Direct interaction between pharmacists with their patients opened the path of easy access to information for the people [36].

14.2.2 Shift in Mode of Purchase of Medicine

COVID-19, a droplet or contact transmitted zoonotic virus first occurred in Wuhan, china, in late December 2019 and then it started to flow all over the globe [75]. Silent emergence, unwarranted therapeutic option and inestimable death have left no option other than 'lock down' to break the contagious infection chain of COVID-19 [29, 72]. Lock down has the ability to restrict the intensity of spread of the disease. People purchase emergency and compulsory goods rather than the needless one during unpredictable situations [11]. COVID-19 created a challenging situation for the patients with chronical illness by making unavailability of proper medications, up surging medicine cost which lead to worse their health condition [10]. Lockdown, curfews, social distancing, border sealing generated drug shortage by evolving panic among consumers with non-medical backgrounds, patients with severe diseases as it led them to make the stockpiling of drugs [4]. Research conducted in Ukraine to identify the factors fuelling people's stockpiling behaviour during COVID-19 found that people increased purchase volume and reduced purchase frequency of drugs [59]. Home isolation measure of Covid-19 leads people to become more internet addicted which resulted in various psychological problems like mental illness, stress, anxiety, depression, etc. [21]. Amidst the phase of COVID-19 pandemic, perceived risk, fear psychosis also compelled people to purchase preventive medicines [35, 24]. Internet saved the period of COVID-19 by giving us the chance of meeting our personal as well as professional needs virtually while maintaining all COVID-19 protocols [68]. Mobile phone usage with upgraded technology became a rescuer of human being as it facilitated them with the help of providing online platform accesses to fulfil their daily needs during pandemic lockdown [14]. In UAE due to the implementation of quarantine measures for breaking the chain of virus spread, purchase of medicine using internet became a prevalent way of adopting medications where marital status, age, gender, level of education were the influencing factors [33]. Due to poor access and scarcity of medicines in local shops, online web-based e-commerce medicine delivery systems were accessed to fight pandemic 24 × 7. The convenience of virtual platform and pleasure given motivations associated with e-pharmacy are also the important factors that are provoking individuals to replace online pharmacy in place of offline one [65]. People are choosing the option e-pharmacy over the offline traditional system due to perceived values, i.e. user friendly, intuitive and through this process they will get the required solution easily and instantly. People considered online pharmacy as the most reliable way of purchasing drugs during pandemic [48]. An inquiry for recognising the factors insisting the population of Silchar, Assam, disclosed that during the pandemic period their purchasing pattern changed significantly which led them to tend more towards the concept of e-pharmacy [20]. A study done in Slovakia revealed that during pandemic Slovak people faced a change in their buying behaviour where consumers' income, age and occupation played pivotal roles [70]. A study made in china during the aftermath of pandemic on patients with different levels of Parkinson disease (PD) exposed that during and after no COVID-19 situation patients experienced severe mental health problems which encourage them to shift their daily activities from offline to online and also it changed the way of managing PD through online drug acquisition and recovery [30]. In Germany Pandemic enforced people to accept the process of Digital Health Application (DHA) in case of Rheumatic diseases [38]. A study examining the obstacles and strategies for the implementation of Medications for Opioid Use Disorders (MOUD) among Medicaid recipients during climate change-related extreme weather events (EWEs), emphasising deployment of Medications for Opioid Use Disorders (MOUD)-specific disaster planning and utilisation of telehealth for providing better and impactful healthcare [64]. During the atrocious situation of New Zealand's COVID-19 lockdown a study on patients experiencing the method of e-prescribing demonstrated that it possesses a promising future and also highlighted the ways to strengthen it [32]. COVID-19 has drastically transformed the purchasing behaviour of US population which led to the change in consumption trend, reduction in buyer's satisfaction and shift towards online purchasing system [8]. Amidst the fast-paced technological growth, pharmacy sector experienced notable metamorphosis from physical to digital with the progress of AI, Distributed Networks, Virtual Media and specially during the COVID-19 period the concept of e-pharmacy emerged rapidly [7]. Pandemic enforced Spain's traditional pharmacy system to adopt telepharmacy for patients counselling to maintain social distancing and to break the chain of virus spread [43]. Initiatives taken by Australian government for ensuring the continuous delivery of lifesaving drugs includes govt. subsidised medicine dispensation, medicine supply for maximum 30 days, home delivery services of medicine for risky population, initiation of telehealth and e-prescription [50].

Even, a post-pandemic period study conducted on 520 registered under 35 years aged young pharmacists in Indonesia revealed the significance of telepharmacy during COVID-19 period for providing incessant medical services and for providing chances to rural areas for accessing better and improved healthcare services [26]. A survey on 28 graduated pharmacy students in New York for understanding the impact of online educational conference unveiled that most of the students agreed that tech-based learning system can be the best choice for the growth of pharmacy curriculum [17]. A cross-sectional study conducted in Uganda on the basis of 174 out of 198 sampled retail pharmacies was to diagnose the obstacles for delivering the Medicare services in rural areas in post-pandemic period exposed that around 58% pharmacist are taking the help of technology for delivering services in remote areas while witnessing the problems like exorbitant internet fees, low internet speed, manpower deficit, regulatory gap in remote pharmaceutical services [23]. E-pharmacies became an easy and important alternative to physical pharmacies [39]. E-patients mostly inquired regarding the cost of treatment, medical insurance and reimbursement of treatment [2]. During pandemic, people purchased more conventional medicine and less modern medicine as compared to pre-pandemic through online mode [74]. For ensuring better healthcare at lower cost different nations are implementing E-health, Telemedicine, M-health but the implementation required high speed internet access, easily accessible electronic gadgets, concern of policymakers and public awareness [25]. By comparing pre- and post-pandemic period e-pharmacy retailers should focus on different supply chain models to enhance their services [44].

Emergency need of medicine, Easy accessibility, Peer recommendation in social media, online advertisement, repeated visibility in digital media, Social media post by e-pharmacy, Easy to interact with the pharmacist, Fear psychosis created in digital media [3], Emotion-based posts, Attractive discount, Celebrity endorsement are the derived factors of adopting online purchase of medicine by the previous researchers. During COVID-19, fake message, Enhanced mobile phone usage, Restriction of movement during pandemic, Enhanced social media awareness, Panic buying [47] have fuelled the online mode of purchasing medicine. Thus, an interesting avenue of research is identifying the influence of digital presence of online medicine delivery APP on subjective motive of purchasing medicine online during COVID-19 pandemic.

14.2.3 Social Media Influence for Shifting the Purchasing Mode of Medicine

Social media played a huge role in trafficking people, which they used mostly to gather disease-related information and medication [5]. For complex product purchase, social media mediates the decision-making process by providing information regarding the brand, associated risks and the high cost of acquisition [71]. Electronic word of mouth and value co-creation creates a trust in message spread through social media sites [61]. Digital media platforms have very influential role in influencing the choice process of medicine online [71]. Various social media are increasingly being used for connecting different health professionals by sharing information regarding details of medicines, use of medicine for diseases, clinical trials, etc., [13]. Digital social media has enhanced the scope of doctor-patients interaction, disease awareness, run-time interaction during medical emergencies [9]. In order to maintain social distancing a study on consumer having active social media account through telephonic interviews revealed that the situation of consumer panic buying was evolved due to unpredictability, lack of product during emergency health crisis, etc., [49]. WhatsApp and Twitter are the most popular media that influenced people maximum to decide the purchase of medicine in online mode [31]. Sharing and posting of personal experiences on social media related to purchases, advice, reviews, issues and problems positively influence in making purchase decisions as it generates awareness among peer groups [51, 6]. The spread of negative emotions via social media fostered the perceived threat of mortality. Fear psychosis led people to use social media to share their emotions [67]. Celebrity endorsement, Hashtags, discounts, promo codes and celebrity face endorsement were designed by Indian online pharmacies with the objective to let people talking about their brand [19, 63]. Evolution of promo codes (sales promotion technique) was a by-product of COVID-19 pandemic [1].

14.2.4 Pandemic Essentials

An exploration by the Jordanian Pharmacists brought to light that pandemic changed the pattern of drug consumption by increasing the demand for prescribed and non-prescribed drugs such as antibiotics, antithrombotic, antiviral, vitamins, minerals and also for medical instruments as well [22]. Excessive and Preposterous use of azithromycin in Karachi during the phase of COVID-19 pandemic facilitated and enhanced antimicrobiotic resistance [69]. Initiation of a multi-professional antibiotic stewardship program (ABS) on medical intensive care unit (MICU) by German university affirmed that multidisciplinary association among doctors, nurses, pharmacist is very important for preserving the usefulness of antibiotics [60]. COVID-19 triggered an increase in supply and demand schedule for complementary and alternative medicines (CAM) as possible preventive medications for eradicating the pandemic and emphasised the need for pharmacists, health professionals to stay informed and updated regarding global CAM practices for rendering better health services [53]. Nonpharmaceutical service providers like small kiosks and minimarkets mainly stocked non-prescribed medicines like drug for common cough and cold, or analgesic-antipyretics [73]. COVID-19 created an upsurge in demand for psychotropic drug by negatively impacting the mental health of mainly the young and weak of French population [42]. A study based on 120 responses from Ichalkaranji discovered that during pandemic people started to prefer over the counter (OTC) allopathic drugs for recovering general illness where factors like fair price, readily available, own decision played important roles [41]. A study carried out in Germany demonstrated that COVID-19 lockdown generated an upsurge in buying preferences for medicines including neurological, cardiovascular, psychotropic, anti-Parkinson, tranquilisers, antiepileptics, antidementive, vitamin-k, beta blokers ACE inhibitors

[40]. For the regeneration and prevention of the spread of SARS-COV-2 virus many chemicals have been identified and have exhibited in vitro efficacy which includes Nitazoxanide, Remdesivir, Chloroquine, etc., but rapid pharmaceutical trials are required to declare the exact remedial efficiency of those drugs [66]. Even though huge clinical experiments are required to announce few categories of drugs including anti-malaria, antiviral and anti-inflammatory, conventional Chinese medicines, some other medicines and treatments are highly effective for the prevention of SARS-CoV-2 virus but they were predominantly recommended for the treatment against of COVID-19 pandemic based on previous pandemic experiences [55].

14.3 Methods

This study analysed the impact of digital media on the purchase of medicine online during the COVID-19 pandemic. The research is based on both primary and secondary data. Secondary data has been collected from several journals, magazines and websites. Primary data has been collected by field survey through a structured questionnaire method from earning members of families. The questionnaire consists of close-ended questions that helped us to collect the reflection of the respondents on a 5-point Likert scale anchored from Strongly Agree—Agree—Maybe—Disagree—Strongly Disagree which were coded as 5, 4, 3, 2 and 1, respectively. Respondents were chosen by applying non-probability convenience methods of sampling. Primary data has been collected in the above-mentioned process and analysed by using MS Excel and other statistical software. To classify the consumer's opinion on choice of medicine, preference of digital medium for delivery of medicine and influence of social media on purchase of medicine online; dimension reduction of the constructs are essential. Therefore, Principle Component Analysis (PCA) has been applied in this study to reduce the dimension of the construct and variables.

14.4 Results and Discussion

According to our analysis we observed that among 107 respondents, 63% are male and 37% are female, with highest qualification degree as Graduation, and most of them being married and belongs to different professions and age group from 28 to 58 years.

The study emanated that, 54% work in service sector, 27% have businesses, 9% are homemaker, 6% are student and the rest 4% are unemployed in which 11% have the monthly household income ranges as Rs. 10,000–30,000; 23% between Rs.30,000 and 50,000; Rs.50,000 and 70,000 by 20%, 23% between 70,000 and 90,000 and the rest 23% having income of Rs.90000 and above. The analysis stated the truth of the study as 97% agreed to know about e-pharmacy whereas 3% disagreed, among which 43% people came across regarding online purchase option of medicine through

Facebook, whereas 32% from Instagram, 2% from Twitter, 15% from WhatsApp and 8% from others.

Since 94% opined that during pandemic, they purchased medicines online more often. From the analysis we can observe that, 45% people agreed to the fact that reviews of e-pharmacy sites forced people to select the most reliable one whereas 35% strongly agreed that during pandemic attractive posts helped to purchase medicines online. The analysis also states that, 91% people were in favour of the statement that purchase of medicines online was an easy and convenient way due to the restrictions during pandemic. 2% respondents disagreed on the statement that during COVID-19 pandemic frequency of e-pharmacy advertisement increased. 81% people opined that that due to increase in advertisement many people visited e-pharmacy. Interesting to observe that, nobody disagreed on the fact that Social media pages of e-pharmacy were an easy option to attract customers. During our interview it has been found that, 65% people were more exposed to social media posts that have enhanced the purchase from e-pharmacy and 1% respondents disagreed that frequent advertisements on social media have made public aware about different e-pharmacy sites. Majority of the people have opined that social media advertisement increased the awareness regarding innovative health products and 48% people also agreed that social media helped in building trust about e-pharmacies among public. Moreover, 74% respondents agreed that panic-driven purchase increased and fear psychosis increased the rate of purchase from e-pharmacy. 78% of the people agreed that Reviews of e-pharmacy sites forced them to select the most reliable one and that attractive discount were the reason that they were provoked and got attracted towards e-pharmacy websites. Further, 67% agreed that Reviews of e-pharmacy sites forced people to select the most reliable one but still 36% of people were uncertain whether attractive discounts of online medicine sites provoked them or not.

To understand the factors influencing the influence of social media on purchase of medicines online during pandemic Principle Component Analysis has been done. The dataset has 100 plus responses which is 5 times more than the variables studied. The Cronbach's alpha value, i.e. scale reliability coefficient is 0.8632. KMO (Kaiser–Meyer–Olkin Measure of Sampling Adequacy) value of the said analysis is 0.6. All the derived values are conforming to the thumb rule. Hence we may proceed for Principle Component Analysis (PCA). The software-based analysis has revealed the following 6 factors with Eigen value greater than 1 (6.03, 2.86, 1.59, 1.23, 1.13 & 1.06). Only, first three factors have variables with loading value equal to or greater than 0.5.

Component I—Emergency need of medicine, Easy accessibility, Peer recommendation in social media, Online advertisement, Frequent visibility in digital media, Social media post by e-pharmacy, Easy to interact with the pharmacist, Fear psychosis created in digital media.

Component II—Emotion-based posts, Fake message.

Component III—Easy to interact with the pharmacist.

Table 14.1 KMO and Bartlett's test

Kaiser–Meyer–Olkin measure of sampling adequacy		0.776
Bartlett's test of sphericity	Approx. Chi-Square	907.543
	df	231
	Sig.	0.000

Therefore we can conclude that Component 1 can be termed as perceived objective value of online medicine purchase, Component 2 can be termed as perceived subjective value of online medicine purchase and Component 3 is perceived consultative value of online medicine purchase (Tables 14.1 and 14.2).

During COVID-19 pandemic, fear psychosis also forced consumers to choose e-pharmacy as not only a medicine service provider but also for interacting with the

Table 14.2 Component matrix[a]

	Component					
	1	2	3	4	5	6
Q4	0.639	−0.414	−0.329	−0.126	0.082	−0.154
Q5	0.605	−0.469	−0.142	−0.058	0.081	−0.186
Q6	0.678	−0.344	−0.242	−0.001	−0.020	0.085
Q7	0.417	−0.399	0.217	0.505	0.231	−0.044
Q8	0.632	−0.200	−0.231	−0.335	−0.003	0.313
Q9	0.464	−0.519	−0.060	0.035	0.135	−0.004
Q10	0.640	−0.145	−0.270	0.015	−0.031	0.009
Q11	0.470	0.139	−0.224	0.311	0.469	0.019
Q12	0.483	0.267	0.002	−0.420	0.260	0.229
Q13	0.392	0.177	0.373	−0.452	−0.007	−0.017
Q14	0.603	0.119	0.382	0.146	0.214	−0.228
Q15	0.571	0.094	0.485	−0.318	0.004	0.008
Q16	0.456	−0.028	0.530	0.062	0.201	−0.225
Q17	0.478	0.587	−0.034	0.050	−0.185	−0.306
Q18	0.487	0.536	−0.295	0.068	−0.247	−0.388
Q19	0.682	0.395	−0.204	−0.138	−0.092	−0.245
Q20	0.481	−0.367	0.107	0.067	−0.357	−0.016
Q21	0.447	0.046	0.360	0.085	−0.055	0.329
Q22	0.507	0.121	0.239	0.359	−0.455	0.173
Q23	0.556	−0.059	−0.062	0.141	−0.381	0.357
Q24	0.229	0.654	−0.196	0.162	0.180	0.200
Q25	0.338	0.586	−0.086	0.180	0.227	0.366

Extraction Method: Principal Component Analysis

paramedical persons. To explore this, the association between 'Social media post by e-pharmacy', 'Easy to interact with the pharmacist' and 'Fear psychosis created in digital media', ANOVA has been applied. The hypothesis is

H0: Approach to an e-pharmacy by panic-driven customer is not associated with the intention of interacting with the pharmacist.

H1: Approach to an e-pharmacy by panic-driven customer is associated with the intention of interacting with the pharmacist.

The one way ANOVA reveals the value of probability is 0.023 which is less than 0.05 at 95% confidence level. Therefore, the alternative hypothesis can't be rejected.

Pharmacists act as an important influencer in this scenario. They are becoming important catalyst that can help for changing the paradigm and in future in absence of doctor, paramedical stuffs are becoming very important choice before the prospective customers to consult and clarify their doubts. Since in India doctor-patients ratio is very weak hence pharmacist can not only become a supplement but can also become a true friend to guide the patients regarding treatment and medicine-related information which will definitely improve the relationship among the patients and healthcare providers. Thus, this can help the online medicine delivery platform to transform and uplift their organisation for making trustworthy among the prospective purchaser who look for consistent and beneficial solution.

14.5 Conclusion

In conclusion COVID-19 expedited people's demand for accepting digital platforms for meeting essential requirements by restricting them with different protocols. As offline meetings were not allowed people were bounded to rely more on online system for daily survival. Thus it ensured dependability on digital media platforms. This practice also transformed the mode of accessing health and medicinal services as an output of emergence of e-pharmacy or online pharmacy. Amidst pandemic e-pharmacy became the reliable and convenient solution facilitating people to get required pharmaceutical services while maintaining social distancing. The analysis has already explored that decision of purchase of medicines was impacted by digital media during the wave of COVID-19 pandemic. The role of social media can't be overlooked. It became an effective factor responsible for the change in consumer's purchasing behaviour. Public were influenced and guided by social media-based promotions or forwarded messages towards purchase of medicine. Emotional posts and or viral messages provoked them to purchase certain drugs in search of health security. Additionally, fake message has potentiated their decisions of approaching e-pharmacy more constructive. During pandemic, panic-driven customers considered e-pharmacy also as a centre of interaction with the pharmacist. From the PCA analysis, it can be concluded that the existing e-pharmacies should focus on these factors to prepare certain marketing strategies to attract customers and keep the sale graph growing even after the pandemic.

The new age in the pharmaceutical market is knocking at the door. The triangular partnership is hitting new heights between physicians, pharmaceutical firms and patients. In order to connect consumers and healthcare professionals (doctors) with their products and services, the pharma industry must follow a 360-degree strategy. This study or analysis observed that majority of the respondents are aware of e-pharmacies through various digital and social media and also that helped them to know more or influenced them towards purchasing medicines from online e-pharmacies more frequently during pandemic. Frequent advertisements on social media helped them connect with the pharmacists more easily without the risk factor of going out during lockdown and create a two way communication process. Here, emotional triggering posts and fear psychosis played a major factor in guiding traffic towards pharmacies. Digital media played a big role for the growth of e-pharmacies and engaging traffic towards online buying of medicines during pandemic. In short, we can say that emergence COVID-19 pandemic not only boost the growth of e-pharmacy but also it opened up a new window for the healthcare sectors to provide convenient, reliable and cost-efficient services. It is evident that e-pharmacy will adopt all upcoming advanced digital innovations and perform as a key controller to reshape the structure of future health system, mainly during another crisis.

References

1. Ackner, R. (2020). *Ecommerce coupon marketing strategies: Give discounts, get a lot more.* https://www.bigcommerce.com/blog/coupon-marketing/
2. Agrawal, S., & Kaur, N. (2015). Influence of social media marketing in indian pharmaceutical industry. *International Journal of Advance Research and Innovation, 3*(4), 735–738.
3. Ahmad, A. R., & Murad, H. R. (2020). The impact of social media on panic during the COVID-19 Pandemic in Iraqi Kurdistan: Online questionnaire study. *Journal of Medical Internet Research, 22*(5), e19556.
4. Al, Zoubi, S., Gharaibeh, L., Jaber, H. M., & Al-Zoubi, Z. (2021). Household drug stockpiling and panic buying of drugs during the COVID-19 pandemic: A study from Jordon. *Frontiers in Pharmacology*. https://doi.org/10.3389/fphar.2021.813405
5. Al-Dmour, H., Masa'deh, R., Salman, A., Abuhashesh, M., & Al-Dmour, R. (2020). Influence of social media platforms on public health protection against the COVID-19 pandemic via the mediating effects of public health awareness and behavioral changes: Integrated model. *Journal of Medical Internet Research, 22*(8), e19996.
6. Ali Taha, V., Pencarelli, T., Skerhacova, V., Fedorko, R., & Kosikova, M. (2021). The use of social media and its impact on shopping behavior of Slovak and Italian consumers during COVID-19 Pandemic. *Sustainability, 13*, 1710. https://doi.org/10.3390/su13041710.
7. Almeman, A. (2024). The digital transformation in pharmacy: Embracing online platforms and the cosmeceutical paradigm shift. *Journal of Health, Population and Nutrition, 43*(1). https://doi.org/10.1186/s41043-024-00550-2

8. Andrew, M., John, N., & Kevin, M. (2020). Changes in consumer decision-making resulting from COVID-19 pandemic. *Journal of Customer Behaviour, 19*(4), 299–321. https://doi.org/10.1362/147539220X16003502334181
9. Arekar, K., Jain, R., & Kumar, S. (2021). Predictive model to study the consumer buying behaviour towards e-pharmacy through social media influence: A cross country study. *Academy of Marketing Studies Journal, 25*(2), 139–159. https://www.abacademies.org/articles/predictive-model-to-study-the-consumer-buying-behaviour-towards-epharmacy-through-social-media-influence-a-cross-country-study-10038.
10. Awucha, N. W., Janefrances, O. C., Meshach, A. C., Henrietta, J. C., Daniel, A. I., & Chidiebere, N. E. (2020). Impact of COVID-19 pandemic on consumers' access to essential medicines in Nigeria. *The American Journal of Tropical Medicine and Hygiene, 103*(4), 1630–1634. https://doi.org/10.4269/ajtmh.20-0838
11. Aydınlıoğlu, Ö., & Gencer, Z. T. (2020). Let me buy before I die! A study on consumers' panic buying behaviours during the Covid-19 pandemic. *Electronic Turkish Studies, 15*(6), 139.
12. Ball, M., & Lillies, J. (2000). E-health: Transforming the physician/patients relationship. *International Journal of Medical Informatics, 61*(1), 1–10. https://doi.org/10.1016/S1386-5056(00)00130-1
13. Bell, M., Douglas, J., & Cutts, C. (2014). How pharmacy's adoption of social media can enhance patient outcomes. *Intregated Pharmacy Research and Practice, 3*, 39–47. https://doi.org/10.2147/IPRP.S42774
14. Bhavya, R., Sambhav, S. (2020). Role of mobile communication with emerging technology in COVID-19. *International Journal of Advanced Trends in Computer Science and Engineering, 9*(3). https://doi.org/10.30534/ijatcse/2020/131932020
15. Chin, R. (2000). The Internet: another facet to the paradigm shift in healthcare. *Singapore Medical Journal, 41*(9).426–429.
16. Cullen, W., Gulati, G., & Kelly, B. D. (2020). Mental health in COVID-19 pandemic. *An International Journal of Medicine, 113*(5), 311–312. https://doi.org/10.1093/qjmed/hcaa110
17. Davydova, M., Divinagracia L.M., Ng K.E., Bradley. N., (2024). Pharmacy Education, Student perspectives on peer education using a virtual platform to enhance advanced pharmacy practice experiences (APPE). *Pharmacy Education, 24*(1), 48–53. https://doi.org/10.46542/pe.2024.241.4853
18. DeLuka, J. M., Enmark, R. (2000). E-health: The Changing Model of Healthcare. *Frontiers of Health Services Management, 17*(1), 3–15.
19. Depoux, A., Martin, S., Karafillakis, E., Preet, R., & Larson, H. (2020). The pandemic of social media panic travels faster than the COVID-19 outbreak. *Journal of Travel Medicine, 27*(3), 4–5. https://doi.org/10.1093/jtm/taaa031
20. Dutta, D., & Bhattacharjee, B. (2021). Consumer preference and buying pattern of medicines through e-pharmacy during the Covid-19 pandemic in Silchar. *ASSAM, 8*(1).
21. Eidi, A., & Delam, H. (2020). Internet addiction is likely to increase in home quarantine caused by coronavirus disease 2019 (COVID 19). *Journal of Health Sciences & Surveillance System, 8*(3), 142–143. https://doi.org/10.30476/jhsss.2020.87015.1104
22. Elayeh, E., Haddadin, R. N., Dawud, R. J., Alsinjlawi, H. O., & Zidan, R. (2024). Navigating changes in patients drug and non-drug item demands in community pharmacies Admist the COVID-19 pandemic. *Jordan Journal of Pharmaceutical Sciences, 17*(1), 31–44. https://doi.org/10.35516/jjps.v17i1.1426
23. Fauzi, I. A., Kintu, J., Luyima, M., Kyeyune, H., Rajab, K., Kutyabami, P., Asiimwe, L. I., Bukenya, I., Kitutu, F. E., & Mwawule, F. W. (2024). The remote provision of pharmaceutical products and services by retail pharmacies in Kampala, Uganda; practices and challenges. *BMC Health Services Research, 24*(1). https://doi.org/10.1186/s12913-024-11537-4
24. Gangopadhyay, S., & Chakrabarty, A. (2022). Will COVID pandemic induced resurge in immunity booster consumption drive an unethical business practice. *Gerantology and Geriatrics Studies, 8*(1), 751–752.
25. Gangopadhyay, S., Ukil, A., & Chatterjee, S. (2022). Emerging paradigm of smart healthcare in the management of COVID-19 pandemic and future health crisis. In C. Chakraborty & M.R.

Khosravi (Eds.), *Intelligent healthcare* (pp. 455–478). Springer. https://doi.org/10.1007/978-981-16-8150-9_20

26. Ghozali. M. T. (2024). Assessment of knowledge, perception, and readiness for telepharmacy-assisted pharmaceutical services among young pharmacists in rural Indonesia. *Exploratory Research in Clinical and Social Pharmacy, 16*. https://doi.org/10.1016/j.rcsop.2024.100513
27. Gordon, R., McDermott, L., Stead, M., & Angus, K. (2006). The effectiveness of social marketing interventions for health improvement: What's the evidence? *Public Health, 120*(12), 1133–1139. https://doi.org/10.1016/j.puhe.2006.10.008
28. Hoek, J., & Jones, S.C. (2011). Regulation, public health and social marketing: A behaviour change trinity. *Journal of Social Marketing, 1*(1), 32–44. https://doi.org/10.1108/20426761111104419. https://www.researchgate.net/publication/342289159_EPHARMACY_IMPACTS_ON_SOCIETY_AND_PHARMA_SECTOR_IN_ECONOMICAL_PANDEMIC_SITUATION_A_REVIEW
29. Huang, C., Wang, Y., Li, X., et al. (2020). Clinical features of patients infected with 2019 novel coronavirus in Wuhan, China. *The Lancet, 395*(10223), 497–506.
30. Huang, P., Tan, Y.-Y., Chen, B., Shang, H.-F., Wang, L.-J., Liu, C.-F., Chen, L., Chang, Y., Wang, H., Wang, X.-L., Lei, X.-G., Yao, L.-F., Yu, Y., Ye, Z., Chen, H.-B., Chen, S.-D. (2024). Life and disease status of patients with Parkinson's disease during and after zero-COVID in China: an online survey. *Translatonal Neurodegeneration, 13*(8). https://doi.org/10.1186/s40035-024-00399-9
31. Ibrahim, O., Ibrahim, R., & Tameemi, N. (2020). The use of social media for pharmacists' professional activities in united Arab emirates: A descriptive study. *Asian Journal of Pharmaceutical and clinical Research, 13*(7), 182–186.
32. Imlach, F., McKinlay, E., Kennedy, J., Morris, C., Pledger, M., Cumming, J., & McBride-Henry, K. (2021). E-prescribing and access to prescription medicines during lockdown: Experience of patients in Aotearoa/New Zealand. *BMC Family Practice*, 22. https://doi.org/10.1186/s12875-021-01490-0
33. Jairoun, A. A., Al-Hemyari, S. S., Abdulla, N. M., El-Dahiyat, F., Jairoun, M., Al-Tamimi, S. K., Babar, Z.U.D. (2021). Online medication purchasing during the Covid-19 pandemic: A pilot study from the United Arab Emirates. *Journal of Pharmaceutical Policy and Practice, 14*. https://doi.org/10.1186/s40545-021-00320-z
34. Kardakis, T., Weinehall, L., Jerden, L., Nystrom, M. E., & Johansson, H. (2013). Lifestyle interventions in primary health care: Professional and organizational challenges. *The European Journal of Public Health, 23*(4), 1–6.
35. Kargar-Soleimanabad, S., Dehbozorgi, R., & Mahdi Mahdavinoor, S. M. (2022). Psychosis due to anxiety related to COVID-19: A case report. *Annals of Medicine & Surgery, 78*(6), 103795. https://doi.org/10.1016/j.amsu.2022.103795
36. Kaur, G., & Quareshi, K. T. (2015). Factors obstructing intentions to trust and purchase products online. *Asia Pacific Journal of Marketing and Logistics, 27*(5), 758–783.
37. Kelly, J. T., Campbell, K. L., Gong, E., & Scuffham, P. (2020). The internet of things: impact and implication s for health care delivery. *Journal of Medical Internet Research, 20*(11). https://preprints.jmir.org/preprint/20135?__hstc=102212634.7e4c5ce11a3410880694b31723070fb9.1733637009561.1733637009561.1734710188941.2&__hssc=102212634.1.1734710188941&__hsfp=1669205919
38. Kernder, A., Morf, H., Klemm, P., Vossen, D., Haase, I., Mucke, J., Meyer, M., Sewerin, P., Bendzuck, G., Eis, S., Knitza, J., & Krusche, M. (2021). Digital rheumatology in the era of COVID-19: Results of a national patient and physician survey. *RMD Open, 7*(1). https://doi.org/10.1136/rmdopen-2020-001548
39. Khan, M., Amin, R., Mamun, A., & Sajib, A. (2021). Development of web based online medicine delivery system for COVID-19 pandemic. *Journal of Software Engineering and Application, 14*(1), 26–43. https://www.scirp.org/journal/paperinformation.aspx?paperid=106664.
40. Kostev, K., & Lauterbatch, S. (2020). Panic buying or good adherence? Incresed pharmacy purchases of drugs from wholesalers in the last week prior to Covid-19 lockdown. *Journal of Psychiatric Research, 130*, 19–21. https://doi.org/10.1016/j.jpsychires.2020.07.005

41. Landge, R. R., & Petare, P. (2021). Study on consumer behaviour in COVD pandemic with respect to OTC medicine. *International Research Journal of Humanities and Interdisciplinary Studies, 2*(5), 104–113. http://doi-ds.org/link/05.2021-98526359/IRJHIS2105014
42. Laurin, A., Bulteau, S., Caillet, P., Artari, P., Sauvaget, A., Gollier-Briant, F., Huon, J.-F., & Bonnot, O. (2023). Psychotropic drugs consumption during 2020 COVID-19 pandemic and lockdowns: Evidence of a surprising resilience of the drugs delivery system in France. *European Neuropsychopharmacology., 73*, 48–61. https://doi.org/10.1016/j.euroneuro.2023.04.004
43. Margusino-Framiñán, L., Illarro-Uranga, A., Lorenzo-Lorenzo, K., Monte-Boquet, E., Márquez-Saavedra, E., Fernández-Bargiela, N., Gómez-Gómez, D., Lago-Rivero, N., Poveda-Andrés, J. L., Díaz-Acedo, R., Hurtado-Bouza, J. L., Sánchez-Gundín, J., Casanova-Martínez, C., Morillo-Verdugo, R. (2020). Pharmaceutical care to hospital outpatients during the COVID-19 pandemic. Telepharmacy. *Farmacia Hoapitalaria. 44*(suppl. 1), 61–65. https://doi.org/10.7399/fh.11498
44. Mariappan, M., Venkataraman, Y., Devi, K., & Wamba, S. (2022). A large-scale real-world comparative study using pre-COVID lockdown and post-COVID lockdown data on predicting shipment times of therapeutics in e-pharmacy supply chains. *International Journal of physical Distribution and logistics Management, 52*(11). https://doi.org/10.1108/IJPDLM-05-2021-0192
45. Martin, S., Yen, D. C., & Tan, J. K. (2002). E-health: Impacts of internet technologies on various healthcare and services sectors. *International Journal of Healthcare Technology and Management*, 71–86. https://doi.org/10.1504/IJHTM.2002.001130
46. Metaxiotis, K., Ptochos, D., & Psarras, J. (2005). E-health in the new millennium: a research and practices agenda. *International Journal of Electronic Healthcare*, 165–175. https://doi.org/10.1504/IJEH.2004.005865
47. Miah, R. M., Hossain, A., Shikder, R., Saha, T., & Neger, M. (2022). Evaluating the impact of social media on online shopping behavior duringCOVID-19 pandemic: A Bangladeshi consumers' perspectives. *Heliyon., 8*(10), 1–13.
48. Moureaud, C., Hertig, J., Dong, Y., Muraro, I. S., & Alhabash, S. (2021). Purchase of prescription medicines via social media: A survey-based study of prevalence, risk perceptions, and motivations. *Health Policy, 125*(11), 1421–1429. https://doi.org/10.1016/j.healthpol.2021.09.007
49. Naeem, M. (2021). Do social media platforms develop consumer panic buying during the fear of Covid-19 pandemic. *Journal of Retailing and Consumer Services, 58*(January), 102226. https://doi.org/10.1016/j.jretconser.2020.102226
50. Parajuli, D. R., Khanal, S., Wechkunanukul, K. H., Ghimire, S., & Poudel, A. (2022). Pharmacy practice in emergency response during COVID-19 pandemic: Lessons from Australia. *Research in Social and Administrative Pharmacy, 18*(8), 3453–3462. https://doi.org/10.1016/j.sapharm.2021.08.013
51. Patarawadee, S. (2013). *Does social media affect consumer decision-making*. Scholars Archive@ JWU. http://scholarsarchive.jwu.edu/mba_student/24
52. Patel, M. P., Kute, V. B., Agarwal, S. K., & behalf of COVID, O. (2020). "Infodemic" COVID 19: more pandemic than the virus. *Indian Journal of Nephrology, 30*(3), 188–191. https://doi.org/10.4103/ijn.IJN_216_20
53. Paudyal, V., Sun, S., Hussain, R., Abutaleb, M. H., & Hedima, E. W. (2022). Complementary and alternative medicines use in COVID-19: A global perspective on practice, policy and research. *Research in Social and Administrative Pharmacy, 18*(3), 2524–2528. https://doi.org/10.1016/j.sapharm.2021.05.004
54. Peattie, K., & Peattie, S. (2009). Social marketing: A pathway to consumption reduction. *Journal of Business Research, 62*(2), 260–268. https://doi.org/10.1016/j.jbusres.2008.01.033
55. Rabby, M. I. I. (2020). Current drugs with potential for treatment of COVID-19: A literature review. *Journal of Pharmacy & Pharmaceutical Sciences, 23*(1), 58–64. https://doi.org/10.18433/jpps31002
56. Rahmqvist, M., & Bara, A. C. (2007). Patients retrieving additional information via the Internet: A trend analysis in a Swedish population, 2000–05. *Scand Journal of Public Health, 35*, 533–539.

57. Rahmqvist, M., & Claudia-Ana, B. (2010). Patient characteristics and quality dimensions related to patient satisfaction. *International Journal for Quality in Health Care, 22*(2), 86–92.
58. Rollnick, S., Kinnersely, P., & Stott, N. (1993). Methods of helping patients with behavior change. *BMJ, 30*(7), 188–190. https://doi.org/10.1136/bmj.307.6897.188
59. Rossolov, A., Kuzkin, O., & Rossolov, H. (2021). Emergency supplies purchase pattern during COVID-19 outbreak in the developing economy: Frequency and stockpiling drivers' assessment. *Journal of Humanitarian Logistics and Supply Chain Management, 12*(1), 106–128. https://doi.org/10.1108/JHLSCM-02-2021-0016
60. Schmid, S., Schlosser, S., Gülow, K., Pavel, V., Müller, M., & Kratzer, A. (2022). Interprofessional collaboration between ICU physicians, staff nurses, and hospital pharmacists optimizes antimicrobial treatment and improves quality of care and economic outcome. *Antimicrobial Agents Used in Intensive Care Unit, 11*(3), 381. https://doi.org/10.3390/antibiotics11030381
61. See-To, E. W. K., Ho, K. K. W. (2014). Value co-creation and purchase intention in social network sites: The role of electronic Word-of-Mouth and trust—A theoretical analysis. *Computers in Human Behaviour, 31*(1), 182–189.
62. Singh, H., Majumder, A., & Malviya. (2020). E-pharmacy impacts on society and pharma sector in economical pandemic situation: A review. *Journal of Drug Delivery and Therapeutics, 10*(3), 335–340.
63. Smiga, J. (2022). *Social media PR: 6 Tactics to improve your social public relations.* https://blog.hubspot.com/marketing/pr-is-now-social-pr, on 13.11.22.
64. Springgate, B., Matta, I., True, G., Doran, H., Torres, W. V., Stevens, E., Holland, E., Mott, K., Ardoin, T. R., Nixdorff, N., Haywood, C., Meyers, D., Johnson, A., Tatum, T., & Palinkas, L. A.(2024). Implementation of medication for opioid use disorder treatment during a natural disaster: The PROUD-LA study. *Journal of Substance Use and Addiction Treatment, 165.* https://doi.org/10.1016/j.josat.2024.209469
65. Srivastava, M., & Raina, M. (2020). Consumers' usage and adoption of e-pharmacy in India. *International Journal of Pharmaceutical and Healthcare Marketing, 15*(2), 235–250. https://doi.org/10.1108/IJPHM-01-2020-0006
66. Stahlmann, R., & Lode, H. (2022). Medication for COVID-19- an overview of approaches currently under study. *Deutsches Ärzteblatt International, 117*(13), 213–219. https://doi.org/10.3238/arztebl.2020.0213
67. Steinert, S. (2020). The role of social media and emotional contagion. *Corona and Value Change, 23*(1), 59–68. https://doi.org/10.1007/s10676-020-09545-z
68. Subudhi, R., & Palai, D. (2020). Impact of Internet use during lockdown. *SSRN, 2*(S), 59–66. https://ssrn.com/abstract=3711177
69. Tanwir, S., Sabah, A., Khatoon, A., & Afridi, F. I. (2024). Increased antimicrobial resistance against azithromycin during COVID: Role of irrational utilization. *National Library of Medicine, 74*(2), 258–271. https://doi.org/10.47391/jpma.8446
70. Valaskova, K., Durana, P., & Adamko, P. (2021). Changes in cosumers' purchase patterns as a consequence of the COVID-19 pandemic. *Mathematics, 9*(15), 1788. https://doi.org/10.3390/math9151788
71. Voramontri, D., & Klieb, L. (2019). Impact of social media on consumer behaviour. *International Journal of Information and Decision Sciences, 11*(3). https://doi.org/10.1504/IJIDS.2019.10014191
72. Wang, M., Cao, R., Zhang, L. (2020). Remdesivir and chloroquine effectively inhibit the recently emerged novel coronavirus (2019-nCoV) in vitro. *Cell Research, 30*(3), 269e271.
73. Wibowo, Y. I., Yana, I. G. A. A. K., Sari, G. A. P. L. P., Brata, C., Sunderland, B., & Setiadi, A. P. (2024). Mapping medication selling practices at nonpharmacy outlets during the COVID-19 Pandemic: An example of Indonesia. *International Journal of Pharmacy Practice, 32*(5), 369–376. https://doi.org/10.1093/ijpp/riae033
74. Wijaya, H., Mustapha, S. M., & Subramaniyan, M. (2023). Impact of Covid-19 pandemic on consumer behaviour and preference: Modern or traditional medicines? *Research Journal of Pharmacy and Technology, 16*(11), 5392–5399.

75. Wu, Y. C., Chen, C. S., & Chan, Y. J. (2020). The outbreak of COVID-19: An overview. *Journal of the Chinese Medical Association, 83*(3), 217–220. https://doi.org/10.1097/JCMA.0000000000000270
76. Wymer, W. (2011). Developing more effective social marketing strategies. *Journal of Social Marketing, 1*(1), 17–31. https://doi.org/10.1108/20426761111104400

Soumik Gangopadhyay is a Professor, Dept. of BBA, IEM-Kolkata (University of Engineering & Mgmt. Kolkata) India. He has 23+ years of combined experience in academics and industry, effectively conducted training sessions extensively in areas of marketing for senior Indian defense personnel, & corporate executives of CESC, Elillily, Indian Oil Corporation, Reserve Bank of India, Shyamali Paribahan, Caplet Pharma etc. He has worked in the pharmaceutical industry for 8+ years. He has received Excellence in Research of 2022–2023 from IEM-UEM Group and Gold medal (2004, 2007) from ZYDUS CADILA Healthcare for best performance. Successfully presented research article in conferences held at IIM-Ahmedabad, IIM-Kozhikode, MAHE, NMIMS, Bangkok (award winner), Indonesia, Colombo (award winner). Published article in UGC approved, ABDC, SCOPUS indexed journals such as Indian Journal of Marketing, JHM (SAGE), IJCMS, The Indian Economic Journal, IJBIR(Inderscience), Springer Nature etc. He is the reviewer of "Health Education" (Emerald), American Journal of Health Promotion. He has reviewed 30+ SCOPUS or WOS indexed journal articles. As an author, he has contributed two chapters in an edited book of Prof. Philip Kotler published by Routledge. He has recently published an edited book "Perspectives in Sustainable Management Practices" (ISBN: 9781032440026) with Routledge.

Peu Das is an Assistant Professor of Dept. of BBA, IEM-Kolkata (University of Engineering & Mgmt. Kolkata) India. She has completed her MA in economics and teaching economics to the undergraduate students for the last two years. She has successfully presented research article in conference held in NMIMS Indore. She is also pursuing her PhD. Her research interest is managerial economics.

Shreosee Kushari has completed her Bachelor of Business Administration (BBA) in 2023 from the Institute of Engineering and Management (IEM) under University of Engineering & Management (UEM), Kolkata where she developed a strong foundation in business and management principles. Currently she is pursuing Master of Business Administration (MBA) from Institute of Management, Nirma University, Ahmedabad, India to enhance the leadership and analytical skill in the field of business.

Shankha Shubhra Sarkar has Completed his Bachelor of Business Administration (BBA) in 2023 from the Institute of Engineering and Management (IEM) under University of Engineering & Management (UEM), Kolkata where he developed a strong foundation in business and management principles. Currently he is pursuing Master of Business Administration (MBA) from Kalinga Institute of Industrial Technology (KIIT), Bhubaneswar, Odisha, India.

Subham Mondal has Completed his Bachelor of Business Administration (BBA) in 2023 from the Institute of Engineering and Management (IEM) under University of Engineering & Management (UEM), Kolkata. Currently he is serving as a CEO of Kargo Motors, Kolkata, India.

Soumojoy Sen has Completed his Bachelor of Business Administration (BBA) in 2023 from the Institute of Engineering and Management (IEM) under University of Engineering & Management (UEM), Kolkata. Currently he is serving as Service Desk Advisor at British Telecom, Kolkata, India.

Siddharth Das has Completed his Bachelor of Business Administration (BBA) in 2023 from the Institute of Engineering and Management (IEM) under University of Engineering & Management (UEM), Kolkata where he developed a strong foundation in business and management principles. Currently he is pursuing Master of Business Administration (MBA) from Kalinga Institute of Industrial Technology (KIIT), Bhubaneswar, Odisha, India.

Chapter 15
Assessing Person-Job Fit Dimensions and Their Impact Over Employee Job Satisfaction: A Study of CBIC Employees in Public Sector in Post-pandemic Timeframe

Manoj Kumar Chaudhary, Rupsha Roy, and Saugat Ghosh

Abstract Person-Job Fit is an important concept associated with organizational psychology and has impact on employee job satisfaction, their commitment toward their organization and growth of organization. The period of pandemic has witnessed major disruptions in organizational demand for necessary skill sets among their employees and employees witnessed harsh consequences for not matching the demand. The new normal post-pandemic period is witnessing new job roles of employees and mounting demand for learning more and advanced skills to work more efficiently. Public sector organizations like Central Board of Indirect Taxes and Customs (CBIC) are not an exception to this situation. The jobs are more demanding now and so are the employees with respect to their personal and professional requirements. Thus, the relevance of Person-Job Fit concept and its constructs, Need-Supply and Demand-Ability concepts are questioned in post-pandemic job scenario. Overcoming such ambiguity of conceptual framework and constructs of Person-Job Fit in the post-pandemic timeframe, is the aim of present study. As pandemic has also forced severe changes in organizational demand even in relatively rigid structure of public sector organizations and as such studies are not much frequent in public sector organizations and almost none in post-pandemic scenario, the focus of the study is fixed on such premises. The findings of this study will help in validation of the theoretical structure of Person-Job Fit and also provide important information to the policymakers to design job accordingly to achieve higher employee satisfaction and thereby securing higher organizational growth and less employee turnover.

M. K. Chaudhary · R. Roy
Amity Business School, Amity University, Kolkata, India
e-mail: manojnaitik09@gmail.com

R. Roy
e-mail: rroy2@kol.amity.edu

S. Ghosh (✉)
Xavier Business School, St. Xavier's University, Kolkata, India
e-mail: saugat.ghosh@sxuk.edu.in

S. Gangopadhyay et al. (eds.), *Pandemic Diaries*,
https://doi.org/10.1007/978-981-96-5415-4_15

Keywords Person-Job fit · Demand-ability fit · Need-supplies fit · Job satisfaction

15.1 Introduction

COVID-19 pandemic has witnessed paradigm shifts in workplace and the new normal has opened floodgates of unprecedented changes in work culture, skill requirement and expectations of both employers and employees. Public sector enterprises, where critical government functions intersect with pre-defined duties and responsibilities, faced havoc during pandemic, particularly because of less flexibility in the bureaucratic structure. The major changes demanded for new skillsets, for which the ability of the employees was mostly insufficient, and training was not adequate. Central Board of Indirect Taxes and Customs (CBIC) is one of the significant government organizations and is no exception from this change. During COVID-19, the organization faced huge alteration in operation. The physical summoning of the defaulters turned online. The cash transactions also changed to electronic transactions. Most importantly, for the purpose of contact free communication, all data and documents were digitalized, and cloud space were bought and used by the organization. As a revenue earning organization the continuation of the operation was essential while the security was a serious concern for remote operation and policies were changing rapidly. Thus, alignment of the required and present skillsets faced serious mismatch. Employees of this organization faced a huge change in their job requirement and acquiring new skill sets were highly demanded. Post-pandemic period also witnessed new inclusion in the demand of the jobs and thus, Person-Job Fit became a critical issue in most government organizations.

Person-Job Fit is an organizational psychological framework where employee skills and the job requirement are evaluated, matched. Higher levels of Person-Job Fit is said to increase productivity, job satisfaction and commitment, while the converse generates a negative outcome. The concept of Person-Job Fit is studied from different aspects. Theoretically the concept is constructed by two dimensions, and they are Need-Supply Fit and Demand-Ability Fit. While the first one speaks of the demand of the required skillsets for performing a particular job, the other one addresses the professional as well as personal requirements and expectations of the employees for doing the job. The proper balance of these two is said to be critical for maintaining a proper Person-Job Fit and thereby maintaining organizational growth and job satisfaction of the employees.

The post-pandemic era is characterized by different requirements from both the employer and employee perspectives. While the organization's priority and required skill sets are modified so is the requirement of the employees. Pandemic has witnessed severe job cuts, reduction in salary and tremendous demand for learning new technology which triggered high level of insecurity related to job, ambiguity related to the job roles and increased the importance employee priorities to the concepts of work-life balance, employee mental health and organization's support to learning and development. During and after the pandemic, there has been significant ambiguity related to specification of the jobs, required skill sets and also about the support provided to the employees. Public sector organizations like CBIC are no exception in this context. The newly defined job scenario thus requires revisiting the theory of Person-Job Fit once more and check whether the constructs are still valid. Besides this, the research works on Person-Job Fit were mostly done on private organizations, while public sector organizations like CBIC are somewhat neglected. The present study aims to bridge this research gaps. The study aims to examine the relevance of Person-Job Fit in post-pandemic scenario and also to explore and identify whether the Need-Supply and Demand-Ability Fit are the constructing components of Person-Job Fit. Besides, it also aims to conduct the study on CBIC, which is one of the most important public sector organizations. Besides this, the study also tries to identify the interrelationship between the dimensions of Person-Job Fit and employee perception of their job satisfaction.

The findings of this study will contribute in theoretical understanding of the composition of Person-Job Fit and also help in establishing its impact on job satisfaction. The empirical findings will use statistical validation of the variables in study, which will provide greater rigor to theoretical conceptualizations. Besides this, the test of interrelationship between job satisfaction and Person-Job Fit will also provide specific information about the impact of specific dimensions of Person-Job Fit on job satisfaction. Findings like this will help the policymakers to identify the specific aspect through which employee satisfaction to their job could be enhanced and higher growth and productivity could be achieved.

15.2 Literature Review

15.2.1 Introduction to Person-Job Fit (P-J Fit)

The concept of P-J Fit is associated with several other fundamental psychological dynamism of an organization and the related fields of human resource management and career planning and growth of employees. According to the theoretical proposition made by Edwards [8], P-J Fit is an analysis of the requirement of a particular job or work and ability of the employee, or their skill sets to address the requirement. From the perspective of organizational productivity and growth, this analysis is crucial, as greater Person-Job Fit ensure higher level of productivity. Kristof [14] was one among the pioneer researchers who explained the role of this fit in delivering several positive employee outcomes which includes job satisfaction, higher levels of employee productivity and positive and high commitment toward the organization. Findings of Cable and DeRue [2] demonstrated that a critically estimated Person-Job Fit ensures enhanced job performance and less amount of employee turnover. Kristof-Brown et al. [15] studied the following topic with the aid of meta-analysis on this topic and identified its impact on job satisfaction, progress of productivity of employees and also their commitment rendered toward their organization. Several Indian studies also identified the interrelationship between Person-Job Fit with increased commitment on organizations and turnover intentions of the employees [3], job satisfaction of the employees [19] and high employee engagement with the given task [13].

Person-Job Fit is studied to have correlation with innovative behavior of employees. Han et al. [10] have identified that both of these organizational psychological components affect the innovative behavior in working situations of the employees. Huang et al. [11] also studied the impact of Person-Job Fit on employee innovation-oriented nature where they introduced involvement of the employees with their job as a mediating component and their commitment toward their career as a component of moderation. The findings show high impact of Person-Job Fit on employee innovation and the impact of both mediator and moderator were statistically significant.

The constructs of the Person-Job Fit were another important avenue and the studies in this field have identified two components as constructs: Demand-Ability and Needs-Supply Fit. Cable and DeRue [2] was the proposer of Demand-Ability Fit theory where they proposed that the proper correlation between job's demand and required employee skills for a job leads to high productivity, satisfaction to the employees and less turnover. The theory of Need-Supply theory was first proposed by Kristof et al. [15] where researchers argued that the job an employee performs needs their expectations from the job to be fulfilled in terms of rewards, benefits and other tangible and non-tangible component. A mismatch in this could trigger a higher amount of employee dissatisfaction and can decrease employee productivity, commitment toward their organization and can also substantially increase turnover ratio.

Although much is said about the constructing components of P-J Fit and Demand-Ability and Need-Supply Fit being the constructing dimensions, the theoretical construct is not empirically substantiated. The importance of study increases a lot in the post-pandemic period as there is a huge shift of demand of the job and the required skill sets of the employees and also in employee needs and process of supplying them by employees.

15.2.2 Constructs of Demand-Ability Fit

Demand-Ability Fit is one of the basic requirements of Job Design. It is an analysis of two things, the requirement for execution of the job and the ability or competence to perform the job [8]. Higher level of Demand-Ability Fit signifies that there is a perfect match between the required skillsets and the availability of the skill among the employees. Such situations lead to high level of job satisfaction and better performance [15, [2]. Edward (1991) also added that high level of Demand-Ability Fit ensures less stress among the employees and reduces turnover intentions among the employees. The ability of Demand-Ability Fit to trigger proactive behavior of employees has been identified by Tims et al. [20] through meta-analysis of the relevant topics. It also ensures innovation in job, among the employees. But, a mismatch in Demand-Ability Fit, according to Lauver et al. [16] and Seong et al. [18], can lead to employee disengagement of the job and high level of turnover. Wright and Davis [24] have conducted similar study on public sector organizations. The findings indicated that the nature of the job in public sector are critical and the job roles are highly defined and structured. The support of the organization for performing the job is also limited. Under such scenario, higher Demand-Ability Fit is crucial as it ensures the alignment between job requirement and employee capabilities. This also leads to higher clarity of the job roles and also ensures high level of job satisfaction. Similar finding is observed in the study of Vigoda [22], where the study identified high level of Demand-Ability Fit in government organizations also lead to high chances of career advancement opportunities for the employees.

15.2.3 Constructs of Needs-Supplies Fit

Need-Supplies fit is concerned with how well the job meets the personal and professional needs of the employee Kristof-Brown et al. [15]. When an employee feels that the job offers sufficient rewards, recognition, career growth, and work-life balance, Need-Supplies fit is said to be high [2]. Studies by Kristof-Brown et al. [15] and Guan et al. [9] suggested that N-S Fit strongly influences job satisfaction, especially in times of economic or social uncertainty, as it was evident during the pandemic.

Furthermore, Guan et al. [9] linked Need-Supply Fit with the cross-cultural setting. Modern organizations are characterized with assimilation of several cultures among employees, and the findings of Guan et al. [9] showcased that such cross-cultural setting compels employees to adjust in their workplace. This adjustment is a demand of the job which is necessary to be met and part of Demand-Ability Fit. Employees are required to be evaluated on this skill before assigning them to such job roles. Besides this, their research also highlighted that in public sector organizations, Need-Supply Fit is the key for job satisfaction. Wright and Davis [24] stated that long-term job satisfaction can only be achieved by suitable judgment of Need-Supply Fit. Failure to ensure this will lead to decreased productivity and increased turnover rate. De Clercq et al. [6] highlighted the impact of Need-Supply Fit on employee psychology. According to them, a well-designed framework of Need-Supply guarantees psychological well-being and also improves rate of employee retention.

15.2.4 The Interrelationship Between Person-Job Fit and Employee Job Satisfaction

The interrelationship between Person-Job Fit and employee satisfaction from their job has been studied by several researchers. The theoretical study of Edwards [8] has identified that greater fit between the Person and Job ensures higher level of employee satisfaction from their jobs. Kristof et al. [15] conducted a meta-analysis on this topic and identified the similar result. Their study has identified that research works have proposed that there could be significant impact of fit from the perspective of demand and ability of job and employees respectively and also of their needs for execution of the job, and supply fit on the satisfaction of the employees derived from their job. According to the study of Jansen and Kristof-Brown [12], a proper balance of Demand-Ability Fit and Need-Supply Fit can help in achieving higher level of job satisfaction. This result was in unison with Cable and DeRue [2], who proposed that higher Person-Job Fit can ensure several employee outcomes such as employee commitment to their organization, higher productivity, higher level of job satisfaction and low turnover rate.

Vigoda-Gadot et al. [22], in their study on public sector organization, over this similar domain, identified that although public sector organizations provide higher job security, provide employees opportunities of making provisions of personal development and growth and clarity of job role needs to be defined and communicated properly for achieving higher level of Person-Job Fit. Studies of Chatman [4] and Vianen [21] underscored dual avenues of job satisfaction. According to their studies, employees feel more satisfied with their job while their ability matches with the requirement of the job. This satisfaction enhances to a higher level when organizations provide necessary support, reward and recognition for successful completion of their jobs. Wright and Davis [24] further elaborated on this and stated that high stress oriented public sector jobs need to have proper match of employee ability and demand of the job, as the mismatch of these two often results in employee dissatisfaction and can lead to low productivity and high turnover rate.

15.2.5 Post-Pandemic Implications for P-J Fit and Job Satisfaction

Like many other things, COVID-19 pandemic has created significant difference in employee perception of job, the parameters that measure the satisfaction derived from the job, the necessary supports required from the organizations to work healthily in job and also the perspective to evaluate Person-Job Fit. According to Bailey and Kaplan [1], pandemic has forced workforce to re-evaluate the demand of their job and the adequacy of the necessary supports provided by the organization. Research works of Parker et al. [17] and Wang et al. [23] have also come up with similar findings. These findings significantly point out that the Need-Supply Fit has gained significant importance as employees seek for higher flexibility in their work mode, enhanced job security and proper work-life balance.

While Need-Supply Fit is gaining popularity among the employees, organizations, according to Bailey and Kaplan [1], should focus on Demand-Ability Fit measures. According to their study, organizations must evaluate the requirement of a particular work to identify the demand of that specific work and describe the necessary abilities to meet that requirement. Wright and Davis [25] voiced the same argument and included that to achieve positive outcomes, organizations must ensure a properly communicated Demand-Ability Fit for a particular Job and also design a thoughtful Need-Supply Fit to maintain job satisfaction and organizational commitment.

Although much is discussed about the constructs of Person-Job Fit and its interrelationship with job satisfaction, certain gray areas are still there which require further contemplation and also justify the present research. The research gaps from other works that justify the present study are as follows:

15.3 Research Gaps, Objectives and Hypothesis

15.3.1 Lack of Consensus on the Dimensions of Person-Job Fit

Several studies conceptualize P-J Fit differently, without reaching a clear consensus on its dimensions. Edwards [8] noted the confusion surrounding whether D-A Fit and N-S Fit should be treated as independent dimensions or part of a broader P-J Fit construct. Some scholars focus on D-A Fit, while others emphasize N-S Fit, leading to confusion regarding whether these dimensions are distinct or interrelated. Thus, the present research works contain the gap which states:

> **RG1:** Despite the extensive body of research on Person-Job Fit (P-J Fit), there remains a lack of consensus on its dimensions, particularly regarding the interplay between **Demand-Abilities Fit (D-A Fit)** and **Needs-Supplies Fit (N-S Fit)**.

15.3.2 Insufficient Examination of the Interplay Between D-A Fit and N-S Fit

While some studies explore either Demand-Ability Fit or needs-supply fit, few investigate on their common impact on both of these dimension interaction and contribution, jointly to employee reactive outcomes such as job satisfaction, performance or retention. Cable and DeRue [2] acknowledged the importance of investigating both D-A Fit and N-S Fit but noted that few studies look at how these dimensions operate together to influence employee outcomes. Thus, the following research gap is constructed from the outcomes of the present studies:

RG2: Previous studies have inconsistently operationalized P-J Fit, often focusing on either D-A Fit or N-S Fit, without fully exploring their combined influence on employee outcomes.

15.3.3 Underrepresentation of Public Sector and Organizational Context in P-J Fit Research

Most P-J Fit studies have been conducted in private-sector settings, leaving a gap in understanding how P-J Fit functions in public sector organizations. Although Wright and Davis [24] pointed out that public sector organizations often have unique job demands and organizational characteristics, which may alter how P-J Fit functions in such settings, much studies are not found to address this avenue in the context of Indian public sector employees particularly in post-pandemic period. Thus, the following research gap is constructed:

RG3: Most P-J Fit studies have been conducted in private-sector settings, leaving a gap in understanding how P-J Fit functions in public sector organizations particularly in post-pandemic period.

15.3.4 Research Objective

From the research gaps identified in the preceding stage, the following objectives of the present research have been constructed:

RO1: To empirically identify and validate the dimensions of Person-Job Fit with respect to post-pandemic scenario in public sector organization like CBIC.
RO2: To check and statistically validate the D-A and N-S Fit as constructs of P-J Fit and their ability to explain the same with respect to post-pandemic situation in public sector organization like CBIC.
RO3: To identify the impact of P-J Fit and its constructs on employee job satisfaction within the public sector like CBIC, specifically in the context of post-pandemic timeframe.

15.3.5 Research Hypotheses

RH1: Significant and statistically validated dimensions of P-J Fit will be identified by the study conducted during post-pandemic period on public sector organization like CBIC.
RH1: Both of D-A and N-S Fit will be identified as distinct but interrelated dimensions of P-J Fit and the same will be evident in the study on public sector employees of CBIC in post-pandemic timeframe.

RH2: The combined influence of D-A and N-S Fit will have a significant positive impact on job satisfaction among CBIC employees in post-pandemic timeframe.

15.4 Research Methodology

15.4.1 Methodology for Identification of Relevant Scale Items of Person-Job Fit

To identify the relevant variables of P-J Fit, the study of Cable et al. [2] and Kristof et al. [15] was considered. Both the studies provided a sketch of the constructs of both D-A and N-S Fit. These items are used as scale items to identify the construct of P-J Fit and also estimating the relevance of both D-A Fit and N-S Fit in constructing P-J Fit. Data were collected on the scale items using a Likert scale of containing five intervals where 1 = Completely Disagree and 5 = Completely Agree. The identified items were primarily checked for reliability and the value of Chronbach's alpha was considered as the measure of the test of reliability. The questionnaire was circulated among 600 employees of CBIC among which 492 were received with complete responses. The sampling scheme used in this study was convenient sampling.

15.4.2 Methodology to Identify the Dimensions of Person-Job Fit

To identify the relevant dimensions of P-J Fit, a PCA was conducted over the collected data and Varimax rotation technique was used. The value of KMO was considered for sample adequacy and the values of Bartlett's test was considered for checking whether there are correlations among the identified variables. The "Total Variance Explained" table was checked for identification of the number of factors and also for the estimation of extent to which variance of the entire P-J Fit was explained by them. The orthogonally rotated component matrix values were used to identify the items associated with each Factor.

15.4.3 Identification of Interrelationship Between P-J Fit and Employee Job Satisfaction

Identification of the interrelationship between P-J Fit and employee satisfaction derived from the job is the final objectives of the study and to estimate this interrelationship primarily the factor scores of the D-A Fit and N-S Fit were estimated. Employee rating on their perception of job satisfaction was also collected through the questionnaire. Finally, a multiple regression analysis was conducted where Employee perception about their job satisfaction was considered as dependent variable and indices of Demand-Ability Fit and Need-Supply Fit were considered as independent variables. The value of Durbin-Watson was considered for autocorrelation and the values of VIF were checked for problems related to multicollinearity. The value of Adjusted R^2 were checked with corresponding p-value to check the validity of the model. The ability of all the independent variables to modulate the dependent variable and their relative strength to create impact on dependent variables is also done through the beta-values associated with the unstandardized coefficients.

15.5 Findings and Interpretations

15.5.1 Test of Reliability

As discussed earlier, the gathered scale items were checked for their reliability and the findings are as follows (Table 15.1).

Table 15.1 List of factors and variables under demand-abilities and needs-supplies fit and the corresponding Chronbach's Alpha values

Name of the Factor	Items under the factor	Chronbach's Alpha
Demand-Abilities Fit (D-A Fit)	1. I have the necessary skills and knowledge to perform my duties effectively in my role at CBIC 2. My qualifications are well-suited to the tasks and responsibilities required in my position at CBIC 3. I am confident in my ability to meet the performance expectations of my job at CBIC 4. The demands of my job are well-matched with my abilities and strengths 5. I am capable of handling the workload and specific tasks required in my role without feeling overwhelmed 6. My role allows me to fully utilize my professional expertise and skills	0.939
Needs-Supplies Fit (N-S Fit)	1. My job at CBIC provides a level of job security that meets my personal needs 2. There are adequate opportunities for career growth and advancement within CBIC 3. My job aligns with my personal values and professional goals 4. The recognition and rewards I receive for my work at CBIC are sufficient to meet my expectations 5. I feel my job at CBIC provides a good balance between work and life	0.890

15.5.2 Identifications of Dimensions of Person-Job Fit

To identify the relevant dimensions under Person-Job Fit, a Principal Component Analysis was administered. The findings are as follows.

15.5.3 Findings Related to KMO and Bartlett's Test of Sphericity

For factor analysis, there are two important things which are required to be considered before estimating other findings, the values estimated for KMO and Bartlett test. The first one speaks about whether the sample size is adequate for the test to run, the Bartlett's test speaks about the interrelationship among the variables. For KMO test, a value exceeding 0.6 or more permits further study, while a lesser value calls for more collection of data. The p value of Bartlett's test shows that whether the H_0 which states that the independent variables have insignificant correlation.

The two main prerequisites for factor analysis are determining the sufficiency of the sample and the presence of a link between the variables. KMO value provides estimation about appropriateness of the samples both for individual variables and for the entire model. The statistic calculates the percentage of variation in each of the model's variables and if it has a value between 0 and 1 and a value we can proceed with the factor analysis if the KMO > 0.6.

To determine whether correlation matrix of the identified variables that have been determined is an identity matrix, Bartlett's test is conducted. While variables' correlation matrix, as per null hypothesis of this test, is proposed to be an identity matrix, thereby uncorrelated with each other, the converse speaks about the correlation among variables to support the continuation of factor analysis (Table 15.2).

Estimated KMO value is found to be 0.881 > 0.6 and confirms the adequacy of the samples size and thereby supports the continuation of factor analysis. The finding also exhibits that estimated p value of Bartlett's test, 0.000 which is less than 0.05 and specifies the correlation among the test variables and support further proceedings with EFA.

Table 15.2 KMO & Bartlett's Test

KMO Result		0.881
Bartlett's Test	Approx.Chi-Square	5129.336
	dof	55
	Sig	0.000

15.5.4 Findings Related to Communality

Communality in factor analysis describes amount of explained variability of the item by the factors. The communality of the variable is between 0 and 1. A variable has a communality of 1 if it has no unique variance, meaning that all of its variance can be explained. In the event that the components fail to account for all variance in the variable, the communality will be zero. As per Child [5], an item must be eliminated from the factor analysis if the extraction value score is less than 0.2. Following is the communality table of the analysis.

Since none of the items in this instance had an extraction value less than 0.2, none of the items were removed from the model (Table 15.3). In addition, the item "My qualifications are well-suited to the tasks and responsibilities required in my position at CBIC" has an extraction value of 0.871, meaning that 87.1% of the variability of this item has been explained by the factors. It's also critical to note that, according to this factor analysis, this item's variance is the maximum that a factor can explain. Conversely, the item "The recognition and rewards I receive for my work at CBIC are sufficient to meet my expectations" had an extraction value of 0.489. This indicates that 48.9% of this item's variability has been explained by the factor. In this factor analysis, this is the least amount of variance that the factor can account for.

Table 15.3 Communalities

	Initial	Extraction
I have the necessary skills and knowledge to perform my duties effectively in my role at CBIC	1.000	0.781
My qualifications are well-suited to the tasks and responsibilities required in my position at CBIC	1.000	0.871
I am confident in my ability to meet the performance expectations of my job at CBIC	1.000	0.798
The demands of my job are well-matched with my abilities and strengths	1.000	0.865
I am capable of handling the workload and specific tasks required in my role without feeling overwhelmed	1.000	0.770
My role allows me to fully utilize my professional expertise and skills	1.000	0.626
My job at CBIC provides a level of job security that meets my personal needs	1.000	0.814
There are adequate opportunities for career growth and advancement within CBIC	1.000	0.836
My job aligns with my personal values and professional goals	1.000	0.822
The recognition and rewards I receive for my work at CBIC are sufficient to meet my expectations	1.000	0.489
I feel my job at CBIC provides a good balance between work and life	1.000	0.660

Table 15.4 Total variance explained

Component	Initial Eigenvalues			Extraction sums of squared loadings			Rotation sums of squared loadings		
	Total	Var. %	Cumm.Var %	Total	% of Var	Cumm. %	Total	% of Var	Cumm.%
1	7.091	64.467	64.467	7.091	64.467	64.467	4.684	42.580	42.580
2	1.240	11.275	75.743	1.240	11.275	75.743	3.648	33.163	75.743
3	0.607	5.515	81.257						
4	0.577	5.242	86.499						
5	0.351	3.189	89.688						
6	0.304	2.767	92.455						
7	0.213	1.939	94.394						
8	0.199	1.809	96.203						
9	0.194	1.760	97.963						
10	0.165	1.498	99.461						
11	0.059	0.539	100.000						

15.5.5 Findings Related to Variance Explanations

Factor analysis derives the number of legitimate factors and the amount of explained variance are two important information which states the success of the test. This is worth mentioning that eigenvalues > 1 are determined the number of factors. The findings related to this are as follows (Table 15.4).

The findings show in the factor analysis performed on the collected data reveals two components, 7.091 and 1.240 with initial eigenvalues > 1. Consequently, two factors have been found via factor analysis. The first component is found to explain 64.467% of the variance and the second component is found to explain 11.275% of the variance. Together, they account for 75.743% of the variance. Consequently, this could be said that these two factors were found via factor analysis, and that these factors account for 75.743% of the variance in the whole.

15.5.6 Findings Related to Rotated Component Matrix

Different elements are found in different components according to factor analysis. One component or factor should contain items with a high association. The rotated component matrix table compiles data regarding the Pearson correlation between the components and the items. We can also determine the type of the component by looking at the objects' characteristics. Here, we employed an orthogonal rotation called the Varimax rotation. Varimax rotation is utilized in this situation because,

Table 15.5 Rotated component matrix[a]

	Component	
	1	2
I have the necessary skills and knowledge to perform my duties effectively in my role at CBIC	0.867	
My qualifications are well-suited to the tasks and responsibilities required in my position at CBIC	0.879	
I am confident in my ability to meet the performance expectations of my job at CBIC	0.761	
The demands of my job are well-matched with my abilities and strengths	0.880	
I am capable of handling the workload and specific tasks required in my role without feeling overwhelmed	0.815	
My role allows me to fully utilize my professional expertise and skills	0.640	
My job at CBIC provides a level of job security that meets my personal needs		0.787
There are adequate opportunities for career growth and advancement within CBIC		0.901
My job aligns with my personal values and professional goals		0.866
The recognition and rewards I receive for my work at CBIC are sufficient to meet my expectations		0.507
I feel my job at CBIC provides a good balance between work and life		0.670

in contrast to oblique rotation, orthogonal rotation does not permit the correlation between the detected components. The findings are as follows (Table 15.5).

Component 1—Demand-Abilities Fit (D-A Fit)

Component 1 comprises of six items that pertain to the way in which respondents express their Demand-Ability Fit attributes. The six elements have an acceptable factor loading range of 0.880 to 0.640, explains 64.467% of the variation of the overall variance on all the components.

Component 2—Needs-Supplies Fit (N-S Fit)

Component 2 comprises of three items that pertain to the way in which respondents express their Needs-Supplies Fit attributes. The five elements have an acceptable factor loading range of 0.901 to 0.507, explains 11.275% of the variation of the overall variance on all the components.

15.5.7 Measuring the Interrelationship Between P-J Fit and Job Satisfaction

Estimation of interrelationship of the P-J Fit measure with employee perception about the satisfaction derived from their job is done through a multiple regression analysis. Here the employee rating about their job satisfaction was considered as dependent variable and factor scores of D-A and N-S Fit as independent variables. The interrelationship measures are as follows (Tables 15.6 and 15.7).

The findings show that the value of Adjusted R^2 is 0.582 with a corresponding p value 0.000. This confirms that value of Adjusted R^2 is statistically significant. The value of Adjusted R^2 signifies that 58.2% of the variability of the dependent variable, job satisfaction, has been explained by both D-A and N-S Fit.

The Durbin-Watson test is a statistical test that detects autocorrelation in the residuals of a regression model and the test statistic of this test is a value between 0 and 4, with an acceptable range of 1.5 to 2.5. A value of 2 or nearly 2 indicates no autocorrelation. Here, the estimated value of this test is 2.060 which is close to 2. This finding indicates that the regression analysis does not suffer the problem of autocorrelation.

The values of VIF are also considered to measure whether the collinearity is posing serious problem to the predictor variables. The estimation of VIF is necessary to check the level of collinearity among them within a multiple regression. According to Norman R. Draper and Harry Smith [7], small values of VIF, such as VIF < 3, indicate significantly low correlation among predictor variables. They estimated the cutoff value of VIF, which is 5. Thus, as per their observation, variables with VIF < 5 should be included in the model. Here the value of VIF is 1 < 3. Thus, this could be stated that the model does not suffer much problem related to multicollinearity (Table 15.8).

All the independent variables of the model are identified as statistically significant as all the p values are 0.000 < 0.05. The value of unstandardized coefficient associated with the factor score of Demand-Ability Fit is 0.739. This means, one unit change in Demand-Ability Fit will affect job satisfaction by 0.739 unit. Similarly, for the factor score of Need-Supply Fit, the value of unstandardized coefficient is 0.196.

Table 15.6 Model summary[b]

Model	R	R^2	AdjustedR^2	Std.Err	Durbin-Watson
1	0.764[a]	0.584	0.582	0.64640377	2.060

a. Predictors: (Constant), FS_NSFit, FS_DAFit
b. DependentVariable: JobSat

Table 15.7 ANOVA[a]

Model		Sum of square	dof	Mean^2	F	Sig
1	Regression	286.677	2	143.339	343.049	0.000[b]
	Residual	204.323	489	0.418		
	Total	491.000	491			

a. Dependent Variable: JobSat
b. Predictors:(Constant), FS_NSFit, FS_DAFit

Table 15.8 Coefficients[a]

Model		Unstd.Coeff		Stand. Coeff	t	Sig	CollinearityStatistics	
		B	Std.Err	β			Tol	V.I.F
1	(Constant)	-4.189E-17	0.029		0.000	0.869		
	FS_DAFit	0.739	0.029	0.739	25.317	0.000	1.000	1.000
	FS_NSFit	0.196	0.029	0.196	6.718	0.000	1.000	1.000

a. Dependent Variable: JobSat

The value of unstandardized coefficient denotes that one unit change in the level of Need-Supply Fit can change job satisfaction by 0.196 unit.

15.6 Conclusion

This research was committed to assess dimensions of Person-Job Fit, particularly focusing on Demand-Ability Fit and Need-Supplies fit and also to estimate their influence on satisfaction of the employees from their job, with respect to the PSU, Central Board of Indirect Taxes and Customs (CBIC) in the post-pandemic period. The findings serve a major role of identification of Person-Job Fit dimensions and also validating the Cable and DeRue [2] given theoretical constructs. Moreover, the study highlights that both Demand-Ability and Need-Supplies play in shaping employee satisfaction, particularly in the public sector, where job demands are often rigid, and organizational support is essential.

The findings revealed that fit of Demand-Ability and Need-Supply, both are statistically significant predictors of job satisfaction. Both of them are found to have statistically significant impact on Person-Job Fit. As the study was conducted on the employees of CBIC, it could be stated that based on the findings, the fit related to Demand-Ability is found to has more impact on the fit of Person-Job than the fit related to the Need-Supply. This finding highlights that public sector organizations needs to check the ability of their employees closely as per the demand of the assigned job to ensure a better outcome.

Need-Supply Fit is also identified to have statistically significant impact on Person-Job Fit. The post-pandemic period has witnessed a major shift in job dynamics where security of their job, opportunities of growth in their career, work-life balance and so on have become more important than the earlier timeframes. These attributes are part of Need-Supply Fit elements and the statistically significant Need-Supply Fit indicates that public sector organizations, such as CBIC has to pay attention on these attributes to ensure a better job satisfaction of their employees.

Thus, to conclude this could be mentioned that the present research has addressed all the identified research gaps, identified the dimensions of Person-Job Fit and thereby validated the theoretical construct given by Cable and DeRue [2] and also identified the relative importance of both constructs of P-J Fit, with respect to the employee job satisfaction of the CBIC employees in post-pandemic period. Findings of this study will provide pragmatic insight to the policymakers while designing a particular job. Further research in this could be conducted where the importance of dimensions of P-J Fit, could be tested in the context of Person-Organization fit. Impact of D-A and N-S Fit could also be tested for other employee outcomes such as employee commitment and turnover intentions, in the new normal timeframe.

References

1. Bailey, D. E., & Kaplan, S. M. (2020). Crisis-driven changes in work: Adjustments in response to COVID-19. *Academy of Management Journal, 63*(5), 1289–1309. https://doi.org/10.5465/amj.2020.4005
2. Cable, D. M., & DeRue, D. S. (2002). The convergent and discriminant validity of subjective fit perceptions. *Journal of Applied Psychology, 87*(5), 875–884. https://doi.org/10.1037/0021-9010.87.5.875
3. Chhabra, B. (2015). Person-job fit: Mediating role of job satisfaction & organizational commitment. *Indian Journal of Industrial Relations, 50*(4). https://doi.org/10.7322/jhgd.119277
4. Chatman, J. A. (1989). Improving interactional organizational research: A model of person-organization fit. *Academy of Management Review, 14*(3), 333–349. https://doi.org/10.5465/amr.1989.4279063
5. Child, D. (2006). *The essentials of factor analysis* (3rd ed.). Continuum International Publishing Group.
6. De Clercq, D., Dimov, D., & Belausteguigoitia, I. (2016). Perceptions of adverse work conditions and innovative behavior: The buffering roles of relational resources. *Entrepreneurship Theory and Practice, 40*(3), 515–542. https://doi.org/10.1111/etap.12128
7. Draper, N. R., & Smith, H. (1998). *Applied regression analysis* (3rd ed.). John Wiley & Sons.
8. Edwards, J. R. (1991). Person-job fit: A conceptual integration, literature review, and methodological critique. *International Review of Industrial and Organizational Psychology, 6*, 283–357.
9. Guan, Y., Deng, H., Bond, M. H., Chen, S. X., Chan, C. C. H., & De La Rama, J. (2010). Person–job fit and work-related attitudes among Chinese employees: Need for cognitive closure as a moderator. *Basic and Applied Social Psychology, 32*(3), 250–260. https://doi.org/10.1080/01973533.2010.495683
10. Han, Y., Yin, J., & Zhang, X. et al. (2015). The impact of person–job fit on innovation behavior: A moderated mediation model. *Journal of Business Research, 68*(5), 955–963. https://doi.org/10.1016/j.jbusres.2014.09.015
11. Huang, Q., Wang, H., & You, X., et al. (2019). Person–job fit and innovation behavior: Roles of job involvement and career commitment. *Frontiers in Psychology, 10*, 1134. https://doi.org/10.3389/fpsyg.2019.01134
12. Jansen, K. J., & Kristof-Brown, A. L. (2006). Toward a multidimensional theory of person–environment fit. *Journal of Managerial Issues, 18*(2), 193–212.
13. Kansal, T., & Sengupta, P. R. (2023). Person-job fit and person-organization fit in educational institutions: Keys to organizational success. *The International Journal of Indian Psychology, 11*(2), April–June. https://doi.org/10.25215/1102.082
14. Kristof, A. L. (1996). Person–organization fit: An integrative review of its conceptualizations, measurement, and implications. *Personnel Psychology, 49*(1), 1–49. https://doi.org/10.1111/j.1744-6570.1996.tb01790.x

15. Kristof-Brown, A. L., Zimmerman, R. D., & Johnson, E. C. (2005). Consequences of individual's fit at work: A meta-analysis of person-job, person-organization, person-group, and person-supervisor fit. *Personnel Psychology, 58*(2), 281–342. https://doi.org/10.1111/j.1744-6570.2005.00672.x
16. Lauver, K. J., & Kristof-Brown, A. (2001). Distinguishing between employees' perceptions of person–job and person–organization fit. *Journal of Vocational Behavior, 59*(3), 454–470. https://doi.org/10.1006/jvbe.2001.1807
17. Parker, S. K., Knight, C., & Keller, A. (2020). Remote managers are having trust issues. *Harvard Business Review, 98*(2), 2–9.
18. Seong, J. Y., Kristof-Brown, A. L., & Park, W. W. (2015). Person-group fit: Diversity antecedents, proximal outcomes, and performance at the group level. *Journal of Organizational Behavior, 36*(3), 389–409. https://doi.org/10.1002/job.1983
19. Saleem, S., Rasheed, M. I., Malik, M., & Okumus, F. (2021). Employee-fit and turnover intentions: The role of job engagement and psychological contract violation in the hospitality industry. *Journal of Hospitality and Tourism Management, 49*, 385–395. https://doi.org/10.1016/j.jhtm.2021.10.009
20. Tims, M., Bakker, A. B., & Derks, D. (2016). Job crafting and job performance: A longitudinal study. *European Journal of Work and Organizational Psychology, 24*(6), 914–928. https://doi.org/10.1080/1359432X.2014.969245
21. Van Vianen, A. E. M. (2000). Person–organization fit: The match between newcomers' and recruiters' preferences for organizational cultures. *Personnel Psychology, 53*(1), 113–149. https://doi.org/10.1111/j.1744-6570.2000.tb00196.x
22. Vigoda-Gadot, E., & Meisler, G. (2010). Emotions in management and the management of emotions: The impact of emotional intelligence and organizational politics on public sector employees. *Public Administration Review, 70*(1), 72–86. https://doi.org/10.1111/j.1540-6210.2009.02112.x
23. Wang, B., Liu, Y., Qian, J., & Parker, S. K. (2021). Achieving effective remote working during the COVID-19 pandemic: A work design perspective. *Applied Psychology, 70*(1), 16–59. https://doi.org/10.1111/apps.12290
24. Wright, P. M., & Davis, G. (2003). Job demands, organizational support, and public sector satisfaction: A comparative study. *Journal of Public Administration Research and Theory, 13*(2), 221–237. https://doi.org/10.1093/jpart/mug016
25. Wright, P. M., & Davis, G. (2020). The importance of job security and demands in public sector settings. *Public Personnel Management, 49*(4), 547–563. https://doi.org/10.1177/0091026020912517

Manoj Kumar Chaudhary is an experienced and dedicated professional with nearly two decades of experience of working in Government sectors. Presently he is associated with Customs and Indirect Taxes as Superintendent of Customs (P) under the Central Board of Indirect Taxes and Customs(CBIC) in Indian Customs. He has a sound academic background with M.Sc. in Physics, and MBA in Human Resource Management. Manoj is currently pursuing Ph.D. in Management at Amity University. His vast experience of working in various government sectors provided him the motivation for research.

Rupsha Roy is a dedicated academician with more than 28 years of experience in academics. She completed her MSc in Ecology and Environment from Sikkim Manipal University in 2003 and MBA from Visva-Bharati University, 2005. She has done her PhD in Management from ICFAI University, Jharkhand in 2017. Presently She is working as an Associate Professor in ICFAI University, Kolkata. Her specialization area is Human Resource Management and Organizational Behaviour. She has keen interest in learning and teaching papers of OB and HR. She has also published research articles of high international repute.

Saugat Ghosh is a passionate academician and researcher in Management and allied fields. He completed her Masters in English from University of Burdwan and MBA from Calcutta University. He gained his PhD degree from the Department of Business Management, University of Calcutta. He has significant contributions in research paper presentations and publications across various international conferences and journals. He has authored more than 13 research articles and also authored a book. His teaching career spans across two decades and also has experience in guiding research scholars in Doctoral Programmes. He is also working in Government and Privately financed projects.

Chapter 16
Behavior of the US Financial Markets During the COVID-19 Crisis

Jayendra S. Gokhale

Abstract The year 2020 saw numerous people across the whole world get infected with a virus never seen before, called COVID-19. The rapid spread of infections and the casualties from the disease forced the global economy to come to a standstill. Consequently economic agents behaved in ways to counter this and overcome the adversity. The government encouraged protection, security and management of hospitals and medical care centers at the same time finding ways to make sure financial and medical relief was being provided to its residents. The Federal Reserve provided liquidity by increasing its open market operations and cutting interest rates. Due to the interest rate cuts, and due to fear of contagion, bond yields dropped as investors sold stocks and resorted to purchase of government debt. Stock markets initially crashed, but regained as better medical management, treatment and vaccination options were found. As investors better understood how different companies were behaving and responding to the pandemic, different stocks behaved differently. Some such as technology and online retail stocks performed very well, while others such as traditional brick and mortar companies did not perform as well. The pandemic saw an advent of a new breed of companies that provided online support for meetings and networking as well as provided online tools for purchase of goods and services. Things slowly began coming back to normal when vaccines were discovered and people developed immunity against the disease due to these vaccinations. By the time all was said and done, the pandemic had left an indelible mark on the global economy.

16.1 Introduction

In 2020, the COVID-19 virus spread all over the world. Many countries around the world imposed restrictions on their residents and some imposed lockdowns curb the spread. Consequently, there was a limit to free movement of people. On the one hand, the lockdown gave time for researchers and the medical community to respond

J. S. Gokhale (✉)
Embry Riddle Aeronautical University, Daytona Beach, FL, USA
e-mail: gokhalej@erau.edu; Jayendra.Gokhale@erau.edu

S. Gangopadhyay et al. (eds.), *Pandemic Diaries*,
https://doi.org/10.1007/978-981-96-5415-4_16

to the virus without (hopefully) causing an explosive spread within communities, thereby limiting the number of infections and casualties due to the virus. On the other hand, this curbed economic activity and growth as observed in several countries worldwide. Companies in nearly all sectors underperformed due to this restriction on free movement. In this chapter, we study the effect of the COVID-19 crisis on the US financial system, focusing on the stock markets during the period of imposition of the restrictions to when the vaccines were announced and released to people. This paper is organized as follows: Sect. 16.1 discusses the state of the overall economic conditions in the US, and Sect. 16.2 discusses the response of different sectors within the US stock markets. Section 16.3 concludes.

16.1.1 COVID-19 Crisis and the US Economy

In the United States, cases began growing in February of 2020. As the virus began spreading across the country, so did fear. As restrictions and closures increased, people began losing jobs. The US economy lost 4 million jobs in April 2020 alone. The decrease in number of jobs due to the pandemic was the greatest since 1929–1934 Great Depression [1]. The unemployment rate increased to nearly 15% during April 2020, a level not seen even during the financial crisis (Fig. 16.1). Figure 16.1 also shows level of employment since the Great Recession of 2008–09. The level of employment (inverse of unemployment) rose steadily after 2009 until the 2020 COVID-19 crisis led to layoffs and employees quitting out of fear of getting infected.

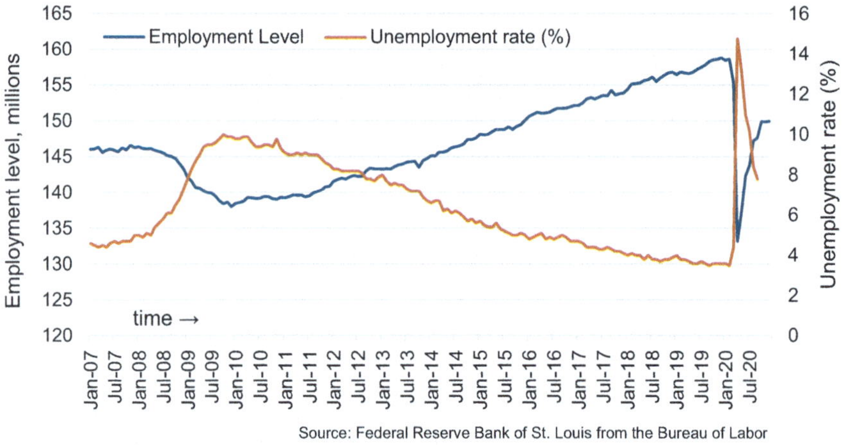

Fig. 16.1 Number of people employed in the United States January 2007–December 2020 and the rate of unemployment (%)
Source: Federal Reserve Bank of St. Louis from the Bureau of Labor Statistics

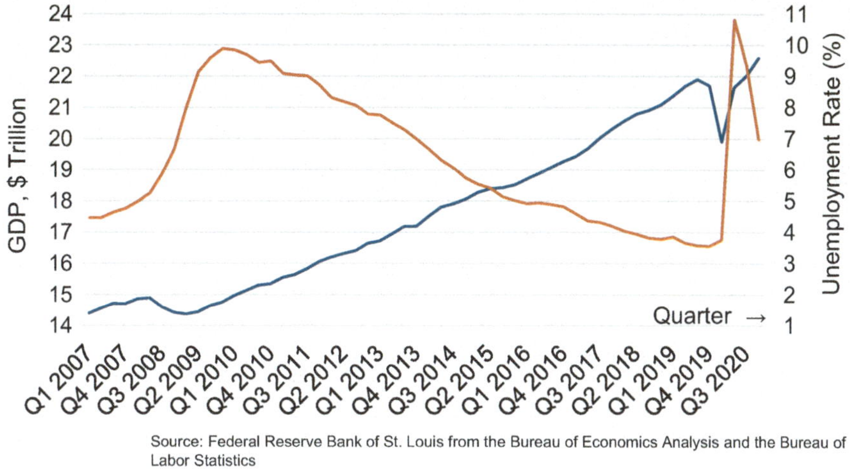

Fig. 16.2 Unemployment and gross domestic product in the United States 2007–2020

Notwithstanding strict measures undertaken by the Federal, state and local governments, as the virus surged in 2020, responding to the safety measures, businesses either offered their services remotely or shut down completely. This caused widespread layoffs. Additionally, out of fear of getting infected, some members of the workforce quit their jobs. Consequently, the levels of economic activity fell (i.e., GDP decreased) as the levels of unemployment (people without a job looking for work) rose. The correlation between GDP and unemployment is shown in Fig. 16.2. One can notice that the negative correlation between these two variables was evident during this period. Slowing economic activity and lower GDP led to a rise in the quarterly average unemployment rate, which increased to a high of 11%.

The US stock market also provided noticeable evidence of a slowdown in economic activity. In March 2020, with significant lockdowns in effect and no signs of vaccines on the horizon, the markets had to price in a deep recession, as shown in Fig. 16.3. The US GDP started slowing down in late 2019 due to a slowdown in Chinese manufacturing. China is a leading exporter of goods to the United States. Eventually stock market followed suit and declined. This decline in the US stock markets started pricing a possible slowdown of future profits for US firms on an account of this adversity. Thus, the medical community was put on notice to come up with treatment options such as vaccines and therapeutics to revive the economic activity. On their part, the Federal Government and the Federal Reserve came up with a plan to rescue the financial markets and the economy with the former providing fiscal stimulus and incentives to taxpayers, while the latter providing greater levels of liquidity and lower interest rates. It seems that the slowdown was relatively short-lived because several state governments started relaxing the lockdown rules. Further, the adoption of technology such as zoom and other web conferencing services provided alternative avenues for employees to get to work remotely. Grocery stores

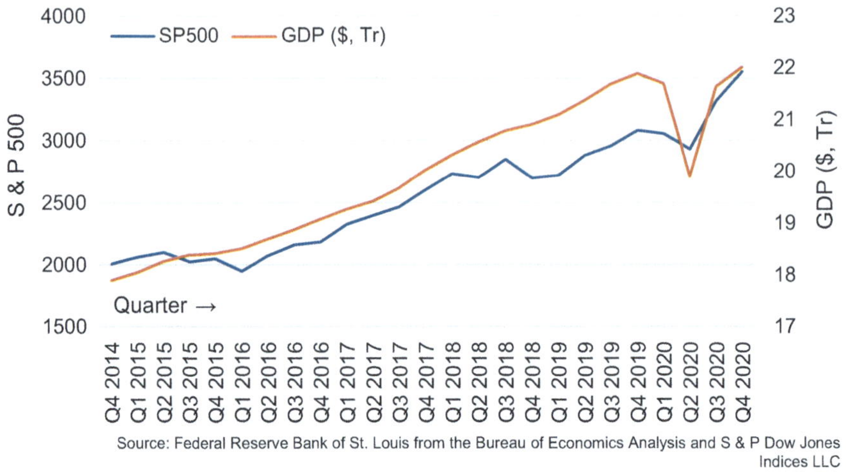

Fig. 16.3 Correlation of stock markets and US GDP

and retailers started offering curbside pickup, so customers did not have to get inside the stores. There was also a slow trickle of employees back to work, particularly those providing essential services such as hospitals and utilities.

The lockdowns imposed by Federal/State/local governments caused consumption spending to drop, leading to lower prices on goods and services. On the one hand, producers wanted to clear their inventory, while on the other, consumers were discovering innovative ways to complete their purchases to meet demands. Sales on online platforms like Amazon.com surged while store delivery service providers such as Instacart and Shipt flourished as well. Due to government and self-imposed restrictions, consumers mostly shopped online and through online delivery services. However, this was not enough to prop up demand to its pre pandemic levels. Consequently, there was a slowdown in demand and this led to lower personal consumption expenditures than the previous years. This can be seen in Fig. 16.4.

16.2 Role of the Federal Reserve and Monetary Policy

The Federal Reserve was established in 1913 by the Federal Reserve Act of Congress. Its primary responsibility is to act as the lender of last resort in the financial markets. It has a dual mandate for ensuring that unemployment rate remains low while inflation is also under control. During the crisis, there was a downward pressure on the price levels, an upward pressure on unemployment and a risk for an overall severe economic outcome for the country. The Federal Reserve (Fed) played an essential role in stabilizing the economy. It cut its target Fed Funds rate (the benchmark interest rate) by 0.5% in February 2020. Thereafter, Fed continued to provide liquidity in the

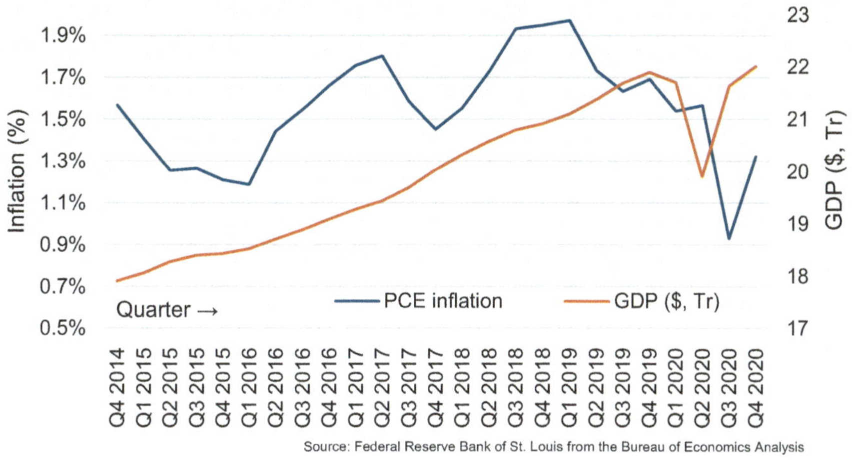

Fig. 16.4 Real gross domestic product and % change in personal consumption index

financial markets as well as decrease its interest rates. This implied that any financial institution facing liquidity issues could easily borrow from financial markets and the Fed at the shorter end of the curve. Fig. 16.5 shows that the personal consumption expenditure (PCE) inflation fell while the Fed funds rate was also cut.

Investors and market participants grew more and more fearful and resorted to purchasing safe-haven assets such as currencies from countries perceived to be safe, such as the United States dollar (USD), measured by Exchange-Traded Fund (ETF): UUP, Switzerland (Swiss Franc (CHF), ETF: FXF) and Japan (Japanese Yen, ETF:

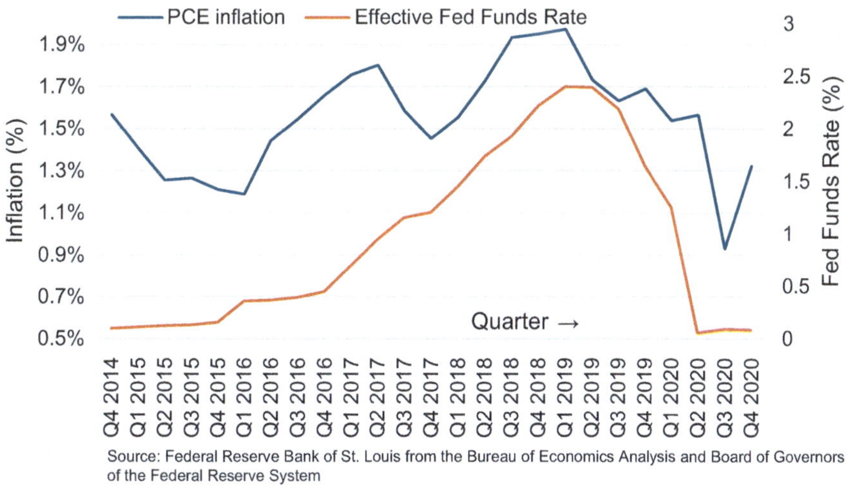

Fig. 16.5 Correlation of fed funds rate and percent change in personal consumption expenditures for the United States 2014–2020

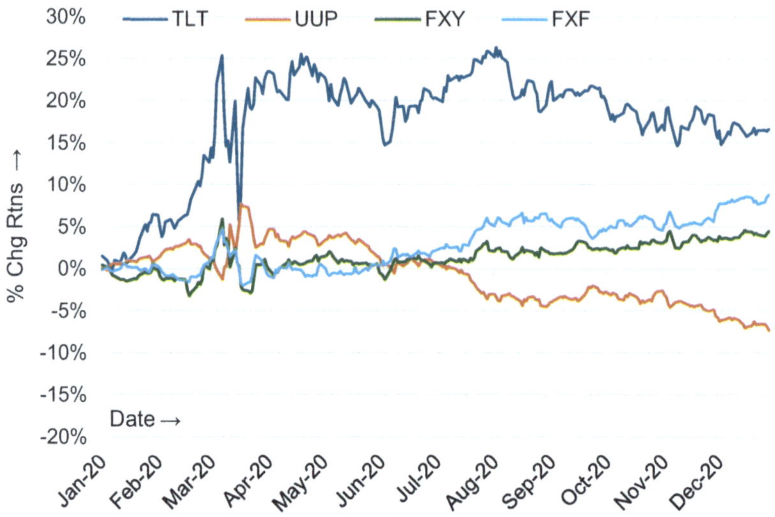

Fig. 16.6a Returns from US dollar (UUP), Swiss Franc (FXF), Japanese Yen (FXY) and Long-term US Government bonds (TLT). *Source* Yahoo! Finance

FXY) as well as US government bonds (ETF: TLT). Consequently, the price of these instruments soared, as seen in Fig. 16.6a. Additionally, a cut in the US interest rates by the Fed exacerbated this intense rise in price levels of the US government debt. The US government 10-year yield bond fell to its lowest 0.54% on March 9, 2020.

As fear due to the pandemic spread, financial market participants also resorted to purchasing certain commodities such as Gold (measured by ETF: GLD) and Silver (ETF: SLV). Since SLV also has industrial applications, it showed greater returns than GLD. However, as the Fed lowered its interest rates to fight the crisis, the USD currency depreciated after its initial surge. Consequently, when the USD initially strengthened, the prices of GLD, SLV, oil (ETF: USO) and its producer-related ETF (XLE) all depreciated due to stronger dollar and weaker commodity prices. However, GLD and SLV (also being safe-haven assets) gradually appreciated in returns, while USO and XLE remained depreciated due to weak demand from energy consumption [4]. This is seen in Figure 16.6b.

As scientists began working on vaccines and therapeutics, pessimism about health and economic conditions gradually started fading and the economy started recovering. Strict lockdowns were replaced with more regulated measures.

Fig. 16.6b Returns from Gold (GLD), Silver (SLV), US Crude Oil price (USO) and the Energy Sector Index (XLE) *Source* Yahoo! Finance

16.3 Stock Market Response

In this section, we examine the performance of stocks in some essential economic sectors that were highly relevant as the COVID-19 crisis unfolded: retail, restaurants, fintech, malls and commercial real estate, technology and stocks related to the travel and tourism sectors. We also look at these stocks from two perspectives: behavior during the peak of the pandemic and after the vaccine announcement in November 2020.

The first group under discussion is restaurant stocks such as Darden Restaurant (DRI), Yum Brands (YUM), Shake Shack (SHAK), Brinker International (EAT) and Chipotle (CMG). Restaurants that adapted quickly and rolled out applications that catered to social distancing by offering customers a chance to order online or pick up orders in-store without dining in fared better. Chipotle's early drive to create innovations such as online delivery service also helped its stock to soar [3]. Figure 16.7 shows the movement of various restaurant stocks in the year 2020.

Another contrasting story is that of retailers. While some discount retailers such as TJ Maxx (TJX) and Ross Stores (ROST) continued to operate on a lean inventory, their sales did not fall drastically despite store closures (primarily in Europe for TJX, [5]). Overall, the companies that traditionally used the brick-and-mortar model for retail such as Kohl's (KSS), Macy's (M) and Nordstrom (JWN) struggled to come to terms with the changed consumer behavior due to the pandemic. Once COVID-19 vaccines were announced, these stocks gradually increased again during the latter part of the year. Curiously, the returns of healthcare-supporting company Chlorox (CLX) surged due to the outsized demand for these products. Figure 16.8 shows stock returns from these retailers over time in 2020. Macy's, Nordstrom and Kohls

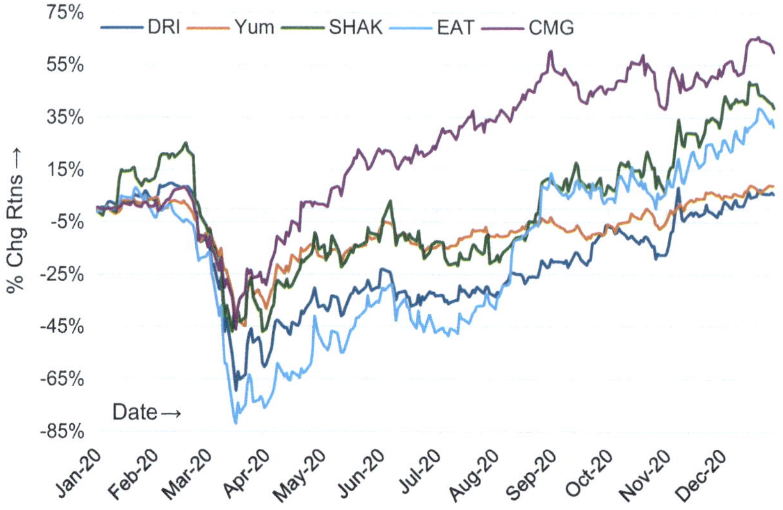

Fig. 16.7 Returns from restaurant stocks

struggled to bring customers to stores and had to act quickly to attract more customers online, which they did not do traditionally.

Customers had to change their shopping habits to protect themselves and their loved ones from the COVID-19 virus. Therefore, even as retailers in brick-and-mortar stores pivoted to online deliveries and pick up outside stores, stocks that provide support to retailers and small businesses and the overall commercial sector,

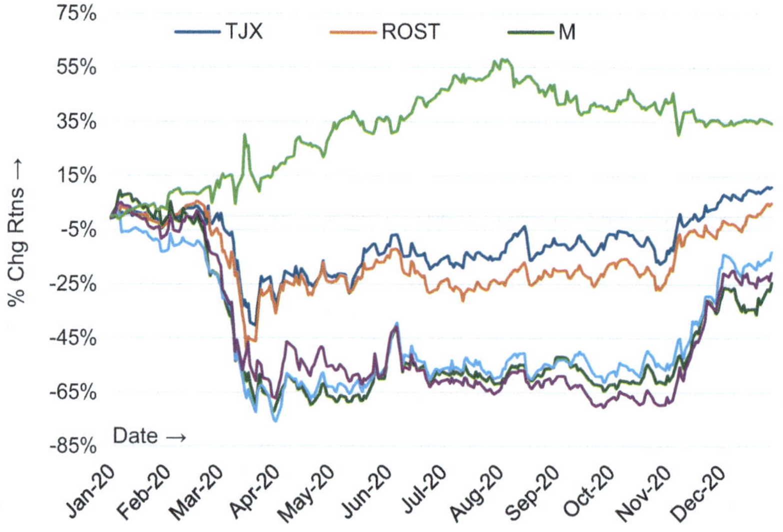

Fig. 16.8 Returns from retailers' stocks

Fig. 16.9 Returns from stocks representing commercial real estate

such as Simon Property Group (SPG), Kimco Realty Corporation (KIM), SL Green Realty (SLG) and Federal Realty Investment Trust (FRT) struggled financially. Even after the announcement of vaccines was made, investors feared a structural shift in consumer behavior due to a lack of desire to return to the mall. Therefore, these stocks did not see a significant bounce even toward the end of the year (Fig. 16.9).

It is important to discuss stocks of payment services companies as no transaction is complete without a financial aspect. There was a modest growth in stock returns of payment system companies such as American Express (AXP, -5%), Mastercard (MA, 15%) and Visa (V, 18%). These stocks which did not grow as significantly in 2020 as compared to the new generation fintech stocks such as PayPal and Square. This was primarily because of the reliance of these former group of companies on credit cards and in-store purchases as well as retail purchases. However, PayPal (PYPL) and Square (SQ) stocks took off [2] & [6]. Of particular interest were Venmo by PayPal and CashApp by Square. These applications provided payment solutions and the ability to make direct person-to-person cash transfers without going directly to the bank. This enabled PayPal to produce returns of 109% for the year. Substantial investment in Bitcoin and the subsequent rise in the price of that asset also helped increase the returns of Square stock significantly (246%) (Fig. 16.10).

One group of stocks deeply hurt by the COVID-19 pandemic was the recreation, travel and transportation industry. Even after the announcement of vaccines was made, investors were nervous about either slow rollout or about people's aversion to travel for recreational purposes in fear of safety from people-to-people transmission of the virus that still lingered. Moreover, there were additional fears for this sector because their fleet had to be grounded, and some of these companies had to borrow from the government to meet their variable operating costs. Consequently, returns from cruise ship company Carnival (CCL) and airlines such as American (AAL),

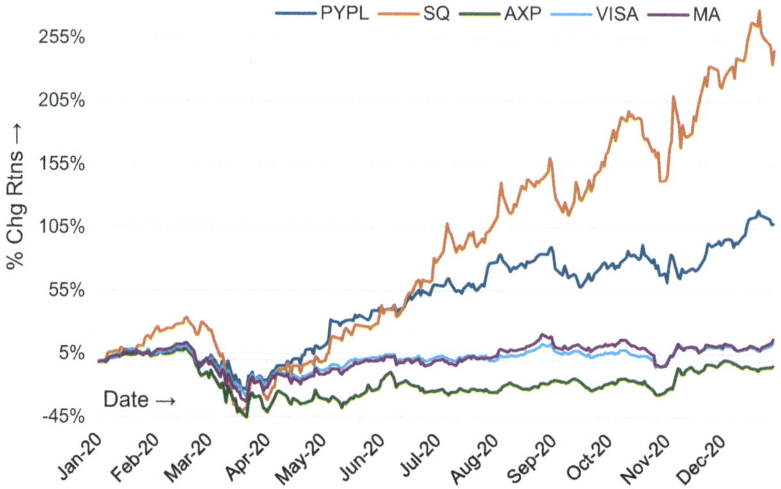

Fig. 16.10 Returns from payment system stocks PayPal, Square, American Express, Visa and Mastercard

United (UAL), Delta (DAL) and Spirit (SAVE), as well as aircraft manufacturing company Boeing (BA) yielded -30 to -57% returns for the year (Fig. 16.11).

In line with the travel, tourism and recreation industry, the hotel industry also showed negative returns at the beginning of the pandemic in fear that guests would avoid visiting places with a greater chance of infection. Hence, the stocks of companies Hilton (HILT), Marriott (MAR) and Wynn (WYNN) initially turned strongly

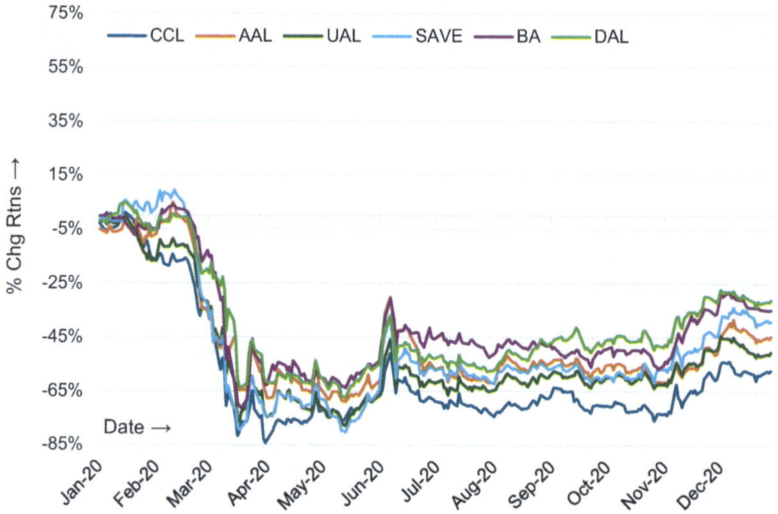

Fig. 16.11 Returns from stocks related to recreation and travel

Fig. 16.12 Returns from hotel stocks

negative during March 2020. However, as the industry adapted to the safety standards and protocols, stock returns began to show signs of recovery by the end of the year. HILT stock showed the most robust rebound (zero percent change during 2020), primarily due to its ability to withstand the opposing pressures of the pandemic [7]. Among all hotels, HILT's owned assets were the least, and most of its income was due to franchise revenues. Other hotels, such as Mar (−13%) and WYNN (−20%), were slower to rebound (Fig. 16.12).

Against this backdrop of the massive hits taken by traditional retailers (Fig. 16.8), the recreation and travel industry (Fig. 16.11) and the hotel industries (Fig. 16.12), there was one industry that was thriving. This consisted of companies that served their customers primarily in the comfort of their homes. This sector is comprised of at-home entertainment (Netflix, NFLX), at-home fitness (Peloton, PTON), online retailers such as Amazon (AMZN) & eBay (EBAY) and online conferencing platform providers (Zoom communications, ZM). There was such a widespread surge in the demand for products of ZM and PTON that their returns went up by more than 400% for the year. NFLX stock also showed returns of 59%. However, these returns paled before ZM and PTON (Fig. 16.13a). Peloton shares surged primarily on account of demand for at-home fitness due to the closure of gyms, so much so that it could not keep up with its backorders. These problems persisted well into the next year [10]. Returns from ZM and PTON in the year 2020 are displayed in Fig. 16.13a for comparison with NFLX. Returns from other stocks are shown in Fig. 16.13b. Per Fig. 16.13b, eBAY and AMZN also provided higher double-digit returns.

While all this was going on, big technology companies Apple (AAPL), Google (GOOG), Microsoft (MSFT) and Facebook (META) provided support to the

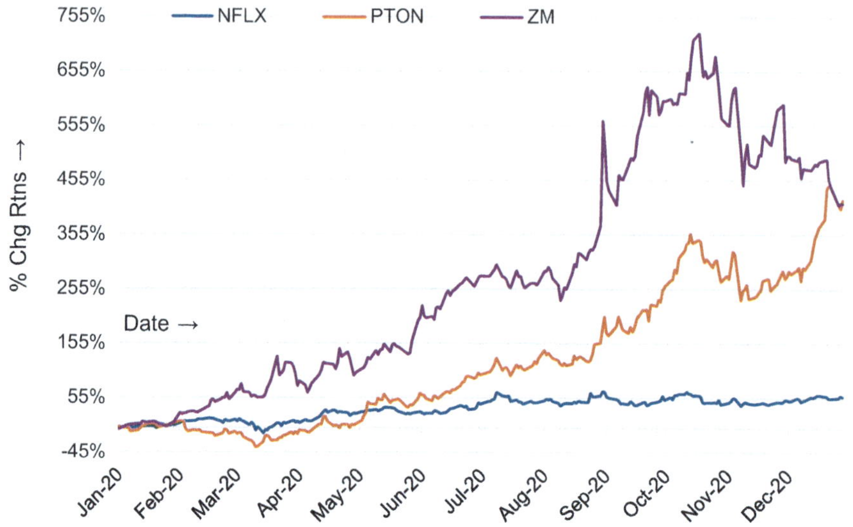

Fig. 16.13a Returns of at-home service provider stocks (Netflix (NFLX), Zoom (ZM) and Peloton (PTON))

Fig. 16.13b Returns of at-home service provider stocks—NFLX, eBay and AMZN

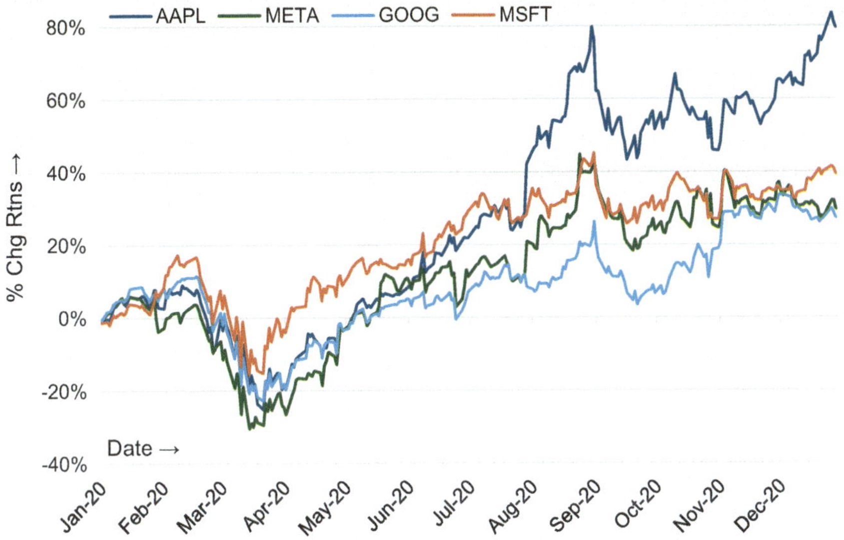

Fig. 16.14 Returns of technology companies Google (GOOG), Apple (AAPL), Microsoft (MSFT) and Facebook (META)

consumer in this new online world, with online interaction via either communications and tools or hardware. Apple was the best-performing stock in this group because of its hardware quality. In a way, Apple was at a sweet spot in 2020, with investors recognizing its hardware quality and an evolving array of software services, with the iOS ecosystem gaining prominence. Lastly, Apple had been a bell weather stock among the big technology companies' stocks in the previous decade [9]. Thus, some investors also viewed it as a safe-haven stock [8]. All these technology companies supported the economy by performing at their optimal levels. In stock markets, their stocks returned 25% or more for the year (Fig. 16.14).

16.4 Conclusion

To conclude, 2020 saw numerous shocks to the global and US economies and stock markets. The returns of various stocks fluctuated meaningfully and predominantly in an orderly fashion. Thanks to the Fed's intervention, the markets had enough liquidity. So, while unemployment ticked up due to health concerns, the economy was primarily disrupted in the more traditional, brick-and-mortar type of industries that required significantly greater people-to-people interaction. Stock markets reflected that by punishing these stocks. However, investors who invested in the stocks of those companies that provided solutions to overcome challenges posed by the pandemic were rewarded with higher returns. These companies ultimately helped emerge from the pandemic successfully.

References

1. Ansell, R., & Mullins, J. P. (2021, June). Covid-19 ends longest employment recovery and expansion in CES history, causing unprecedented job losses in 2020 : Monthly Labor Review. *U.S. Bureau of Labor Statistics*. https://www.bls.gov/opub/mlr/2021/article/covid-19-ends-longest-employment-expansion-in-ces-history.htm
2. Ballard, J. (2021, January 7). Why PayPal stocks climbed 116% in 2020. *The Motley Fool*. https://www.fool.com/investing/2021/01/07/why-paypal-stock-climbed-116-in-2020/
3. Fernando, J. (2022, July 22). *Chipotle earnings: What happened. Investopedia*. https://www.investopedia.com/chipotle-earnings-4773164
4. Johnston, M. (2022, February 22). What happened to oil prices in 2020. *Investopedia*. https://www.investopedia.com/articles/investing/100615/will-oil-prices-go-2017.asp
5. McConnell, D. (2021, February 24). *The TJX Companies, Inc.. reports Q4 and full year FY21 results; reports above-plan Q4 overall open-only comp store sales of down 3%; plans to declare quarterly dividend at current rate in the first quarter of fiscal 2022*. The TJX Companies, Inc. https://investor.tjx.com/news-releases/news-release-details/tjx-companies-inc-reports-q4-and-full-year-fy21-results-reports
6. Roberts, D. (2020, December 29). Square soared in 2020, even as the pandemic hit its core customers. *Finance Yahoo*. https://finance.yahoo.com/news/square-soared-in-2020-even-as-covid-19-pandemic-hit-its-core-customers-201148563.html
7. Strauss, L. C. (2021, February 7). Why hilton stock is primed for postpandemic upside. *Barrons*. https://www.barrons.com/articles/hiltons-asset-light-focus-primes-shares-for-postpandemic-upside-51612558801
8. Sun, W. (2021). The changes of apple's stock price during the pandemic. *Atlantis Press International B.V.* https://www.atlantis-press.com/proceedings/icemci-21/125966048
9. Team, T. (2020, August 20). Apple stock is up 50% this year. What's going on? *Forbes*. https://www.forbes.com/sites/greatspeculations/2020/08/20/apple-stock-is-up-50-this-year-whats-going-on/
10. Thomas, I. (2021, October 22). *How Peloton is trying to stay ahead of supply chain issues*. CNBC. https://www.cnbc.com/2021/10/22/how-peloton-is-trying-to-stay-ahead-of-supply-chain-issues.html

Jayendra Gokhale is Associate Professor at the College of Business at the Daytona Beach Campus of Embry Riddle Aeronautical University. He received his Ph.D. in Economics at Oregon State University. His areas of research are Financial Economics, Industrial Organization and applied Micro-econometrics. He has taught Principles of Microeconomics and Macroeconomics, Money and Banking, International Economics, Corporate Finance and Econometrics. He has worked for the Finance and Treasury Divisions of General Electric and Axis Bank (India). He completed his MBA in Finance at the Indian Institute of Management, Kozhikode. His undergraduate major is in Biochemical Engineering from the Indian Institute of Technology, Delhi. Jayendra has published in International Journal of Economics and Accounting, Journal of International Finance and Economics, Journal of Wine Economics, International Journal of Economics and Finance, Quarterly Review of Economics and Finance and the Journal of Behavioral Finance. His research interests are in industry research, event studies in aviation and non-aviation, new techniques to determine noise trading in financial markets and motion pictures.

Part IV
Turmoil and Pacification: Impact of COVID-19 on Social Change and Management

Chapter 17
Exploring the Connection Between Happiness, Resilience, and Death Anxiety in College Students With and Without Psychological Distress

Jhilli Das Tewary, Debopriya Ghatak, and Shweta Sah

Abstract Increase in the level of stressors during COVID-19 had challenged mental health across all ages. The most affected were the youth who had been isolated and cut off from all the richness of experience which was due to them. In this scenario, the study attempts to investigate the association between Happiness, Resilience, and Death Anxiety in College Students. Data was collected from 70 college students grouped into Psychiatric Morbidity High Risk group (n = 42) and Psychiatric Morbidity Low Risk group (n = 28). The sample was grouped using General Health Questionnaire (Goldberg 1978), a screening instrument for psychological morbidity. Data was collected through Google form by administering a detailed information schedule, death anxiety scale (Templer, D. I.) (1970), resilience appraisal scale (Johnson et al., 2008), and subjective happiness scale (Lyubomirsky, S. & Lepper, H. S.) (1999). Statistical analysis of data was done using SPSS. Descriptive Statistics, t-test, correlation, and regression were computed from the data. Results showed significant differences between Psychiatric Morbidity High Risk Group and Psychiatric Morbidity Low Risk Group with respect to emotional coping, situational coping, and happiness. Low risk group were found to be happier than the high risk group and were more effective in the use of emotional and situational coping mechanisms.

Keywords Psychiatric morbidity · Happiness · Resilience · Death anxiety

J. D. Tewary (✉)
Department of Psychology, St. Xavier's University, Kolkata, India
e-mail: jhilli.das@sxuk.edu.in

D. Ghatak
Department of Psychology, Adamas University, Barasat, India

S. Sah
Department of Psychology, Sister Nivedita University, Kolkata, India

S. Gangopadhyay et al. (eds.), *Pandemic Diaries*,
https://doi.org/10.1007/978-981-96-5415-4_17

17.1 Introduction

COVID-19 came like a tsunami; no one was prepared for it. News of death and infection started coming sporadically until, suddenly, we found ourselves completely engulfed by it. News of death and suffering started pouring in from all parts of the world. In a few months the virus had infected more than 100 million people throughout the world, the death toll rising to over 2 million causing widespread disruption and distress in nearly every country. The consequences of this unprecedented situation had far-reaching effects.

Confronted with this pandemic the government tried its best to contain the spread of the rapidly advancing virus by adopting different strategies. Physical contact between individuals were reduced, in-person business transactions were limited, traditional classrooms shifted to a digital mode, and restrictions were imposed on social gatherings. As per the Centers for Disease Control and Prevention [12], certain groups such as older adults and individuals with serious health conditions were recognized as "being at a greater risk of severe illness from SARS-CoV-2 infection". To safeguard their health, it was advised that they adopt additional safety measures, such as minimizing their contacts with persons outside of their home and engaging in social distancing [18]. The pandemic had also led to significant social isolation, altering daily routines and revealing deep-seated existential fears. Life has changed dramatically, leaving a profound impact on individuals and communities alike. Suffering manifested in various forms.

17.2 Survey of Literature

17.2.1 Suffering During the Pandemic

Suffering has multiple origins. It could be physically accompanied by intense pain or it could be psychological such as mental illness or diminishing hope. It could be triggered by social causes, i.e. losing a loved one or a breakup in a relationship. Systemic causes such as social inequalities and poverty have also been identified as possible causes of suffering [18]. According to Van Tongeren and Showalter Van Tongeren [65], suffering can be characterized as "(a) cognitively threatening, as it challenges deeply held beliefs about the world, (b) chronic, meaning it is persistent or long-lasting, and (c) consequential, in that it fundamentally or profoundly changes people". "The COVID-19 pandemic led to a wide range of suffering, which in turn contributed to chaos in multiple areas within the healthcare system, education, social and economic and business world" [65].

The characteristics of suffering frequently highlighted existential issues that could cause a great deal of distress. The widespread terror and actions of self-isolation began even before the government declared the lockdown and social isolation. During the first few months of the epidemic, everyone was riveted to their TVs, following the number of deaths and the hospitals' deteriorating state. When medical science struggled, preventive, non-scientific alternatives were sought. Some individuals in India began ingesting cow dung and urine under the pseudonym "cow dung therapy" for the treatment of COVID-19 [32].

17.2.2 Effect of Pandemic on Youth

The pandemic had both direct and indirect effects on the youth of India. Constant bombardment of negative news of innumerable lives being lost globally with no immediate solution to the situation has led to an escalation of daily stresses in their lives. Added to this, prolonged home confinement, excessive internet usage, constant exposure to fake news in social media, financial worries, career-related worries, shifting to an online mode of education with its related problems have led to widespread emotional distress and increase in psychiatric problems. People with a COVID-19 diagnosis, those in quarantine and even those who are not physically afflicted by the virus have all been shown to experience psychological discomfort, worry, along with depression [51, 74].

During this period there was also a rise in anxiety issues among the youth especially among the students. The causes identified were related to adverse economic conditions, everyday life circumstances, and difficulties with their academic work [10]. Indians too experienced high levels of anxiety [52].

17.2.3 Death Anxiety and Maladjustment

Death anxiety is described as the distress or fear that arises from both conscious and unconscious concerns about death and dying [31]. There is evidence linking psychopathology, maladjustment, and neurotic symptoms to death anxiety [45, 44]. Various studies found a favorable correlation between general anxiety with death anxiety scales [1, 38, 66]. Similarly, White and Handal [71] discovered a strong link between "mental distress" and "death anxiety" [71]. Greater death anxiety was also reported by Gilliland and Templer (1985–86) in the psychiatric population as compared with the normal [25]. Existentialist Sartre concludes that death limits restrict a person's capacity for self-actualization; nonetheless, a person who is aware of their purpose in life will experience less fear of dying [57].

Terror Management Theory (TMT), which draws partially from Becker's ideas, has significantly advanced our understanding of the psychology of death anxiety. Both Becker and TMT propose that "positive illusions" can help reduce anxiety about death [16, 27, 63].

Resilience was found to be one of the buffers against death anxiety. Individuals with resilience are more effective at managing stress and difficult situations, which in turn helps lower anxiety. This suggests that resilience is another vital element for overall health and well-being.

17.2.4 The Power of Resilience

Resilience has been defined as "a dynamic process encompassing positive adaptation within the context of significant adversity. Implicit within this notion are two critical conditions: (1) exposure to significant threat or severe adversity; and (2) the achievement of positive adaptation despite major assaults on the developmental process" [24, 39, 42, 54, 70]. Resilience has been viewed as a vital aspect of personality that supports good health and serves as a "key" to it. It may also be viewed as a "meta-source" with exceptional regulatory power, supporting the activation of other resources required to manage difficult life circumstances [46].

Resilience played a critical role as a protective factor during the pandemic, helping individuals cope with the uncertainties and stress caused by widespread health crises. According to recent studies, the pandemic made the youth more vulnerable due to isolation policies, uncertainty, and volatility in their life. They also had a greater prevalence of "stress, anxiety, and depression symptoms" [48]. According to Zhang et al., a study done in Wuhan, China, on 296 patients who had mild COVID-19 symptoms revealed that patients with greater levels of resilience also had lesser levels of anxiety and sadness [74]. This pattern of inverse relationship suggests that resilience served as a buffer against depression and anxiety [22]. Ego resilience partly works by producing positive emotions. When confronted with stress, individuals with high ego resilience tend to feel more positive emotions compared to those with lower resilience, even though both groups experience similar levels of negative emotions. The variation in their positive emotions helps account for their greater ability to bounce back from adversity and stress, prevent depression, and continue personal growth [23, 47, 67].

17.2.5 Happiness and Mental Health

Resilience fosters happiness and good health. "Happiness is a complex but highly valued phenomenon at the intersection of the individual and the contextual". It is characterized as "frequent positive emotions and high life satisfaction" [8, 53]. According to Everly's (2009) theory, a person's capacity for either regret or happiness may be correlated with their level of resilience [20]. Everybody will face adversity at some point in their lives, thus developing resilience requires the ability to manage stress well.

17.2.6 Psychiatric Morbidity and Stress

The COVID-19 Crisis had its long-lasting aftereffects. The responsibility of learning through online mode of education was shifted more toward the student. This resulted in poor quality of learning in unmotivated students. As there is less opportunity for more interactive and hands-on learning, online learners frequently faced feelings of isolation or disconnection from the course, the teacher, or other students. This resulted in a loss in learner motivation and engagement [28, 33]. According to WHO, "in the first year of the COVID-19 epidemic, there was a discernible 25% increase in the prevalence of worry and stress worldwide" [72]. Research shows that COVID-19 had a wide range of psychological effect which were likely to last longer than six months [62]. Numerous factors may contribute to mental health problems, such as dread of dying, isolation and loneliness, hospital admission stress, bereavement from a personal loss, and even financial loss during the epidemic [72].

Psychiatric morbidity refers to the presence of psychological disorders or symptoms, such as anxiety, depression, post-traumatic stress disorder (PTSD), and other mental health issues. Psychological distress is caused by the emotional suffering which accompanies these symptoms. Studies show that 37% of young adults in the US, who are primarily college students between the ages of 15 and 24, possess a minimum of one diagnosable mental disorder. Similar findings were observed among university students in Britain [69]. In his study, Phippen discovered that of the 152 adolescents receiving counseling services, 64% had major mental health conditions [50]. Resilience operates in two ways. First it combats the development of mental illness and second it encourages "thriving and well-being". Following natural catastrophes like hurricanes and earthquakes, resilience has been demonstrated to act as a mediator between stress and mental health morbidity [73].

The current investigation has been conducted with the following objectives in mind. The study aims to examine the differences between high-risk and low-risk groups for psychiatric morbidity with respect to the variables of emotional coping, situational coping, social support, death anxiety, and happiness. Additionally, it seeks to investigate the associations among these variables—emotional coping, situational coping, social support, death anxiety, and happiness—within both the low-risk and high-risk groups separately. Furthermore, the research intends to assess whether death anxiety can be predicted by emotional coping, situational coping, social support, and happiness within the low-risk group and the high-risk group, respectively.

17.3 Methods

17.3.1 Participants

The sample included 70 undergraduate students drawn from some colleges in Kolkata. Their ages ranged from 18 to 22 years. Out of the sample 50.2% were female students and 49.8% were male students. 25 (35.7%) students were from 2nd year, 28 (40%) students were from 1st year, and 17 (24.2%) students were from 3rd year. 52.9% students belonged to high socioeconomic status and 30% students belong to moderately high socioeconomic status and the other 17.2% belonged to moderate socioeconomic status. 53% students have reported that they were diagnosed with psychiatric disorder and 30.2% have reported they were not diagnosed but have psychiatric symptoms and the rest have reported they have no psychiatric symptoms.

17.3.2 Procedure

The study was conducted near the end of COVID-19 phase when the schools and colleges started to reopen after a long time of being closed down in the month of January 2022. The students were very much apprehensive about coming to schools and colleges since the third wave was on the verge of hitting the country at that time. Purposive sampling method was used to collect data from undergraduate students of Kolkata. Prior to commencing the study, participants received an explanation of the objectives and the students' consent was obtained from the students. The ones who did not give the consent were not included in the study. The General Health Questionnaire (GHQ) was administered to the students online through Google form. On the basis of the cut-off scores of GHQ the sample has been divided into two groups—high risk group (HRG: Score above the cut-off point in GHQ) and the (LRG: Score below the cut-off point in GHQ).

17.3.3 Measures

General Health Questionnaire:

It consists of 28 questions which is a self-report screening measure which gives a measure of psychological disorder. Sub-scores should not be used to identify a particular psychological diagnosis because subscales are not independent of one another. Instead, the measurement is employed to determine whether symptoms are present in relation to the individual's normal range [56]. According to Goldberg and Hiller [26], the scale highlights two primary concerns: (i) the subject's ability to perform regular functions, and (ii) the manner in which the "new" and "distressing" phenomenon manifests. A higher score suggests a higher risk of psychological disturbance [26]. According to Anderson et al. [2], a score of ≥ 5 is thought to suggest a probability of psychiatric disorder. Four subscales that were found by factor analysis of the GHQ-28 were: somatic symptoms, anxiety/insomnia, social dysfunction, and severe depression [2].

Death Anxiety Scale:

It is a 15-item self-report questionnaire. "It was developed on a two factor model of death which involved both internal and external psychological factors" [55]. Cronbach's alpha coefficient ($\alpha = 0.839$ and 0.831), composite reliability (CR = 0.845 and.832), Theta ($\theta = 0.893$ and.867), and McDonald Omega ($\Omega = 0.796$ and.743) were used to evaluate dependability. Two factors were involved, according to the analysis. The factors' discriminant validity and reliability were verified [59].

Resilience Appraisal Scale:

These are evaluations of the person's capacity for managing emotions, finding solutions to problems, and obtaining social assistance. It was designed by Johnson et.al 2008 where it used to design the appraisals [34]. The appraisals are the individual's ability to cope with emotions, the ability to solve problematic situations to gain social support. The scale consists of 12 items. Four of these evaluated emotion coping appraisals, four evaluated situation coping appraisals, and four evaluated social support appraisals. Responses were scored on a five-point rating system ranging from "strongly disagree" to "strongly agree". Using information from a pilot sample, a confirmatory factor analysis was carried out to evaluate the suggested three-component structure of the scale. "Alpha reliabilities were 0.88 for the overall scale, 0.93 for the social support sub-scale, 0.92 for the problem solving sub-scale and 0.92 for the emotion coping sub-scale" [35].

Subjective Happiness Scale:

The 14-item subjective happiness scale is a self-report tool that evaluates a person's overall level of happiness. Here a seven-point Likert scale is used where higher scores indicate greater happiness; the range of scores is from 1.0 to 7.0 [40]. The scale has been tested and validated among 14 groups of students, involving a total of 2,732 participants. The results revealed that the "scale has high internal consistency" has been proven to be "stable" across samples. "Test retest and self-peer correlations have suggested good to excellent reliability and construct validation studies of convergent and discriminant validity have confirmed that the scale can be used to measure the construct of subjective happiness" [40].

Table 17.1 The mean, standard deviations, and t ratio of the subdomains of resilience appraisal

Dependent variables	Mean (HRG)	Standard deviation (HRG)	Mean (LRG)	Standard deviation (LRG)	t value	Sig. Value
Resilience sub-domains Emotional coping	13.17	3.19	14.96	3.12	2.339*	0.023
Social support	13.64	3.17	14.00	3.92	0.402	0.690
Situational coping	13.81	2.72	15.14	2.30	2.202*	0.031
Happiness	3.97	1.10	4.67	1.15	2.543*	0.014
Death anxiety	7.52	2.45	6.64	2.94	1.309	0.197

17.4 Data Analysis

General Health Questionnaire was administered to the participants and according to the cut-off point score ≥ 5 of the scale, participants were grouped into two categories which includes psychiatric morbidity and participants not having psychiatric morbidity. SPSS for Windows (Version 26.0) was used to analyze the data. Descriptive statistics was calculated. Independent sample t test has been conducted to see if there is any significance between the two groups with respect to death anxiety, resilience, and subjective happiness. Pearson's correlation was run to explore the relationships between the variables. Correlation was run to see the relationship between each domain of GHQ (somatic symptom, anxiety/insomnia, social dysfunction and severe depression) along with the domains of resilience appraisal (emotional coping, situational coping, social support), death anxiety, and subjective happiness. Linear regression was also computed to predict how death anxiety predicts the other variables.

17.5 Results

Table 17.1 presents the Mean and SD of HRG and LRG along with their t ratios. From the results it may be said that the two groups differ significantly in the variable of emotional coping in which the LRG group ($M = 14.97$) has a higher emotional coping than the HRG group ($M = 13.17$). The LRG group ($M = 15.14$) differs significantly from the HRG group ($M = 13.81$) in the variable of situational coping. The two groups also differ significantly in the variable of Happiness. The LRG ($M = 4.67$) are happier than the HRG group ($M = 3.97$). There were no significant difference between the two groups with respect to death anxiety.

Table 17.2 presents the correlation coefficient of emotional support, situational support, social support, happiness, and death anxiety in the high risk group (HRG:N = 42) and low risk group (LRG:N = 28). As can be seen from this table there is significant correlations between the variables of emotional coping, social support, situational coping, and happiness among the low risk group (LRG).

In the high risk group (HRG) there is significant correlation between the variables of emotional coping with situational coping and happiness. There is also significant correlation between social support and happiness (Table 17.4).

A multiple regression analysis was run to predict death anxiety from the variables of happiness, situational coping, social support, and emotional coping for both the groups. The analysis (Table 17.3) showed a good fit model for the low risk group, $F(4,23) = 3.238$, $P < 0.05$, $R2 = 0.360$. The individual predictors were examined further and indicated that emotion coping ($t = -2.611$) $P < 0.05$, Happiness ($t = 2.776$) $P < 0.05$ were significant predictors of death anxiety in LRG. The model for HRG was not significant. Examination of the individual predictors revealed that none of the variables significantly predicted death anxiety.

17.6 Discussion

The study investigated how the low risk group for psychiatric morbidity differed from the high risk group for psychiatric morbidity with respect to the variables of resilience appraisal and its subdomains of emotional support, situational support, social support along with happiness and death anxiety. An attempt was also made to investigate the association within these variables in each of these groups.

Results from the statistical analysis revealed a significant difference between both the groups within the variables of emotional coping, situational coping, and happiness. The low risk group had better emotional and situational coping than the high risk group. As a result of which they were found to be happier. According to Uchida et al. [68] happiness is the experience of a low arousal state, such as harmony, tranquility, and interpersonal connectedness [68]. Furthermore, harmonious relationships and other people's emotional support are the best indicators of an individual's happiness. A fundamental component of positive emotion and bonding is intimate social engagement [3]. Social relationships are crucial in preventing mental illnesses, physical ailments, and other forms of distress. A well-known study conducted in California by Berkman and Syme in 1979 involved interviewing 7,000 participants to assess the strength of their supportive social networks. Nine years later, the researchers followed up with the participants to determine who was still living. It was found that individuals with strong social networks had better survival rates compared to those with weaker social connections [7]. There was a relationship between happiness and social support in both groups. According to Stanton et al. (2000), emotional processing and expression as a coping strategy work best in emotionally responsive social contexts, when the stressful event is viewed as mostly uncontrollable, and incorporates interpersonal content rather than achievement-related material [61]. Further studies have stated that high correlation exists between mental health and social support [30].

Table 17.2 Correlations between emotional support, situational support, social support, happiness, and death anxiety in the high risk group (HRG: N = 42) and low risk group (LRG:N = 28)

PSYMORB			Emcoping	Socsupport	Sitcoping	Happiness	DAS
LRG (N = 28)	Emcoping	Pearson correlation	1	0.414*	0.624**	0.651**	−0.332
		Sig. (2-tailed)		0.028	0.000	0.000	0.085
	Socsupport	Pearson correlation	0.414*	1	0.127	0.452*	0.022
		Sig. (2-tailed)	0.028		0.520	0.016	0.910
	Sitcoping	Pearson correlation	0.624**	0.127	1	0.377*	−0.287
		Sig. (2-tailed)	0.000	0.520		0.048	0.139
	Happiness	Pearson correlation	0.651**	0.452*	0.377*	1	0.159
		Sig. (2-tailed)	0.000	0.016	0.048		0.420
	DAS	Pearson correlation	−0.332	0.022	−0.287	0.159	1
		Sig. (2-tailed)	0.085	0.910	0.139	0.420	
HRG (N = 42)	Emcoping	Pearson correlation	1	0.218	0.621**	0.495**	0.241
		Sig. (2-tailed)		0.166	0.000	0.001	0.124
	Socsupport	Pearson correlation	0.218	1	0.138	0.330*	0.000
		Sig. (2-tailed)	0.166		0.382	0.033	0.998
	Sitcoping	Pearson correlation	0.621**	0.138	1	0.086	0.191
		Sig. (2-tailed)	0.000	0.382		0.589	0.227
	Happiness	Pearson correlation	0.495**	0.330*	0.086	1	−0.029
		Sig. (2-tailed)	0.001	0.033	0.589		0.855
	DAS	Pearson ccorrelation	0.241	0.000	0.191	−0.029	1
		Sig. (2-tailed)	0.124	0.998	0.227	0.855	

*. Correlation is significant at the 0.05 level (2-tailed)
**. Correlation is significant at the 0.01 level (2-tailed)

Table 17.3 Model summary of multiple regression analysis of LRG and HRG group

PSYMORB	Model	R	R Square	Adjusted R square	Std. error of the estimate
LRG	1	0.600[a]	0.360	0.249	2.554
HRG	1	0.296[a]	0.087	−0.011	2.466

Predictors: (Constant), Happiness, Sitcoping, Socsupport, Emcoping

Table 17.4 ANOVA summary table for multiple regression analysis of LRG and HRG group

PSYMORB	Model		Sum of squares	df	Mean square	F	Sig
LRG	1	Regression	84.459	4	21.115	3.238	0.030[b]
		Residual	149.970	23	6.520		
		Total	234.429	27			
HRG	1	Regression	21.545	4	5.386	0.886	0.482[b]
		Residual	224.931	37	6.079		
		Total	246.476	41			

a. Dependent Variable: DAS
b. Predictors: (Constant), Happiness, Sitcoping, Socsupport, Emcoping

Table 17.5 Multiple linear regression analysis results predicting death anxiety from the variables of happiness, situational coping, social support, and emotional coping of LRG and HRG group

PSY MORB	Model		Unstandardized coefficients		Standardized coefficients	t	Sig
			B	Std. Error	Beta		
LRG	1	(Constant)	10.373	3.560		2.914	0.008
		Emcoping	−0.662	0.254	−0.701	−2.611*	0.016
		Socsupport	0.029	0.145	0.039	0.200	0.843
		Sitcoping	−0.118	0.278	−0.092	−0.426	0.674
		Happiness	1.617	0.582	0.632	2.776*	0.011
HRG	1	(Constant)	5.941	2.576		2.306	0.027
		Emcoping	0.263	0.187	0.342	1.407	0.168
		Socsupport	−0.008	0.129	−0.010	−0.059	0.953
		Sitcoping	−0.003	0.191	−0.004	−0.017	0.987
		Happiness	−0.434	0.443	−0.195	−0.979	0.334

a.Dependent Variable: DAS
b. Predictors: (Constant), Happiness, Sitcoping, Socsupport, Emcoping

From Table 17.1 we can see that there is no noticeable difference between the two groups with respect to death anxiety. Happiness and social support could have acted as protective factors against death anxiety. Emotional coping has been found to reduce the effect of stress. A study on Kuwaiti college students suggest that, rather than being mutually exclusive, death distress and happiness are distinct characteristics. As a result, the cheerful individual could or could not be bothered by death and vice versa [1]. Even when they are reminded of mortality on a regular basis, many strong people manage. For other people, however, adapting to stressful situations may not come naturally, particularly during periods of extreme stress. These people may exhibit an unhealthy method of handling stress or controlling their emotions, which can lead to the development of psychopathological symptoms. Kasi et al., [36], have suggested that the most popular coping style used by the patients of anxiety and depression is religious coping and not emotional coping [36].

From Table 17.2 it is seen that happiness is correlated with emotional coping. This is supported by a study which stated that "positive emotions predicted increase in both resilience and life satisfaction" [15]. Positive emotions help to become ego resilient and hence people high on positive emotions display high ego resilience. It helps them to rebound from adversity and helps them to grow. Thus, happiness helps one to adopt emotional and situational coping strategies to adapt to stressful situations. Situational coping is a transactional process whereby people adapt their coping mechanism to match the demands of particular stressors in their lives [37].

The results shown in Table 17.2 indicate a strong positive correlation between situational coping and emotional coping strategies across both groups in our study. This suggests that individuals who frequently use situational coping strategies—such as actively addressing a problem or changing their environment—are also likely to engage in emotional coping strategies, like managing their emotions and seeking emotional support The results of correlation found among the high risk group in relation to emotional coping, social support, and death anxiety are justified by another study which stated that emotional regulation leads to perceived social support which results in decrease of fear of death through death acceptance but not for externally generated death anxiety [14].

From Table 17.5 it is also seen that emotional coping is inversely proportional to death anxiety in the low risk group. "Emotion-focused coping seeks to actively regulate one's emotions, and can include talking about grief-related stressors, disclosing feelings, and interacting with others" [11]. Thus by managing their emotional responses, they are better equipped to handle thoughts of mortality, reducing the associated anxiety. The results show that happiness is also a predictor of death anxiety in the low risk group. High levels of happiness is associated with increased fear of death. This is supported by a study by Cahill [9] which was conducted on "nursing population exposed to death and dying" [9].

Table 17.2 shows that social support is found to be positively correlated with happiness. Perceived social support from peers and families helps in developing resources for coping with adversities and increasing happiness. Some scholars have argued that "friendship is an important source of happiness" suggesting a causal link and a great single cause for happiness [5]. Another study suggested that young people emphasize "meaningful sources of social support (peers, residential care, and family)" and consider it as an enhancer of positive psychological functioning [21]. It is also suggested that for LRG, death anxiety could have been reduced with situational and emotional coping. Coping helps the individual to become happier and hence it results in low death anxiety.

17.7 Conclusion

The results imply a close relationship between situational and emotional coping mechanisms and their frequent use to reduce stress. By enhancing an individual's capacity to manage stress-related emotional responses as well as external obligations, the complementary application of coping methods promotes overall well-being. While emotional coping in HRG makes it harder for members of the group to control their own emotions because they already experience emotional disturbances, emotional coping in HRG groups facilitates situational coping by increasing acceptance of death. For LRG individuals, emotional problems are not a major concern; instead, they wish to efficiently manage stressful situations because doing so increases their sense of satisfaction and contentment and reduces their fear of dying.

References

1. Abdel-Khalek, A. M. (2005). Death anxiety in clinical and non-clinical groups. *Death Studies, 29*(3), 251–259. https://doi.org/10.1080/07481180590916371
2. Anderson, C., Laubscher, S., & Burns, R. (1996). Validation of the Short Form 36 (SF-36) health survey questionnaire among stroke patients. *Stroke, 27*, 1812–1816.
3. Argyle, M. (1988). Social cognition and social interaction. *The Psychologist, 1*, 177–183.
4. Argyle, M. (1997). Is happiness a cause of health? *Psychology & Health, 12*(6), 769–781. https://doi.org/10.1080/08870449708406738
5. Argyle, M. (2001). *The psychology of happiness* (2nd ed.). Routledge.
6. Bergsma, A., & Veenhoven, R. (2011). The happiness of people with a mental disorder in modern society. *Psychology of Well-Being: Theory, Research and Practice, 1*(1), 2. https://doi.org/10.1186/2211-1522-1-2
7. Berkman, L. F., & Syme, S. L. (1979). Social networks, host resistance, and mortality: A nine-year follow-up study of Alameda county residents. *American Journal of Epidemiology, 109*, 186–204.
8. Boehm, J. K., Ruberton P. M., & Lyubomirsky S. (2017). The promise of fostering greater happiness. In C. R. Snyder (Ed.), *The Oxford handbook of positive psychology* (3rd ed., pp. 944–958). Oxford Academic.
9. Cahill, H. (2022). *The effect of happiness on death anxiety in the nursing population*. Master's thesis, Rhode Island College.
10. Cao, W., Fang, Z., Hou, G., Han, M., Xu, X., Dong, J., & Zheng, J. (2020). The psychological impact of the COVID-19 epidemic on college students in China. *Psychiatry Research, 287*, Article 112934.
11. Carver, C. S., Scheier, M. F., & Weintraub, J. K. (1989). Assesing coping strategies: A theoretically based approach. *Journal of Personality and Social Psychology, 56*, 267–283. https://doi.org/10.1037/0022-3514.56.2.267.[DOI][PubMed][GoogleScholar]

12. Centers for Disease Control and Prevention. (2020). Coronavirus (COVID 19). https://www.cdc.gov/coronavirus/2019-ncov
13. Chu, C., Buchman-Schmitt, J. M., Moberg, F. B., & Joiner, T. E. (2015). Thwarted belongingness mediates the relationship between fear of negative evaluation and suicidal ideation. *Cognitive Therapy and Research, 40*(1), 31–37. https://doi.org/10.1007/s10608-015-9715-3
14. Chukwuorji, J. C., Uzuegbu, C. N., Chukwu, C. V., Ifeagwazi, C. M., & Ugwu, C. (2019). Social support serves emotion regulation function in death anxiety among people living with HIV/AIDS. *South African Journal of Psychology, 50*(3), 395–410. https://doi.org/10.1177/0081246319894700
15. Cohn, M. A., Fredrickson, B. L., Brown, S. L., Mikels, J. A., & Conway, A. M. (2009). Happiness unpacked: Positive emotions increase life satisfaction by building resilience. *Emotion, 9*(3), 361–368. https://doi.org/10.1037/a0015952
16. Collins, R., Taylor, S., & Skokan, L. (1990). A better world or a shattered vision? Changes in perspectives following victimization. *Social Cognition, 8*, 263–285.
17. Coronavirus disease 2019 (COVID-19). *People at increased risk and other people who need to take extra precautions*. https://www.cdc.gov/coronavirus/2019-ncov/need-extra-precautions/index.html
18. Cowden, R. G., Counted, V., & Ramkissoon, H. (2021a) in press. Place attachment and suffering during a pandemic. In V. Counted, R.G. Cowden, H. Ramkissoon, (Eds.), *Place and post-pandemic flourishing: Disruption, adjustment, and healthy behaviors*. Springer. https://www.springer.com/us/book/9783030825799
19. Dursun, P., Alyagut, P., & Yılmaz, I. (2022). *Meaning in life, psychological hardiness and death anxiety: individuals with or without generalized anxiety disorder* (GADh https://doi.org/10.1007/s12144-021-02695-3
20. Everly, G. (2009). *The resilient child: Seven essential lessons for your child's happiness and success*. DiaMedica Publishing.
21. Ferreira, S., Magalhães, E., & Prioste, A. (2020). Social support and mental health of young people in residential care: A qualitative study. *Anuario de Psicología Jurídica, 30*(1), 29–34. https://doi.org/10.5093/apj2019a12
22. Fredrickson, B. L., & Joiner, T. (2002). Positive emotions trigger upward spirals toward emotional well-being. *Psychological Science, 13*, 172–175.
23. Fredrickson, B. L., Tugade, M. M., Waugh, C. E., & Larkin, G. R. (2003, February). What good are positive emotions in crises? A prospective study of resilience and emotions following the terrorist attacks on the United States on September 11th, 2001. *Journal of Personality and Social Psychology, 84*(2), 365–376. https://doi.org/10.1037//0022-3514.84.2.365.PMID:12585810;PMCID:PMC2755263
24. Garmezy, N. A. (1990). closing note: Reflections on the future. In J. Rolf, A. Masten, D. Cicchetti, K. Nuechterlein, & S. Weintraub (Eds.), *Risk and protective factors in the development of psychopathology* (pp. 527–534). Cambridge University Press.
25. Gilliland, J. C., & Templer, D. I. (1986). Relationship of death anxiety scale factors to subjective states. *Omega: Journal of Death and Dying, 16*(2), 155–167.
26. Goldberg, D. P., & Hillier, V. F. (1979, February). A scaled version of the General Health Questionnaire. *Psychological Medicine, 9*(1), 139–145. https://doi.org/10.1017/s0033291700021644. PMID: 424481.
27. Greenberg, J., Pyszczynski, T., Rosenblatt, S., Veeder, M., & Kirkland, S. (1990). Evidence for terror management theory; the effects of mortality salience on reactions to those who threaten or bolster the cultural worldview. *Journal of Personality and Social Psychology, 58*(2), 308–318. https://doi.org/10.1037/0022-3514.58.2.308
28. Grieve, R., Padgett, C. R., & Mofitt, R. L. (2016). Assignments 2.0: the role of social presence and computer attitudes in student preferences for online versus offline marking. *Internet High Educ. 28*(4), 8–16. https://doi.org/10.1016/j.iheduc.2015.08.002
29. Guessoum, S. B., Lachal, J., Radjack, R., et al. (2020). Adolescent psychiatric disorders during the COVID-19 pandemic and lockdown. *Psychiatry Research, 291*, Article 113264.

30. Harandi, T. F., Taghinasab, M. M., & Nayeri, T. D. (2017, September 25). The correlation of social support with mental health: A meta-analysis. *Electronic Physician, 9*(9), 5212–5222. https://doi.org/10.19082/5212.PMID:29038699;PMCID:PMC5633215
31. Harmon-Jones, E., Simon, L., Greenberg, J., Pyszczynski, T., Solomon, S., & McGregor, H. (1997). Terror management theory and self-esteem: Evidence that increased self-esteem reduces morality salience effects. *Journal of Personality and Social Psychology, 72*, 24–36.
32. Herald National. Cow Urine a Rare Gift for Mankind: Enhances Longetivity, Cures Diseases: Govt Body's Reference Document. 2021.
33. Huang, X., Chandra, A., DePaolo, C. A., & Simmons, L. L. (2015). Understanding transactional distance in web-based learning environments: An empirical study. *British Journal of Educational Technology, 47*(4), 734–747. https://doi.org/10.1111/bjet.12263
34. Johnson, J., Gooding, P., & Tarrier, N. (2008). Suicide risk in schizophrenia: Explanatory models and clinical implications, the Schematic Appraisal Model of Suicide (SAMS). *Psychology and Psychotherapy: Theory, Research and Practice, 81*, 55–77.
35. Johnson, J., Gooding, P. A., Wood, A. M., & Tarrier, N. (2010, March). Resilience as positive coping appraisals: Testing the schematic appraisals model of suicide (SAMS). *Behaviour Research and Therapy, 48*(3), 179–186. https://doi.org/10.1016/j.brat.2009.10.007. Epub 2009 Oct 22 PMID: 19906364.
36. Kasi, P. M., Naqvi, H. A., Afghan, A. K., Khawar, T., Khan, F. H., Khan, U. Z., Khuwaja, U. B., Kiani, J., & Khan, H. M. (2012). Coping styles in patients with anxiety and depression. *ISRN Psychiatry, 2012*, 1–7. https://doi.org/10.5402/2012/128672
37. Lazarus, R. S., & Folkman, S. (1984). *Stress, Appraisal, and Coping*. Springer.
38. Lonetto, R., & Templer, D. I. (1986). *Death anxiety*. Hemisphere Publishing Corp.
39. Luthar, S. S., & Zigler, E. (1991). Vulnerability and competence: A review of research on resilience in childhood. *American Journal of Orthopsychiatry., 61*, 6–22.
40. Lyubomirsky, S., & Lepper, H. S. (1999). A measure of subjective happiness: Preliminary reliability and construct validation. *Social Indicators Research, 46*, 137–155.
41. Machado, L., Tavares, H., Petribú, K., Zilberman, M., Torres, R. F., & Cantilino, A. (2015). Happiness and health in psychiatry: What are their implications? *Archives of Clinical Psychiatry (São Paulo), 42*(4), 100–110. https://doi.org/10.1590/0101-60830000000058
42. Masten, A., Best, K., & Garmezy, N. (1990). Resilience and development: Contributions from the study of children who overcome adversity. *Development and Psychopathology., 2*, 425–444.
43. Moore, S. L., Metcalf, B., & Schow, E. (2006). The quest for meaning in aging. *Geriatric Nursing, 27*(5), 293–299. https://doi.org/10.1016/j.gerinurse.2006.08.012
44. Neimeyer, R. A., Wittkowski, J., & Moser, R. P. (2004). Psychological research on death attitudes: An overview and evaluation. *Death studies, 28*(4), 309–340.
45. Neimeyer, R. A., & Van Brunt, D. (1995). Death anxiety. In H. Wass & R. A. Neimeyer (Eds.), *Dying: Facing the facts* (3rd ed., pp. 49–88). Taylor & Francis.
46. Ogińska-Bulik, N., & Juczyński, Z. (2011). Prężność u dzieci i młodzieży; charakterystyka i pomiar—polska skala SPP-18. *Polskie Forum Psychologiczne, t., 16*(1), 7–28.
47. Ong, A. D., Bergeman, C. S., Bisconti, T. L., & Wallace, K. A. (2006). Psychological resilience, positive emotions, and successful adaptation to stress in later life. *Journal of Personality and Social Psychology, 91*, 730–749.
48. Pappa, S., Ntella, V., Giannakas, T., Giannakoulis, V. G., Papoutsi, E., & Katsaounou, P. (2020, August). Prevalence of depression, anxiety, and insomnia among healthcare workers during the COVID-19 pandemic: A systematic review and meta-analysis. *Brain Behav Immun, 88*, 901–907. https://doi.org/10.1016/j.bbi.2020.05.026. Epub 2020 May 8. Erratum in: Brain Behav Immun. 2021 Feb;92:247. https://doi.org/10.1016/j.bbi.2020.11.023. PMID: 32437915; PMCID: PMC7206431.
49. Parker, M., Clifton, E., & Thomsen, C. (2013). The role of emotional coping in reducing death anxiety. *Journal of Psychological Research, 12*(4), 55–69.
50. Phippen, M. (1995). The 1993–4 survey of counseling services in further and higher education. *Newsletter, Association for Student Counseling*, 26–36.

51. Qiu, J., Shen, B., Zhao, M., Wang, Z., Xie, B., & Xu, Y. (2020). A nationwide survey of psychological distress among Chinese people in the COVID-19 epidemic: Implications and policy recommendations. *General Psychiatry, 33*(2), e100213.https://doi.org/10.1136/gpsych-2020-100213
52. Roy, D., Tripathy, S., Kar, S., Sharma, N., Verma, S., & Kaushal, V. (2020). Study of knowledge, attitude, anxiety & perceived mental healthcare need in Indian population during COVID-19 pandemic. *Asian Journal of Psychiatry., 51*, Article 102083. https://doi.org/10.1016/j.ajp.2020.102083
53. Ruberton, P. M., Kruse, E., & Lyubomirsky, S. (in press). Boosting state humility via gratitude, self-affirmation, and awe: Theoretical and empirical perspectives. In E. Worthington, D. Davis, & J. Hook (Eds.), *Handbook of humility*. Rout ledge
54. Rutter, M. (1990). Psychosocial resilience and protective mechanisms. In J. Rolf, A. S. Masten, D. Cicchetti, K. H. Nuechterlein, & S. Weintraub (Eds.), *Risk and protective factors in the development of psychopathology* (pp. 181–214). Cambridge.
55. Saleem, T., Gul, S., & Saleem, S. (2015). Death anxiety scale: Translation and validation in patients with CVD. *Professional Med J, 22*, 723–732.
56. Salter, K., Campbell, N., Richardson, M., Mehta, S., Jutai, J., Zettler, L., Moses, M., McClure, A., Mays, R., Foley, N., & Teasell, R. (2013). *Outcome measures in stroke rehabilitation*. http://www.ebrsr.com/sites/default/files/Chapter21_Outcome-Measures_FINAL_16ed.pdf
57. Sartre, J. (1956). *Being and nothingness: An essay on phenomenological ontology*. Philosophical Library.
58. Sasaki, M. (2005). Dispositional and situational coping and mental health status of university students. *Psychological Reports, 97*(7), 797. https://doi.org/10.2466/pr0.97.7.797-809
59. Sharif-Nia, H., Sharif, P., & Saeed & Koocher, Gerald & Yaghoobzadeh, Ameneh & Haghdoost, AliAkbar & Win, Ma Thin & Soleimani, Mohammad. (2017). Psychometric properties of the death anxiety scale-extended among patients with end-stage renal disease. *OMEGA—Journal of Death and Dying, 80*. https://doi.org/10.1177/0030222817733220
60. Skapinakis, P., Bellos, S., Oikonomou, A., Dimitriadis, G., Gkikas, P., Perdikari, E., & Mavreas, V. (2020). Depression and its relationship with coping strategies and illness perceptions during the covid-19 lockdown in Greece: A cross-sectional survey of the population. *Depression Research and Treatment, 2020*, 1–11. https://doi.org/10.1155/2020/3158954
61. Stanton, A. L., Danoff-Burg, S., Sworowski, L. A., Collins, C. A., Branstetter, A. D., Rodriguez-Hanley, A., et al. (2002). Randomized, controlled trial of written emotional expression and benefit finding in breast cancer patients. *Journal of Clinical Oncology, 20*, 4160–4168.
62. Taquet, M., Geddes, J. R., Husain, M., et al. (2021). 6-month neurological and psychiatric outcomes in 236 379 survivors of COVID-19: A retrospective cohort study using electronic health records. *Lancet Psychiatry, 8*, 416–427.
63. Taylor, S. E., & Brown, J. D. (1988). Illusions and well-being: A social psychological perspective on mental health. *Psychological Bulletin, 103*, 193–210.
64. Templer, D. I. (1970). The construction and validation of a death anxiety scale. *Journal of General Psychology, 82*(2), 165–177. https://doi.org/10.1080/00221309.1970.9920634
65. Van Tongeren, D. R., & Showalter Van Tongeren, S. A. (2020). *The Courage to suffer: A new clinical framework for life's greatest crises*. Templeton Foundation Press.
66. Tramill, J. L., Davis, S. F., Bremer, S., Dudeck, M. M., & Elsbury, D. L. (1982). A proposed relationship between the unidimensional short form of the TMAS and the DAS: The effects of embedding vs. separate administration. *Bulletin of the Psychonomic Society, 19*, 209–211.
67. Tugade, M. M., & Fredrickson, B. L. (2004). Resilient individuals use positive emotions to bounce back from negative emotional experiences. *Journal of Personality and Social Psychology, 86*, 320–333.
68. Uchida, Y., & Ogihara, Y. (2012). Personal or interpersonal construal of happiness: A cultural psychological perspective. *International Journal of Wellbeing., 2*, 354–369. https://doi.org/10.5502/ijw.v2.i4.5
69. Uner, S., Ozcebe, H., Telatar, T. G., & Tezcan. S. (2008). Assessment of mental health of university students with GHQ-12. *Turkish Journal of Medical Science, 38*, 437–46. [Google Scholar]

70. Werner, E., & Smith, R. (1982). *Vulnerable but invincible: A study of resilient children.* McGraw-Hill.
71. White, W., & Handal, P. J. (1991). The relationship between death anxiety and mental health/distress. *OMEGA—Journal of Death and Dying, 22*(1), 13–24. https://doi.org/10.2190/YBTU-176L-D48C-YQK8
72. World Health Organization: Mental health and psychosocial considerations during COVID-19 outbreak. World Health Organization. 2020 (January). pp. 1–6.
73. Xi, Y., Hui, Y., Yao, Y., Peng, K., Wang, Y., & Chen R. (2020). Post-traumatic stress disorder and the role of resilience, social support, anxiety and depression after the Jiuzhaigou earthquake: a structural equation model. *Asian J Psychiatr, 49*, 101958. https://doi.org/10.1016/j.ajp.2020.101958
74. Zhang, J., Lu, H., Zeng, H., Zhang, S., Du, Q., Jiang, T., & Du, B. (2020). The differential psychological distress of populations affected by the COVID-19 pandemic . *Brain, Behavior, and Immunity, 87*, 49–50. https://doi.org/10.1016/j.bbi.2020.04.031

Jhilli Das (Tewary) is an Assistant Professor in the Department of Psychology at St. Xavier's University. She holds a PhD in Psychology from the University of Calcutta and a Postgraduate Diploma in Rehabilitation Psychology. Her research interests focus on Positive Psychology, Personality, and Social Psychology. With over a decade of teaching experience, including 6 years as a guest lecturer and 6 years and 9 months as a full-time Assistant Professor, Dr. Das has made significant contributions to academia. She has guided one research thesis in Positive Psychology and published six research articles in national and international journals, along with two book chapters with Routledge and Nova Publishers. Dr. Das is a life member of the Indian School Association.

Debopriya Ghatak is a distinguished academic and researcher with expertise in developmental psychology, forensic psychology, and positive psychology. Known for her interdisciplinary approach, she has designed and taught diverse courses, including Sports Psychology, Forensic Psychology-II, Criminal Profiling, and Psychology of Violence. Her research spans multiple domains, focusing on human behavior in contexts such as athletic performance, mental well-being, and criminal behavior. As a dedicated educator, she is deeply committed to curriculum development and mentoring students, fostering intellectual and professional growth. Debopriya's work blends academic rigor with practical applications, offering valuable insights into both theoretical and real-world challenges. Her impactful contributions to teaching, research, and mentorship have made her a respected figure in the field of psychology.

Shweta Sah is an Assistant Professor in the Department of Psychology at Sister Nivedita University. She earned her PhD in Psychology from the University of Calcutta in 2016 and holds a professional registration under the Rehabilitation Council of India. With over a decade of industry experience, Dr. Sah has worked extensively in diagnosis and therapy for children and adults, specializing in Rehabilitation Psychology. She has actively participated in numerous national and international conferences, where her contributions have been recognized with multiple Best Paper awards. Combining academic expertise with practical experience, Dr. Sah is dedicated to advancing psychological research and practice, particularly in the field of rehabilitation and therapeutic interventions.

Chapter 18
Gendered and Regional Impact of COVID-19 on the Government Transfer Income of Members of Self-Help Groups in India

Nishi Malhotra

Abstract Approximately 1.7 billion people are below the poverty line and they do not have access to collateral or collateral to access bank finance or credit. In this regard, social finance and transfer payments by the government play an extremely important role in promoting equality and ensuring the inclusion of the poorest of the poor within the financial system. During COVID-19 the plight of the poor people deteriorated manifold as they could not access the finance. In this regard, various schemes under which the direct benefit transfer were provided to the people at the bottom of the pyramid played an extremely important role in reducing poverty. This study aims to explore the impact of COVID-19, on the Government transfer income of the people. In this study, the theoretical lens of Dynamic Punctuated Equilibrium was used to provide transfer payments to the people at the bottom of the pyramid. It aims to study the regional and gendered effects of COVID-19 on the Government transfer income of the members of the self-help groups in India. The difference in different methods has been used for analysis. The study shows that rural regions have higher Government transfer income during COVID-19. However, there is no visible gendered impact of COVID-19 on the Government transfer income of the members of self-help groups in India. This study provided great insights to the policymakers and academicians about relevance of transfer payments, especially during crisis to providing financial aid and support to the people at bottom of the pyramid.

Keywords COVID-19 · Difference in difference · Government transfer income

N. Malhotra (✉)
Indian Institute of Management, Sambalpur, India
e-mail: nishim@iimsambalpur.ac.in

S. Gangopadhyay et al. (eds.), *Pandemic Diaries*,
https://doi.org/10.1007/978-981-96-5415-4_18

18.1 Introduction

Under the SDGs (sustainable development goals) given by the UN, poverty reduction has been adopted as the most important sustainable development goal. Goal 1 is to end poverty in all forms and by 2030 eradicate poverty for all people, reduce by half the proportion of men living below the poverty line, and implement nationally appropriate social protection systems. Thus, the social protection systems have become an extremely important policy decision, which helps to promote social inclusiveness and economic welfare among the poor. In the post-pandemic era, the reduction of poverty has become the most important goal for emerging economies like India. It becomes extremely important for emerging countries to work toward the reduction of inequalities and poverty. The number of poor people living below poverty is declining steadily till 2020. But it was in 2020, with the onset of COVID-19 that the proportion of people living below the level of poverty started increasing steadily. Poverty in the context of developing economies has been defined as living below $2.15 per person per day. The main reason for the increasing economic crisis was the disruption in work, working from home, and the loss of income. As per various reports, more than half a billion people came below the poverty line. Sustainable livelihood emerged as a challenge for most people living below the poverty line. The basic indicators of poverty are extreme hunger, limited access to education, social discrimination, and other social evils. By 2030, nations all over the world are aiming to achieve the sustainable development goal of poverty reduction. The global pandemic has threatened global efforts to tackle poverty and reduce inequality. Before the pandemic, poverty levels declined drastically from 36 to 10% in 2015. Our vision for 2030 is to reduce poverty to less than 3%, i.e., the number of people living below 2.15 $ is to be reduced drastically. Despite all efforts, the problem of extreme poverty is very acute in the African nations and other underdeveloped nations of the world. More than 700 million people or 10% of the world's population live below the poverty line. They struggle to fulfill their basic needs such as food and in the post-pandemic period, these economic losses have increased with people having no social protection. Most of the poor people live in underdeveloped countries such as sub-Saharan Africa and Southern Asia with an additional 23 to 36 million people being added to the status of extreme poverty in a post-pandemic era. Globally, an additional 70 million people have been added to the status of extreme poverty, which means that by 2030 still 6 million people will be living below the poverty, missing out on the target of 100% poverty elimination. The progress of more than 4 years has been erased by COVID-19. Rising inflation and the war in Ukraine further derail the process of economic development. Due to the pandemic, unemployment and cash benefits have increased manifold in developing countries, which has led to the derailing of the process of development. The problem of poverty has many dimensions which include unemployment, social exclusion, and a decline in the life expectancy of the population. An increase in poverty levels leads to an increase in social inequalities which disrupts social cohesion. It further leads to an increase in the level of political and social tension. In the modern-day world, both the public and private sectors help in propelling the engine of growth by creating an enabling environment. Social innovation has an important role to play in reducing inequalities and promoting harmony and development through poverty reduction. Government transfers have emerged as an important tool to promote equality and reduce poverty. During the period of the pandemic, the Government of India introduced various social security schemes including cash transfer for the poor to promote equity and reduce the inequalities in the system. This scheme was rolled out by the Government of India on January 1, 2013, and was aimed at providing access to subsidies in the form of cash and kind. There is a lack of research on the status of this Government transfer scheme in India and this regard, this study aims to measure the income shock of this Government transfer scheme during COVID-19 in India.

18.2 Survey of Literature

During the period of the pandemic, social protection is the key to promoting economic welfare and effectiveness. Social protection is extended by the Government in the form of unconditional and conditional transfers. COVID-19 supplements the income of the poor sections of society during the pandemic. The existing literature draws a clear demarcation between conditional and unconditional cash transfers. According to [16] conditional cash transfers, are important when the key stakeholders suffer from information asymmetry regarding the investments that are considered key to the economic development of the nation. [10] earlier research articles highlight, that unconditional cash transfers lead to better health outcomes, better access to food, and diversity in diet. During the crisis period, various developing nations such as Nepal and India have implemented unconditional cash transfers to supplement the economic resources of the people at the bottom of the pyramid [4]. Most of the cash transfers aid in improving consumption levels and also promote psychological well-being among the members of the community. [8] propagates that unconditional cash transfer helps to improve health services and also has a positive impact on the health expenditure of the members of the community. A case study of the m-Pesa program in Kenya is a case in point, where the Government agencies transferred a fixed sum of amount to the members of the community. The studies revealed that this cash transfer has a positive impact on the economic and psychological health of the members of the community [6]. [1] report highlights the importance of the Bantuan Langsung Tunai (BLT) program as an unconditional cash transfer program that provides access to cash to households that were affected by the economic shocks. As per the report, though this program is not free from demerits, it provides temporary protection to poor households. It helped to address the cash difficulties of the poorer households. This program provided consumer protection and also promoted a higher standard of living through better consumption expenditures. [17] report highlights that the Government of Zambia (Child Grant Program) promotes savings among the members of the group. From the study, it is highlighted that the grant program brings about sustainable improvement in the well-being of the members of the community. The study highlights that the poverty-targeted unconditional cash transfer has significantly increased the level of protection among the members of the group. This microfinance initiative promotes higher savings and smoothens the level of consumption, thus improving the well-being of the members of the community. [7] research study highlights that cash transfer programs have the potential to improve the well-being of children and lead to social development including health and economic development of the members of the community. [15] research has highlighted that cash transfer programs do not need to be wasteful expenditures but indeed reduce poverty and lead to inclusive growth. [18] their article mentioned that cash transfers led to better jobs, improved human capital, and raised the skills of the human capital. The article highlights that there are three kinds of inequalities, i.e., inequalities of opportunity, inequality of living conditions, and inequality of human capital. As per the author, in the short-term policy measures should be targeted to redistribute the wealth, but in the long run, measures should be taken to reduce the inequalities of human capital, and opportunities to promote inclusive growth. [13] in their study through empirical study highlights that the unconditional cash transfer program does not lead to any undesirable behavior or risky behavior. [12] study highlights that cash transfer programs lead to better human capital and minimum threshold income. Thus, it leads to the economic welfare of the households. The research study highlights that the timing of the cash transfers has an impact on household expenditure. [9] in their study highlight that cash transfers reduce poverty, increase food security, and do not lead to undesirable expenditure on tobacco and alcohol. [3] the research study highlights that cash transfers are an effective way of promoting social welfare and do not lead to misuse of funds. During the COVID-19 pandemic, the income and consumption of the people at large decreased significantly. To mitigate the impact of the reduction in consumption for the people at the bottom of the pyramid, the Government of India introduced cash transfers and additional food transfer schemes. It was estimated that during COVID-19, the number of people below the pyramid can rise from a mere 135 million to 260 million, [11]. In January 2020, the Coronavirus pandemic spread across the globe, and to combat the negative impact of COVID-19 on March 24 2020 the Government declared a lockdown. During COVID-19, the GDP of India contracted tremendously and the year-on-year unemployment increased to 23.5%. During COVID-19 due to scarcity of food, the consumption of the households came down drastically. On March 26, 2020, the Government announced the Pradhan Mantri Garib Kalyan Yojana. This initiative was mainly focused on providing support to the rural communities, laborers, craftsmen migrant workers, and farmers to name a few. PMGKY (Pradhan Mantri Jan Kalyan Yojana) mainly aimed at providing conditional cash transfers to people from below the poverty line to buy cooking gas and free rations for the vulnerable. Further on May 17, 2020, an additional 1 lac crore was allocated to various infrastructural bottlenecks. COVID-19 induced significant economic and social disruptions in India. The social assistance namely PM GKY (Pradhan Mantri Grib Kalyan Yojana) provided cash direct benefit transfers and was introduced to combat the negative influences of the increasing COVID-19 during the period. The COVID-19 pandemic badly impacted the credit requirement and liquidity needs of the people at the bottom of the pyramid. The cash transfer programs implemented during the period of crisis helped the members of the society to earn a livelihood and maintain their salaries. Within the ambit of PMGKY (Pradhan Mantri Garib Kalyan Yojana) several different programs such as *Pradhan Mantri Kisan Samman Nidhi* (PM-KISAN), *Pradhan Mantri Ujjwala Yojana* (PM-UY), *Pradhan Mantri Jan Dhan Yojana* (PM-JDY), and *Pradhan Mantri Ann Vitran Yojana* (PM-AVY) to provide access to cash to the people at the bottom of the pyramid. The various Government schemes were aimed at providing access to free cooking gas, regulators, and cash to people who do not have access to financial resources. Through the direct benefit transfer, the Indian Government established the direct benefit transfer and cash transfer scheme for the people at the bottom of the pyramid. During COVID-19 the Government also gave push to the MNREGA (Mahatma Gandhi National Rural Employment Guarantee) program to provide financial support, to combat the pandemic and financial crisis that erupted due to COVID-19. Though the literature highlights the disparities that exist among the different sections of society such as females and males, rural and semi-rural areas, many studies to tried to evaluate the role of the various cash transfer schemes on the income of the household during the pandemic. This study will aim to explore the conditional impact of cash transfer schemes on the income of the household given the gender and region of domicile of the recipients.

18.2.1 Policy Changes During COVID-19 in India

As per the punctuated equilibrium model, the policies of the Government stay the same for a certain period and then they change suddenly, quickly, and dramatically. This paper aims to evaluate the sudden policy changes introduced by Governments all over the world during COVID-19. A case in point is the various Government transfer schemes introduced by the Government of India during the period of epidemic or COVID-19. During the pandemic, the Government of India under the leadership of the Honorable Prime Minister Narendra Modi introduced the direct benefit transfer schemes. The data for analysis was taken from (https://dbtbharat.gov.in/, 2022) [5]. For the number of beneficiaries under the direct benefit transfer scheme in India. From 2013 to 14 during the Prime Ministership of Dr. Manmohan Singh from the UPA party, the number of beneficiaries from the direct benefit transfer scheme increased incrementally from 10.8 Crore in 2013–14 to 22.8 Crore in 2015–16. In the year 2015–16, the NDA Government came to power and they initiated the direct benefit transfer (DBT) in kind. In the domain of the welfarist state, the policy of equal distribution of wealth, which was incremental for years, saw a profound change in the year 2017–18, which set a new direction for the policy future. For the first time in the year 2014, the Honorable Prime Minister Sh. Narendra Modi started the Mann Ki Baat, a novel democratic program to ensure the setting up of the policy communities and agenda setting. The policy communities comprise the citizen groups, interest groups, and public officials from the various Government officials. This electronic platform provides a set of relationships that provide a forum for consensus and social agreement to resolve the policy problem of financial exclusion. This facilitates the social network of the state, democratic institutions, financial players, and interest groups. Thus, during the period 2020–21 and 2021–22, the Indian state witnessed an incremental shift and increase in the number of beneficiaries under the direct benefit transfer scheme. The policy measures adopted by the NDA led to the successful adoption of a new policy image successfully. This facilitated the involvement of the excluded groups. This involved the creation of competitive pluralism, policy communities, and close relationship between certain interest groups (Figs. 18.1 and 18.2).

Ref. [11] suggested that the Government of India initiated a social contract in the form of social relationships among the different members of the group to facilitate discussion on important policy issues. Through this "Policy Community," the Government of India facilitated consensual relationships between the members of

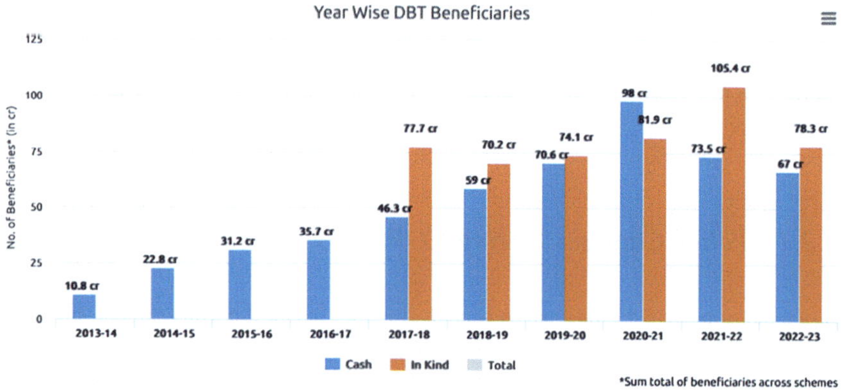

Fig. 18.1 Year-wise DBT beneficiaries

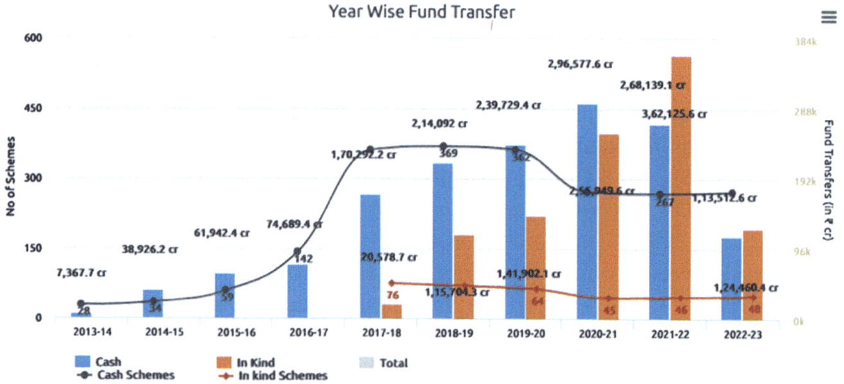

Fig. 18.2 Year-wise fund transfer

the group and the Government of India. The stability of this process of community policy formulation through democratic participation depends on socialization, inclusion within the policy community depends on gaining the personal trust of through awareness and involves immersion within a common culture. Thus, from the policy perspective, the study concludes that a policy of democratic participation ensures the formulation of the *Direct Benefit Transfer* scheme that facilitates two-way communication and receiving feedback from interest groups and media. Thus, this two-way communication network facilitates the measurement of policy changes in a systematic way. Keeping this in mind, the year 2020–21 and 2021–22 represent the punctuation with sudden bursts and increments in the number of direct benefit beneficiaries under the fund transfers. Thus, from the previous study it becomes apparent, that participatory governance has proved to be important in improving the quality of public administration and poverty reduction. Exclusion of the poor leads to a negative impact on the life of people and various Government transfer schemes help in improving the standard of living of the people in the rural region. The direct benefit transfer is a social contract between the state and the beneficiaries that helps in improving the social welfare and food intake of the members of society. Cash transfers help in reducing inequalities and provide a level playing field for the rich and the poor. Thus, the cash transfers result in an income effect. Another dimension that can be studied is whether the cash transfers actually lead to an increase in the amount of consumption and financial outcomes. This certainly depends on the income elasticities of the commodities that are being consumed. Redistribution of the transfer income helps in better social risk management. The Indian economy has very well adapted the transfer scheme to provide cash benefits that help to achieve synergies through administrative reforms. The cash transfer scheme helps in achieving the minimum level of consumption and it also helps to break the vicious circle of poverty by transmitting the benefits across the generations. Cash transfers can be of two kinds conditional cash transfer and unconditional cash transfer. In conditional cash transfers, the payments are made conditional on certain events. Conditional cash transfer is more relevant where there is political stability. However, conditional cash transfer suffers from two main issues such as information asymmetry, i.e., the recipient might know less about the schemes and secondly, it might exclude some people. There have also been many arguments against cash transfer, i.e., it might dissuade the people from entering the labor supply, and secondly, it might reduce gross capital formation in the nation. But the favorable argument for the cash transfer is that it leads to direct benefit transfer and in the case of public expenditure, there might not be a trickle-down effect. The unconditional cash transfer helps to smoothen consumption through the trickle-down effect. Certain inequalities are not inherited from the parents but certain inequalities are inherited such as background, race, etc. In these cases, it becomes the moral responsibility of the state to provide economic assistance through the direct cash benefit transfer. The existing literature highlights that the existing cash transfer schemes suffer from private procrastination and self-control problem. The investments from the private cash transfers involve some form of myopia, lack of information, and incomplete altruism. In case of market failure and lack of redistributive justice, the unconditional cash transfer might be a good strategy. Another aspect of the problem is that if the potential beneficiaries are poorly informed about future decisions, it might lead to the misutilization of conditional cash transfers. In a country like India, the Government of India has adopted the direct benefit transfer scheme which is unconditional in nature.

18.2.2 Theoretical Framework

Within the complex macroenvironment, the banks, the democratic institutions, and the Government work together and they indulge in a continuous exchange of information depending on the existence of the social networks. [2], earlier research has highlighted that the Government's response to social inputs changes proportionally across the period of policy punctuations. Thus, during the stable period of stability, the institutional frictions and bounded rationality limit the level of policy change. In a tumultuous environment, the existing level of social change and the external environmental causes lead to dramatic policy shifts. This study uses the theoretical framework of Dynamic Punctuated Equilibrium to study the impact of pandemics such as COVID-19 on policy changes such as direct benefit transfer, employment, and bank accounts under the (Pradhan Mantri Jan Dhan Yojana) by the Government in India. And how these policy measures have helped to reduce poverty in India. Through the theoretical lens of Social Theory, the Punctuated Equilibrium Paradigm, the concept of sustainability seems to evolve gradually through relatively long periods of stability. It is punctuated by more fundamental and revolutionary bursts of change. The terms "sustainability" and "periods" allude to the phases of equilibrium that are followed by periods of short and sudden bursts. Through the lens of Dynamic Equilibrium, self-help groups have been defined as a complex system that maintains a balance between the outflows and inflows of funds and between social goals and financial rate of return. Through the theoretical lens of punctuated equilibrium, this diagram depicts the various factors that impact the effectiveness and efficiency of group lending.

18.3 Data and Methodology

The study uses the data from CMIE (Centre for Monitoring Indian Economy) Consumer Pyramid Household Survey (CPHS). The data was taken from two waves. One of the waves is the pre-pandemic wave (Jan—Dec 2019) and the post-pandemic wave (Jan—Dec 2020). This data provides us with a panel of the pre-pandemic and post-pandemic panels of individuals, which allows us to track the changes using the difference in different methods or approaches. From all the empirical results, a panel of individuals was observed across the different waves.

18.3.1 Descriptive Statistics

The summary of the descriptive statistics is given below (Table 18.1).

From the analysis of data, it becomes clear that there are 10,46,430 households with a mean income of Rs. 129.13 and a standard deviation of Rs. 829.68. And 66.7% of the observations are from the COVID-19 period. Moreover, 78.3% of the households are headed by male members.

Figure 18.3 shows the difference in the difference graph for the impact of the region on the income of the households and Fig. 18.4 shows the difference in the difference graph for the impact of gender (males) on the income of the households.

Table 18.1 Summary of descriptive statistics

	Observations	Mean Income	Standard Deviation
Income of households	10,46,430	129.13	829.68
Covid	10,46,430	0.667	0.471
Gender	10,46,430	0.783	0.411

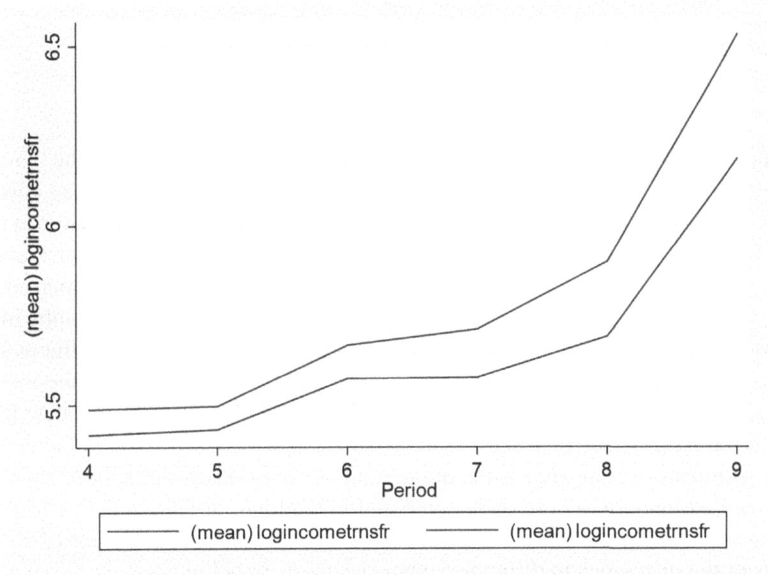

Fig. 18.3 Impact of region on income of households

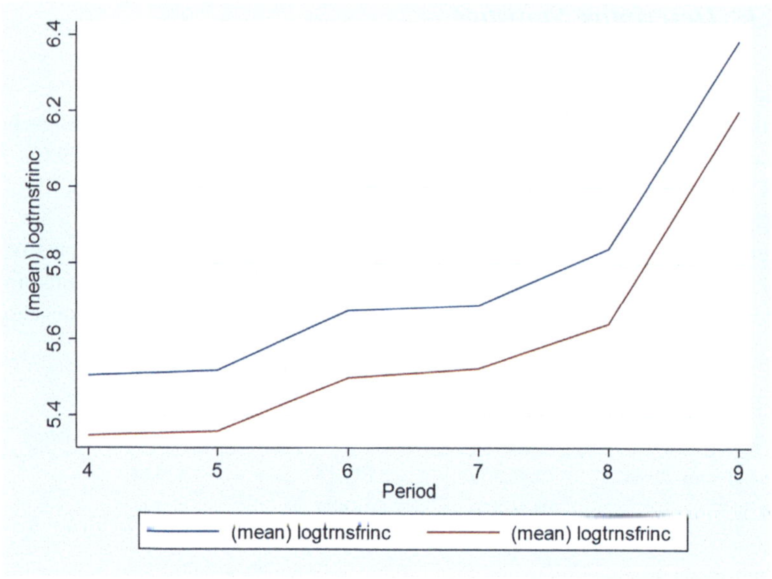

Fig. 18.4 Impact of gender on income of households

18.4 Methodology

This paper examines the regional and gendered impact of COVID-19 on the Government transfer income of the households in different parts of the nation as covered by the CMIE database. We have used the difference in difference estimator to estimate the impact of COVID-19 on the growth of the Government transfer incomes in the nation. The period (Jan—Dec 2019) is taken as the pre-pandemic period and the period (Jan—Dec 2020) is taken as the post-pandemic period. The results of the data further show that during the COVID-19 period in the rural region, the income of the households increased drastically. During the COVID-19 period, the income of the households headed by males increased more as compared to females. Rural regions during COVID-19 witnessed approximately a 14.4% increase in the income of the members as compared to the urban regions. The households headed by males witnessed an increase of 15.5% as compared to the households headed by females. During the COVID-19 period, the income of households increased by 30.2%. The results of the difference in difference analysis are given below.

Table 18.2 highlights the results of the regression analysis to identify the impact of COVID-19 and gender on the access to the Government transfer income. From time to time, the Government brings about various transfer incomes to support the livelihood and employment of the people at the bottom of the pyramid. The analysis of data highlights that female members in a household have higher probability to access the Government transfer funds and the people from the rural region have higher access to the Government transfer incomes. Recently in the year 2019, on

Table 18.2 Regression results for the impact of rural regions on the Government transfer income of households

	Coefficient
Rural COVID-19	0.144*** (0.005)
COVID-19	0.306*** (0.003)
Region **Gender** **Constant** **R Square**	0.070*** (0.004) 0.155*** (0.003) 5.337*** (0.003) 0.283

March, 2019, COVID-19 was declared as an emergency. In wake of this pandemic, the Government transfer to support the livelihood of the people at the bottom of pyramid increased tremendously. The Government of India has started various initiatives to provide subsidy and Government transfer support to the marginalized. And during the COVID-19, the disbursement of the financial support to the poor households increased manifold. The government subsidy further helped the microentrepreneurs to combat the operational risk.

Table 18.3 shows the impact of the male gender on the income of the households through the transfer income. From the analysis of results, it becomes apparent that households headed by males as compared to females generally get higher income through Government transfer. During COVID-19, there was no major increase in the income of the households headed by males as compared to females. The rural regions showed an increase in income through Government transfers. Earlier the Government of India has launched the *Pradhan Mantri Jan Dhan Yojana*, under which access to the banking accounts was provided to the people at the bottom of the pyramid. During the COVID-19 pandemic, the Government introduced cash transfers which enabled the people at the bottom of the pyramid to develop resilience and combat the challenges posed by the pandemic.

Table 18.3 Regression results for the impact of gender on the Government transfer income of households

	Coefficient
Male. COVID-19	0.0002 (0.005)
COVID-19 **Males**	0.349*** (0.004) 0.155*** (0.004)
Rural region **Constant** **R Square**	0.141*** (0.003) 5.31*** (0.003) 0.282

18.5 Conclusions

This study analyses the gendered effect of COVID-19 on the Government transfer income of the people of India. More specifically, it discusses the regional impact of COVID-19 on the Government transfer income of the people of India. As per the study, the income of the rural regions increased drastically during COVID-19 and there was no visible disparity in the transfer of subsidy income during COVID-19. Thus, results show that the Government transfer of incomes reduces the income disparity and mitigates the gender disparity, thus reducing the difference between men and women. Hence, it can be concluded that the income of the Government transfer and subsidies have an important effect upon the income of the households. Thus, the subsidy plays an important role in promoting the welfare and increasing the income of the households. Government transfers plays an important role in promoting the sustainable livelihood for the people from the bottom of the pyramid. This study will be the seminal study in the domain of impact of the Government transfers on sustainable livelihood and in achievement of the sustainable development goal of poverty reduction.

18.6 Contribution to Literature

Organizational effectiveness is emerging as an area of focus to study the impact of inclusive systems on the sustainability of the groups or organizational structure. Organizational change and adaptability have become imperatives to ensure the organization's effectiveness. This research paper only discusses the impact of the Government transfer income on the income of the household, but it does not mention the role of social intermediation and financial training in the success of micro-enterprises. Though further research can be conducted on importance of habitus and organizational learning in promoting cooperation and solidarity. In current context, the Government transfers play an important role in promoting the higher income and sustainable livelihood for the members of the household. Thus, in future more research can be conducted around the impact of the organizational learning and social intermediation on the income of the households. Further for policymakers, the study can be expanded to cover the role of financial literacy and digital literacy in promoting the welfare of the people at the bottom of pyramid. Moreover, during the COVID-19 the importance of social subsidies and Government transfer payments increased manifold.

References

1. Bank, T. W. (2012). *BLT temporary unconditional cash transfers*. The World Bank.

2. Baumgartner, F. R., & Jones, B. D. (1991). Agenda dynamics and policy subsystems. *The Journal of Politics, 53*(4), 1044–1074.
3. Evidence. (2020, December 20). *Research on cash transfers.* Retrieved from Research on cash transfers: https://www.givedirectly.org/research-on-cash-transfers/
4. Haushofer, J. (2016). The short-term impact of unconditional cash transfers to the poor: Experimental evidence from Kenya. *The Quarterly Journal of Economics, 131*(4).
5. https://dbtbharat.gov.in/. (2022). https://dbtbharat.gov.in/. Retrieved from Direct Benefit Transfer: https://dbtbharat.gov.in/
6. Late Latif Jameel Poverty Action Lab. (2022). *Poverty action lab.* Retrieved from JPAL Poverty Action Lab: https://www.povertyactionlab.org/evaluation/improving-economic-and-psychological-well-being-through-unconditional-cash-transfer
7. Laura Robertson et al. (2013). *Effects of unconditional and conditional cash transfers on child health and development in Zimbabwe: A cluster-randomised trial.* Lancet.
8. Management, O. P. (2022). *A Cochrane review.* Retrieved from Unconditional cash transfers for reducing poverty and vulnerabilities: https://www.opml.co.uk/blog/unconditional-cash-transfers-reducing-poverty-vulnerabilities
9. Maxwell Young. (2019, March 8). *Busting six common myths about unconditional cash transfers in Africa.* Retrieved from https://www.ifpri.org/blog/busting-six-common-myths-about-unconditional-cash-transfers-africa: https://www.ifpri.org/blog/busting-six-common-myths-about-unconditional-cash-transfers-africa
10. Pega F et al. (2022, May). https://www.cochrane.org/CD011135/PUBHLTH_does-giving-money-people-low-and-middle-income-countries-without-conditions-attached-lead-better. Retrieved from Cochrane: https://www.cochrane.org/CD011135/PUBHLTH_does-giving-money-people-low-and-middle-income-countries-without-conditions-attached-lead-better
11. Rhodes, R. A. W., & Marsh, D. (1992). New directions in the study of policy networks. *European Journal of Political Research, 21*(1–2), 181–205.
12. Ricardo et al. (2019). Schooling responses to income changes: Evidence from unconditional cash transfers in Rwanda. *International Journal of Educational Research*, 177–189.
13. Ridho et al. (2020). The behaviourial effects of unconditional cash transfers : Evidence from Indonesia. *SMERU Working Paper Series* .
14. Samuel et al. (2012). Evaluating Indonesia's unconditional cash transfer program, 2005–6. *3IE Evaluation Report*.
15. Sudhanshu Handa et al. (2018). Myth-Busting? Confronting six common perceptions about the unconditional cash transfers as a poverty reduction strategy in Africa. *The World Bank Researcher Observer*, 259–298.
16. UNICEF. (2016). *Making money work :Unconditional cash transfers allow women to save and reinvest in rural Zambia.* UNICEF.
17. UNICEF. (2016). *Conditionality in cash transfers: UNICEF's approach.* UNICEF.
18. Kumarawamy .V. (2019). *Are unconditional cash transfers desirable?* https://www.thehindubusinessline.com/opinion/are-unconditional-cash-transfers-desirable/article26762792.ece: https://www.thehindubusinessline.com/opinion/are-unconditional-cash-transfers-desirable/article26762792.ece
19. Wambogo, E. A., Ghattas, H., Leonard, K. L., & Sahyoun, N. R. (2018). Validity of the food insecurity experience scale for use in Sub-Saharan Africa and characteristics of food-insecure individuals. *Current development in nutrition*, 1–10.

Nishi Malhotra is a highly accomplished professional in the field of finance. She holds a Ph.D. in Finance from IIM Kozhikode, an MBA from MDI Gurgaon, a B Com (Hons) from Sri Ram College of Commerce, and an MCOM from IGNOU. Additionally, she has actively pursued courses in financial markets from the Bombay Stock Exchange and corporate finance from Coursera, demonstrating a committed effort to enhance her expertise. Throughout her academic journey, Dr. Malhotra has contributed significantly to the field of finance by teaching an array of subjects such as Financial Accounting, Digital Finance, and other finance-related topics at esteemed institutions like OP Jindal Global University, Rajagiri College of Social Sciences, Dr DY Patil Vidyapeeth, Banasthali Vidyapeeth, and Guru Gobind Singh Indraprastha University. Combining her academic achievements with industry experience spanning over 10 years, she has gained invaluable insights and made substantial contributions to the finance sector.

Chapter 19
COVID-19's Implications on Sustainability and Digital Growth in Small and Medium-Sized Businesses: A Theoretical Framework

Sayani Saha and Soma Sur

Abstract Online platforms are instruments for carrying out an enormous number of duties. Instead of being process-driven, it primarily makes possible a data-driven world. Digital platforms find widespread application in marketplaces, media outlets, information platforms, and other domains. The increasing popularity of digital platforms can be attributed to technological improvements like cutting-edge machine learning and the most popular artificial intelligence. The COVID-19 pandemic has acted as a stimulant for the meteoric rise of online retail platforms. Thus, there is an urgent need for adaptation to volatility from a pandemic like COVID-19. Small- and medium-sized enterprises (SMEs) are especially susceptible to pandemic. SMEs do, already employ some coping mechanisms, though there is very little data on the nature and efficacy of these tactics. Therefore, it is still unknown how SMEs can become more resilient to pandemic risk and less vulnerable to it by implementing documented coping strategies that are effective and long-lasting. This paper examines scholarly literature to find proof concerning pandemic risk effects on SMEs, adaptive strategies developed to address these impacts, and the extent to which these strategies strengthen and lessen SMEs' fragility in India. Several key pandemic risk impacts for SMEs were found in the literature review. Several crucial coping mechanisms were also described in the study. Analysis of the study shows that there is little empirical support for most of the conceptual findings in resilience research among SMEs. Considering the complexity of SMEs, this study recommends further adaptation research aimed at creating fresh approaches to pandemic risk education.

Keywords SMEs · COVID-19 pandemic · Coping mechanisms · Risk factors · Technology · Online · Adaptive strategies · Sustainability

---

S. Saha (✉)
Faculty of Commerce and Management, St. Xavier's University, Kolkata, India
e-mail: sayani.saha@sxuk.edu.in

S. Sur
Xavier Business School, St. Xavier's University, Kolkata, India

S. Gangopadhyay et al. (eds.), *Pandemic Diaries*,
https://doi.org/10.1007/978-981-96-5415-4_19

19.1 Introduction

The global epidemic of COVID-19 has altered the lives of humans. The tactics that were used included lockdowns, seclusion from the outside world, and social disengagement. The quantity of internet-based media that people consume to keep themselves entertained, involved in society, and in touch with friends and family is greatly impacted by these measures [2]. Digital media was the best option for individuals to remain in touch. The pandemic has highlighted the basic necessity of bridging digital functionality within and between countries; digital platforms have offered answers to a variety of problems. Technology has proven to be a useful and crucial tool for helping people in times of need. In this specific context, technology plays a significant role in people's lives by providing them with access to electronic meeting platforms, social networks, electronic commerce, internet-based data, and online educational materials. Since they are electronically self-sufficient, individuals feel secure and independent while employing online resources for a range of tasks. The necessities were delivered right to the door via e-commerce platforms. Groceries, masks, sanitisers, and other necessities were among the commodities that were mostly delivered via electronic commerce platforms that were utilised for the purchasing and selling of goods and services. With the machine learning and artificial intelligence technologies usage, delivery windows for product delivery were established, and routes were mapped out to ensure prompt and secure product delivery. People were ready to explore more online purchasing choices for offerings, both non-essential and essential, from a safety viewpoint, considering the COVID-19 outbreak since there was very little physical contact required to uphold security and safety norms, the main issue was whether or not the standards of social distancing would be supported in their implementation. The majority of consumers including senior citizens now conduct their online transactions using digital wallets. Not many people object to transferring money using digital wallets due to the enhanced security of UPI payments. Electronic commerce is the most effective assistance framework available to residents in regions that have been placed under lockdown in order to combat the pandemic. Researchers believe that overcoming the many obstacles that e-commerce faces will allow it to reach its full potential both during and after the pandemic. Although using technology has many advantages, there are unquestionably disadvantages as well.

Online platforms are regarded to be the instruments for carrying out an enormous number of duties. Instead of being process-driven, it primarily makes possible a data-driven world. Internet-based platforms are widely used in media outlets, data platforms, marketplaces, and other areas. The rising popularity of shopping via the internet and the use of extra services that are either free or require a subscription on a number of internet-based sites have resulted in an enormous spike in the number of people utilising online resources over time. The COVID-19 pandemic on a global scale has spurred the explosive growth of e-commerce platforms. COVID-19 has affected consumer purchasing patterns across a variety of digital channels [5]. The chain of supply in its entirety, sales volume, and customer behaviour have all been significantly impacted by the pandemic. Online shoppers often feel vulnerable, which has caused a significant fluctuation in the volume of orders placed for a variety of products, including luxury, fashion, groceries, and healthcare. There have been several lockdown stages thus far, all in India, with different patterns in the online shopping behaviour of the customers. Advances in technology are responsible for the growing acceptance of digital platforms. Additionally, by using artificial intelligence and machine learning technologies, all organisations are trying to shift their operations to digital platforms. The ability of digital platforms to generate new models, methods, or processes for operation has sparked interest across many disciplines. Technology and digitalisation alone have been able to make up for the loss and have completely changed how people think about working while living at home. The digital age is responsible for the transformation of office meetings into video conferences and board room conversations, real cash into digital wallets, and offline shopping into online shopping on e-commerce.

19.1.1 *Objectives*

The prime objective of the study is to identify the effect of pandemic risk on SMEs, to analyse the coping mechanisms of SMEs post the COVID-19 pandemic, and to determine the impact of coping strategies on SMEs' increased susceptibility to pandemic risk variables.

19.2 Shift in Traditional Mode of Purchase to Online Format

According to Statista, five percent of Canadian and French consumers had made their first-ever online purchase since the coronavirus pandemic was initially announced in 2020. Just 2% of adult German and Japanese respondents said they had never made an internet transaction. Prior to COVID-19, the great majority of individuals worldwide were already familiar with making purchases online. Although the trend towards online shopping was already underway, the COVID-19 outbreak accelerated the change and led to developments in electronic shopping behaviours that will have long-term effects. Consumers started purchasing online which increased the number of product categories that can be purchased on the online platform.

According to Statista, an estimated 346 million Indians participate in digital transactions, such as e-commerce and online payments. One of United Payment Interface's primary benefits is its higher degree of safety. UPI runs on an extremely robust infrastructure and consistently provides end-to-end protection of data and privacy. The RBI's KYC mandate fortifies the platform even further. Thus, this financial revolution in India hinged on the success of UPI has led to increased trust on online payments among consumers. The emergence of financial technology firms and the integration of technology based on blockchain into payment methods have both expanded the market for electronic payments. The marketplace for electronic payments is expanding due to several factors. First, because digital payments are quick and convenient, consumers—especially younger ones—prefers them. Second, there is a growing demand for digital payment solutions that can reduce costs and streamline transactions as organisations become more digitally integrated. Thirdly, the need for secure and efficient payment solutions that facilitate global trade has increased with the rise of online marketplaces and e-commerce. Eventually, the COVID-19 pandemic has increased the usage of contactless financial transactions as businesses and consumers place a greater value on hygiene and safety. The chief of RBI remarked that UPI transactions currently make up 80% of all digital payments made in India, and that India accounts for 46% of global digital payments. There were 162 crore digital payments made in 2012–2013. Till February 2024, this amount increased to 14,726 crores in 2023–2024. Consequently, the shift to online activities predates the COVID-19 pandemic. The global online shopping boom, according to UNCTAD, 2020, demonstrates the importance of ensuring that everyone can take advantage of the opportunity provided by digitalisation as the world transitions from pandemic response to restoration. As a result, COVID-19 caused a shift in consumer behaviour that affected all product categories and also extended to other high involvement goods. The most recent generation of consumers can make decisions about expensive products in real time because they have access to more sophisticated information technology.

Sixty-two percent of Indian respondents to a June 2020 consumer sentiment survey expressed excitement about going shopping in the first three months following the lifting of the COVID-19 shutdown in India. More than 80% of participants said they would rather purchase shoes from physical retailers. In addition, sixty percent of the male respondents preferred to visit physical stores, whereas seventy percent of the female respondents chose to purchase electronics and consumer durables online.

19.2.1 Growth of E-Commerce and M-Commerce Across the Globe

E-commerce has become a vital facet of global shopping in recent years. The advent of the internet has brought about a substantial change in the buying and selling of commodities, similar to a number of other industries. The ongoing automation of modern life has made it possible for consumers across the world to benefit from online transactions. With more than five billion internet users worldwide, Statista reports that as internet availability and usage increase rapidly, more people are transacting online. Global e-commerce sales are expected to reach over $6.3 trillion in 2024, and these figures are expected to increase in future. Several websites allow visitors to explore, assess, and purchase goods and services. By 2023, online marketplaces will account for the majority of worldwide internet sales. Based on traffic figures,

Amazon is the world's most popular online shop. The Seattle-based e-commerce giant, which offers consumer goods, computer services, digital content, and e-retail, saw 6.1 billion direct hits on its website in December 2023. However, Amazon trails behind Chinese competitors Taobao and Tmall run by Alibaba Group with respect to gross merchandise value, ranking third. In 2023, projections show that the e-commerce markets in Asia and America will experience substantial development. Top of the list are Mexico and the Philippines, where internet sales are expected to rise by about 25% and 24%, respectively. With an 18% growth rate, Malaysia takes third place, not far behind. With e-retail sales predicted to increase by more than 15%, Argentina and Brazil were predicted to outperform other countries. Turkey will lead the world in retail e-commerce expansion from 2024 to 2029, with a CAGR of 11.6 percent, according to current industry estimations. The e-commerce market in Turkey is presently valued at 3.4 trillion Turkish Lira. The e-commerce markets in India and Brazil have the fastest growth rates in the world, with compound annual growth rates of 11%. The global retail e-commerce CAGR for the same period was predicted to be 9.5%. Globally, e-commerce makes up 20.1% of total retail sales.

Mobile supplements e-commerce traffic at the global level. The most expected trend is the increasing utilisation of mobile devices in e-commerce. At the start of 2024, smartphones accounted for around 80% of all retailing internet traffic worldwide, generating more online orders than desktops and tablets. As the use of mobile devices increases, especially in places without additional digital infrastructure, mobile usage will continue to influence future shopping experiences. Over seventy percent of all online sales conducted using mobile devices take place in countries like China and South Korea, making Asia a leader in M-commerce.

Even though online buyers' characteristics differ greatly between nations, there are general patterns that may be utilised to understand the core elements of digital consumer behaviour. Well-known online marketplaces serve as both a source of inspiration and the primary location to search for goods prior to making a purchase. Despite being less interactive than in-store purchasing, online buyers like the ease and versatility of e-commerce. Regarding delivery preferences, consumers are beginning to favour having their goods delivered to parcel stations as opposed to their houses, as the latter is typically a less environmentally friendly delivery option. Consumer electronics and fashion make up most online retail sales and are two of the biggest e-commerce categories in the business-to-consumer sector. Sales and revenue from internet purchasing have increased recently. By 2023, e-commerce has generated $6.151 trillion in global sales according to a recent study in 2024.

19.2.2 Rise of E-Commerce in India

India's internet retail business has expanded rapidly from its infancy in the mid-2000s to its current USD19.5 billion sector gross number of transactions prior to rejections and refunds. New consumers are adopting the internet at a higher rate owing to aggressive governmental initiatives and telecommunication service providers' efforts to lower the cost of cell phones and data packages. The way the consumers behave has evolved as well. While e-commerce was formerly a drastically discounted marketplace, many users now prefer the ease of ordering goods online from the comfort of their homes over going to real businesses. This idea is supported by the Technology Acceptance Model (TAM), where the behavioural intentions of a user based on the users' perceptions of the technology's usefulness and ease of use.

According to a latest report on Statista, an estimated 250 million Indians have purchase online in 2023. By 2027, in India 425 million shoppers are projected to be operational. A market with 250 million online consumers is expanding naturally because of the expanding workplace ecosystem and new customers who are already among India's internet user's category, most of whom are from tier II and lower cities and towns. The latter are joining the online buying trend because of several digital initiatives by e-retailers. However, the lack of regional language support, poor logistics and delivery infrastructure, and low internet penetration are the main reasons why these prospective clients are now untapped. Moreover, these clients live in remote areas and have no idea how to use technology, which makes logistics for just one purchase difficult.

According to a recent report of KPMG in 2024, post-pandemic, India's e-commerce retail business has grown significantly; with an estimated value of $57–60 billion in 2023 and is predicted to surpass $160 billion by 2028. India's e-commerce retail sector has expanded significantly due to rising internet penetration, smartphone usage, and language choice on online shopping sites. Furthermore, the general move towards digital payments has helped to increase the popularity of digital wallets. As a result, the retail logistics e-commerce sector has experienced rapid growth. Despite the rapid expansion, the industry is still facing some innate difficulties that it is working to resolve. The industry faces several challenges, including significant returns, severe cost pressures, and inadequate physical infrastructure.

Rising customer expectations, changes in e-commerce shopping, and an increase in the need for same-day deliveries are all influencing the methods in which the retail supply chain for e-commerce operates. Innovative business strategies for last mile delivery, such as delivery through local businesses, and multi-channel retailing are also becoming more prevalent in this industry. The logistics needs of the retail e-commerce industry are always changing due to shifting business requirements. The sector's growth is still anticipated to continue its current rising trajectory.

19.3 Methods

A qualitative descriptive technique was employed. The secondary data used in this study was obtained from international journal articles. Rather than conducting fieldwork to gather research data, we gathered data via a review of pertinent literature. The analysis method involved gathering and examining data sources in order to draw conclusions and findings that aligned with the research theme regarding the COVID-19 pandemic's impact on SMEs in India. This study examined the ways in which the COVID-19 pandemic affected Indian SMEs and offered potential remedies.

19.4 Results and Discussion

Before going over the total results in the section on discussion, each of the three research questions is addressed in turn in this section.

19.4.1 Impact of Pandemic Risk on SMEs

SMEs primarily operate on marginal profit for in-person customers because of their constrained finances and smaller distribution infrastructure. Due to lockdowns and other limitations, such as prohibitions on travelling, social estrangement, and isolation from others, residents lacked the ability to travel, which lead to weaker supply chain services and poor business ties taking over [4]. Financial resources were strained because of these shutdowns, which resulted in disappointing sales, cancellations of various events, hotel, and travel reservations, revenue declines, and payment contribution defaults for service industries, which caused logistical and operational problems [17]. The significance of the local environment, which includes things like shutdowns and restrictions on traffic, to the survival and resilience of SMEs is also highlighted in 36% of the studies that topped the list of pandemic hazards [10]. The primary objective of logistics is to coordinate the flow of commodities between production and consuming sites to meet the demands of customers and businesses. Pandemic risks exacerbate productivity problems, delay the reopening of shuttered enterprises, and worsen layoffs [11].

The government's short-term assistance programmes for SMEs within the framework of policy directives was another prominent risk factor for the SME pandemic. These assistance measures were initially hurried and instinctive, emphasising immediate goals over structural and long-range goals. Despite criticism, these policies with a short-term focus had a direct positive impact.

The United Nations Conference on Trade and Development (UNCTAD) claims that COVID-19 has had disastrous effects on people all over the world. SMEs were resource-constrained by nature, having not adopted technological innovations, and thus were unable to alter their operations at the pace at which the pandemic proliferated. On the other hand, the "new normal" would necessitate major adjustments from businesses, such as operational and business innovation and online transformation. For instance, SMEs in the manufacturing sector, although heavily utilising digital channels, primarily social media [13], did not typically adopt digital technologies of Industry 4.0 [6]. The main causes of this are a dearth of cutting-edge technologies, funding for investments, a lacklustre management team, unclear returns on investment, and, most importantly, a shortage of knowledgeable labour. The Industry 4.0 adoptions were motivated by anticipated benefits, potential markets, employee issues, consumer demands, competition, and an excellent image.

Consequently, a "just-in-time" strategy was used by the majority of SMEs in the production and manufacturing sector in place of innovative aggressive risk management and risk reduction. For instance, the agri-food industry lost some of its resilience to the pandemic's realities. Due to this, SMEs were unable to take advantage of any digital technology advantages, which created demand and logistical issues [18].

19.4.2 SMEs Coping Mechanisms Post the COVID-19 Pandemic

The companies' main concern is the ability to create plans to deal with and react to the changing circumstances brought about by the pandemic. Resilience and coping strategies are frequently linked to human psychology. On the one hand, coping has been defined as the behavioural and cognitive strategies used by humans to deal with and manage traumatic events [9]. Conversely, resilience refers to a person's capacity to bounce back from difficult or traumatic circumstances as well as misfortune. In the broadest sense, the meanings of these two concepts are not mutually exclusive [15]. Considering their close relationship, they have therefore been used synonymously to denote the same thing [20]. For instance, when discussing "coping or that resilience as the ability of a firm to be alert to, adapt to and quickly respond to the changes brought about by supply chain disruption," Ambulkar et al. used both concepts concurrently. Businesses must adopt what is known as organisational behaviour—which involves considering a timely, deliberate change of capacity—in order to deal with the swift declines in the economy, health, and social repercussions that follow pandemics. Important elements included the adoption and adaptation of digital and technological tools, like robust mobile applications. It implies that small and medium-sized enterprises can maintain their competitive edge and adaptability in the rapidly evolving business landscape, even with limited expertise and technological resources. The path to SMEs adopting digital technology starts with digital awareness, inquiry, cooperation, and transformation [14]. For instance, Germany's "Digital in NRW" programme provides a range of services, from educating the public by showcasing remedies in smart factories to debating and creating digital solutions that make up the roadmap for digital transformation and strategies of SMEs.

On the other hand, not many studies address "strategy-as-practice" while taking a wider variety of pandemic risks into account. Only the dynamic capability is documented among the reviewed studies. Evidence was gathered for this review to demonstrate that, in addition to dealing with the effects of risks associated with pandemics, a thorough, in-depth analysis of the manner in which strategy is developed, planned, and executed—making it a crucial business function—should also receive attention. A practice perspective also helps SMEs comprehend how their capabilities grow, alter, and progress because it is a pragmatic approach, which makes strategy-as-practice a crucial component of their global epidemic approach.

The remaining coping mechanisms include financial assistance from the government, novel business models and adjustments to business operations, enhanced internal as well as operational competency, restructuring business strategies to firm strategies, supply chain and business agility, and crucial transformation of the business [1]. The ability of businesses to endure in a highly unstable surrounding is a prerequisite for an operational business transformation.

19.4.3 Effect of Coping Strategies on SMEs' Increased Susceptibility to Pandemic Risk Variables

When it comes to businesses, resilience, however, is a shared outcome of SMEs' vulnerabilities, coping mechanisms, and capabilities in the face of unfavourable business environments. Even better, companies or organisations are able to withstand external shocks and carry on with business as usual [16]. It is possible to become resilient through cooperation, transparency, success, ingenuity, and longevity. When a company is more open, it may be able to enhance its active competencies and withstand economic downturns like the one caused by COVID-19. It included using digital technologies like social media to maintain a positive business relationship through end-to-end transparency in all-channel communication with key stakeholders. A high degree of agility enabled by the digital revolution is also necessary for this coping strategy in order to actively involve stakeholders in order to meet unpredictable and shifting business environments. Even though the road appeared difficult for SMEs, gaining a competitive edge requires meeting current needs and laying the groundwork for future innovation.

Coping mechanisms used by SMEs are typically independent and reactive. Even though the latter may warrant a specific intervention, they tend to concentrate more on short-term pandemic difficulties rather than enduring resilience, focusing instead on behavioural changes and internal business structures. Many of the strategies under review demonstrate strong coping abilities and stress the significance of sharing and spreading risk, particularly through online platforms and diversification of revenue, at least in the landscape of pandemic risk [11]. Nonetheless, there is proof that coping mechanisms, such as using savings or selling assets to rebuild, can lessen SME resilience and regain operations in the aftermath of the pandemic.

Making an investment in innovation entails transferring resources and skills from the individual to the institutional level. Ali and Suleiman also discovered that the norms of risk management, flexibility, and engineering in supply chain are prerequisites for true resilience [21]. This review makes it clear that resilience-promoting factors are not evenly distributed. The ex-ante literature provided a pitifully poor definition and explanation of the concept of "resilience." In order to effectively navigate this cycle, SMEs need to make the most of their easily accessible resources in order to react swiftly to shifts in crisis situations and to turn these resources into goods and services that can eliminate customer issues. SMEs can also mobilise new and additional resources and knowledge from their network to help implement significant changes to their business models after eliminating customer issues [3].

The company has a competitive advantage because of its ability to innovate, adapt to change, and bring about change those benefits both customers and rivals. Operations that enable the deliberate and expert alteration of a firm's strategic value yield flexible capabilities. They are important forces behind a company's ability to create value, maintain a competitive edge, and perform better than average in dynamic environments [7]. Adaptive strategies require internal insight to detect and react to changes in external business conditions. As a result, the strategy must be adaptive and flexible by nature. Systems' inherent ability to adapt to changes within predetermined boundaries is commonly associated with their flexibility. SMEs' pandemic coping strategies include, but are not limited to, employee training, strategy-as-practice "through collective sensing" during a downturn in the economy, and healthy operations. The term "strategy-as practice" refers to the routine activities that employees and organisations take that lead to strategic outcomes. The practice of strategy flourishes in business and organisational settings that value knowledge, trust, and creativity. This leads to a dynamic interaction between stability, adaptation, and sustainability, which is influenced by the contexts of the individual, the organisation, and the market [8].

19.5 Conclusion

SMEs remain the cornerstone of any economy and are essential to job generation and economic growth. However, out of all the enterprise-size classifications, SMEs remain the most vulnerable as evidenced by the pandemic. Like other institutions, SMEs were taken by surprise, creating several vulnerabilities.

The study's conclusions supported three main themes: the pandemic risk factors that SMEs must consider, the coping mechanisms they developed to mitigate those risks, and the coping interventions that strengthen their resistance to those risks. The most remarkable pandemic risk faced by SMEs was regarding logistics, which included organising the commercial activity of sending goods to customers. The most popular coping strategies were digital transformation and technology adoption, indicating how important it is for SMEs to start considering the unavoidable role that digital technologies will play in the survival and continuity of modern businesses. This is notably important, as the pandemic has demonstrated, if SMEs are to withstand any future shocks to the world economy. SMEs in particular need to digitally transform their operations in order to take advantage of technology opportunities. The adoption of digital technologies helps to improve employee satisfaction, lower costs by introducing automation, and optimise internal business processes. Digitalisation, according to Winarsih and Fuad in 2021, will make business processes more efficient and help entrepreneurs make strategic decisions. The pandemic demonstrated the significance of SMEs' digital transformation and acted as a catalyst. SMEs can reevaluate their decision-making processes and use technology in valuable and sustainable ways because of digital transformation. Thus, digitalisation is crucial for creating an efficient competitive strategy. The changing nature of the new digital economy, the enormous influence of technology, and the steadily growing impact of Software as a Service on businesses are the main drivers of the transformation in the SME sector, according to the "Digitalisation of Indian SMEs survey," which was carried out by Capterra India. Globally, the COVID-19 pandemic has changed consumer attitudes and behaviour. The global economy has been transformed by technologies like blockchain, IoT, 5G, artificial intelligence, cloud-based applications, and data analytics. As a result of these developments, small- and medium-sized enterprises (SMEs) have come to understand the necessity of digitising their business models. 40% began the process of digitalisation within the last three to five years: A fraction of the respondents indicated that their companies began digitalising processes after the onset of COVID-19, but 88% of firms began their efforts within the last five years. 88% of respondents claim to have implemented a digitalisation strategy at their company: The majority of decision-makers in an organisation actively participate in choosing how to digitalise processes and place a high value on digitalisation inside the organisation. Next, businesses intend to digitise marketing and accounting procedures: In addition to digitalising the processes, 43% of respondents remarked they will concentrate on accounting, and 42% believed they will concentrate on marketing.

This was witnessed during the post-COVID-19 era [19].

Furthermore, these factors—which are grouped under the two themes and are as wide-ranging and comprehensive as they are—will function as the foundation for an investigative path that deepens our understanding of SMEs during pandemics. One important lesson to be learnt from the study's findings is that, following pandemics and/or disasters, SMEs, legislators, and business think tanks should start having dialogues about coping mechanisms and business continuity. SMEs are becoming more interested in the broader adoption and greater embrace of digital technology due to COVID-19's operational effects. When assessing the extent of expressed enthusiasm for different technological developments between 2019 and 2021, it became evident that most SMEs found an enormous rise in the appeal of the different digital methods. The shift to remote work may help to partially explain this. These technologies are essential for facilitating data collection, monitoring, and cooperation between workers while working remotely from the factory or workplace [12]. Table 19.1 exhibits the comparative analysis of SME inclination in the adoption of digital technology before and after the pandemic.

Since SMEs are essential to a larger economic system, more will be needed to restore pre-pandemic regularity than just remedial measures. Flexibility and adaptability are key components of SMEs' crisis management and adjustment capabilities. The study's conclusions indicate that SMEs' capacity to use network resources and expertise is essential to enhancing their crisis-resilience. In a demanding and unpredictable environment, SMEs can innovate and expand by creating opportunities. The

Table 19.1 Comparison of SME interest in digital adoption pre- and post-COVID-19 pandemic

Types of digital technologies	Pre-COVID-19 (2019) (%)	Post-COVID-19 (2021) (%)
IoT	4	19
Cloud Technologies	7	16
Big Data	4	16
Artificial Intelligence	4	12
ERP, MES	10	11
Additive Manufacturing	7	11
Digital Twins	4	10
Wearables	7	9
Virtual and Augmented Reality	4	8

Source World Economic Forum Report, December 2021

final thoughts of our analysis point out that coexisting strategies like survival, adaptability, and inventiveness can help SMEs become more resilient and less dependent on external assistance.

19.5.1 Implication of the Study

A meta-analysis of the results across reviews was not possible due to the variability of the systematic analyses and the small percentage of studies reporting meta-analysis results. Further research will be required to establish a causal relationship between pandemic risks and resilience of SME, as well as to investigate other variables that may interact with the relationship outlined above. Similar to this, we argue that as more data on COVID-19's effects develop, a rise in long-term qualitative research will spur the emergence of new trends that will advance our knowledge of the pandemic's actual effects and enable focused development of policy interventions in the years to come.

Moreover, even though SMEs have been the subject of much research, in order to advance the current debate beyond its conceptual stage, a more reliable and broadly applicable model of pro-SME pandemic resilience needs to be explored. Consequently, even in cases where additional conceptual research is still required, particularly for understudied research contexts, the study suggests conducting cross-sectional and extensive studies. The ongoing discussion about SMEs and pandemics could be influenced by comparing the resilience stance of SMEs in various regions, analysing the effect of the pandemic on SMEs in various locations, or comparing coping mechanisms among SMEs in other areas. The new findings from these thorough empirical studies will spark fresh arguments and conversations about the subject, which may direct further research in this direction.

SMEs faced the risk of going insolvent during the pandemic as a result of growing labour and manufacturing costs as well as challenges in obtaining clients. During pandemics, there are various ways in which the government can assist SMEs in order to reduce their risk of bankruptcy. For example, a government could provide early warning systems to support SMEs through its entrepreneurial ecosystem and strengthen institutional capacity. Nonetheless, there are not many studies that look at how differently sector-specific SMEs respond to bankruptcy hazards. Which SMEs are most likely to go bankrupt during and after a pandemic could be the subject of intra-sectoral research. Changes in governance or style of management, along with qualitative data about the traits of managers, could also be considered.

This review reveals that the field is still centred around theoretical approaches that are typically related with the assessment of SMEs, impacts of COVID-19, and coping strategies, like institutional theory and expectancy theory. However, in order to assess how top managers' emotional responses shape an SME's strategy to cope, ideas and social science theories from disciplines like social psychology must be considered. Furthermore, the literature demonstrating that small businesses can react to an emergency in different ways suggests that cognitive appraisal theory offers a novel means of evaluating the cognitive-emotional interpretations of events and their influence on coping strategies. Therefore, applying multidisciplinary viewpoints that transcend the boundaries of one school of thought will be necessary for advancing the field.

References

1. Abhari et al. (2022). The impact of Covid-19 pandemic on small businesses in tourism and hospitality industry in Malaysia. *Journal of Research in Marketing and Entrepreneurship*, 75–91.
2. Bartik, A. W., et al. (2020). How are small businesses adjusting to COVID-19? Early evidence from a survey (No. w26989). *National Bureau of Economic Research.*
3. Bivona, E., & Cruz, M. (2021). Can business model innovation help SMEs in the food and beverage industry to respond to crises? Findings from a Swiss brewery during COVID-19. *British Food Journal*, 3638–3660.
4. Caballero-Morales, S. O. (2021). Innovation as recovery strategy for SMEs in emerging economies during the COVID-19 pandemic. *Research in International Business and Finance.*
5. Cowling et al. (2020). <? covid19?> Did you save some cash for a rainy COVID-19 day? The crisis and SMEs. *International Small Business Journal*, 593–604.
6. Dilyard, J., et al. (2021). Thunderbird International Business Review. *Digital innovation and Industry 4.0 for global value chain resilience: Lessons learned and ways forward*, 577–584.
7. Dyduch et al. (2021). Dynamic capabilities, value creation and value capture: Evidence from SMEs under Covid-19 lockdown in Poland. *Plos One*, e0252423.
8. Erdiaw-Kwasie et al. (2002). Does circular economy knowledge matter in sustainable service provision? A moderation analysis. *Journal of Cleaner Production*, 135429.
9. Folkman, S., & Moskowitz, J. T. (2004). Coping: Pitfalls and promise. *Annual Review of Psychology*, 745–774.
10. Lu et al. (2021). Perceived impact of the Covid-19 crisis on SMEs in different industry sectors: Evidence from Sichuan, China. *International Journal of Disaster Risk Reduction*, 102085.

11. Manyati et al. (2021). Leveraging green skills in response to the COVID-19 crisis: A case study of small and medium enterprises in Harare, Zimbabwe. *Journal of Entrepreneurship in Emerging Economies*, 673–697.
12. Merritt, J. (2021). COVID-19 and technology adoption in small and medium-sized enterprises: The impact and the way forward. *World Economic Forum,* 19–20.
13. Moeuf et al. (2018). The industrial management of SMEs in the era of Industry 4.0. *International Journal of Production Research*, 1118–1136.
14. Ratten, V. (2021). Coronavirus (Covid-19) and entrepreneurship: Cultural, lifestyle and societal changes. *Journal of Entrepreneurship in Emerging Economies*, 747–761.
15. Steinhardt, M., & Dolbier, C. (2008). Evaluation of a resilience intervention to enhance coping strategies and protective factors and decrease symptomatology. *Journal of American College Health*, 445–453.
16. Stich et al. (2020). Measures for a successful digital transformation of SMEs. *Procedia CIRP*, 286–291.
17. Sun et al. (2022). The impact of COVID-19 on SMEs in China: Textual analysis and empirical evidence. *Finance Research Letters*, 102211.
18. Taiminen, H. M., & Karjaluoto, H. (2015). The usage of digital marketing channels in SMEs. *Journal of Small Business and Enterprise Development*, 633–651.
19. *The Economic Times*. (2023, August 2). https://economictimes.indiatimes.com/small-biz/sme-sector/new-survey-reveals-how-digitalisation-has-paved-way-for-a-new-era-of-growth-for-smbs-in-india/articleshow/102341070.cms?from=mdr
20. Zamirinejad et al. (2014). Effectiveness of resilience training versus cognitive therapy on reduction of depression in female Iranian college students. *Issues in Mental Health Nursing*, 480–488.
21. Zutshi et al. (2021). From challenges to creativity: enhancing SMEs' resilience in the context of COVID-19. *Sustainability*, 6542.

Sayani Saha is an Assistant Professor of Management, St. Xavier's University, Kolkata. She has about 10 years of teaching experience in Marketing Management, Advertising, Brand Management and Consumer Behaviour. She was the Head of the Department for bachelor's in management studies at St. Xavier's University, Kolkata. She holds a master's in business administration from MAKAUT. Ms. Saha has published several book chapters and scholarly articles in Scopus-indexed and peer-reviewed journals. Her innovative teaching methods and commitment to student success has earned her great admiration. She also has an impressive corporate experience of 3 years in the field of banking.

Soma Sur Professor and Former Dean of Xavier Business School of St. Xavier's University, Kolkata and is presently the Hony. Director of Father Lafont Centre for Excellence in Research and Innovation at the University. She has published widely in national and international journals and book chapters in countries like USA, Malaysia, Philippines, Sri Lanka, Indonesia, and India. She has also presented papers in various national and international conferences. She has been widely appreciated in international forums for her lecture deliberations and invited talks in Loyola Marymount University, Los Angles, California, USA, Jiatong University, Bejing, China, and Keio University, Japan.

Chapter 20
COVID-19 and Tuberculosis Notifications in India: The Dilemma in Public Health Management

Paramita Barman Sen and Arijita Dutta

Abstract While the recent COVID-19 pandemic struck a severe blow to the Indian healthcare system and economy at large within a short span of time claiming lives and livelihood, tuberculosis (TB)—an age-old infection, continues to plague the country, despite a dedicated public TB program functioning across the nation for more than two decades now. Diversion of resources from TB to COVID-19 management threatened to negate any recent progress in reducing the burden of TB in India as well as globally. Specifically, challenges in access to healthcare during the pandemic led to a spiralling in both underdiagnosis and underreporting of TB cases. The current research attempts to explore the status of TB notifications in India in the wake of the recent pandemic. The study uses data relating to notified TB cases from the NIKSHAY portal of the Government of India and employs simple descriptive statistics and tools for graphical representation. The study reveals that overall TB notification rates in India dipped significantly during the first and second waves of the COVID-19 pandemic at all-India level. Also, the share of public notification in total TB notification in the country suffered a setback during the pandemic. Select states recorded public notification shares lower than both all-India level shares during COVID-19 and respective state level shares pre-COVID-19. The government should implement stricter surveillance and ensure greater inclusiveness of the national TB program particularly in the post-pandemic situation to prevent the existing burden of TB from taking a heavier toll on the economy.

Keywords COVID-19 · Pandemic · Tuberculosis · Healthcare service · National TB program · TB notification · NIKSHAY portal · Underreporting · Departures of TB notifications

P. B. Sen (✉)
Faculty of Commerce and Management, St. Xavier's University, Kolkata, India
e-mail: pb.sen@sxuk.edu.in

A. Dutta
Department of Economics, University of Calcutta, Kolkata, India

S. Gangopadhyay et al. (eds.), *Pandemic Diaries*,
https://doi.org/10.1007/978-981-96-5415-4_20

20.1 Introduction

The emergence of the COVID-19 pandemic hurled a number of challenges at the Indian economy, healthcare delivery being one of the most prominent among them. As the pandemic caught the most advanced of health systems across the world unawares, it was natural that the Indian healthcare system was also shaken to its core. Even amidst the recent COVID-19 pandemic, higher number of deaths per quarter from tuberculosis (TB) than from COVID-19 were recorded in India [9] despite TB being an ancient disease. The national TB control program in India (currently known as the National TB Elimination Programme or NTEP) has been providing free of cost diagnostic and treatment services to patients for over two decades now. However, India remains the country with the highest TB burden in the world. Further, with underdiagnosis and underreporting of TB, India accounts for the highest share in the global gap between the annually estimated and notified[1] TB cases [26] that constitute the "***missing millions***".[2] With the advent of the pandemic, diversion of economic and political resources from TB towards COVID-19 further threatened to reverse the fruits of any recent progress in arresting TB burden worldwide and in India. Lost earnings during the phasic lockdowns following the pandemic badly hit the marginalized communities, fuelling poverty and malnutrition, both of which happen to be major risk factors for TB [12]. In India, lockdowns resulted in a drastic fall in TB notification rates following shut-down of out-patient departments in hospitals and scarcity of medical personnel due to lack of conveyance and general apprehension about COVID-19 infection. As India continued to fight the pandemic, lockdowns crippled TB services and limited access to healthcare and anti-TB drugs as well as disrupted treatment services [13]. It has been predicted that short-term strict responses like lockdowns to arrest COVID-19 would have a trailing impact on TB particularly in high burden economies like India [24]. With the increasing deployment of human resources to COVID-19 response on an emergency basis, active case finding for TB was deferred by some Indian states. This reversed the pace in active screening of TB among the vulnerable populations [5]. The national TB program was distressed and overburdened with increased attention to COVID-19 care, overutilization of laboratories meant for TB work, restricted availability of TB care workers and use of drug resistant TB centres and CB-NAAT machines for COVID-19-related work. Shift of focus from TB, an already existing malady, to COVID-19 without adequate backup to contain the former, caused progress in TB control to take backseat [2]. Some states like West Bengal, however, foresaw the need to adopt strategies like home delivery of anti-TB drugs for patients to ensure delivery of uninterrupted TB services under the

[1] Reporting diagnosis and/or treatment of Tuberculosis cases to the nodal public health authority or officials by every healthcare provider, public (establishments run by the Government) or private (NGO sectors and/or individual practitioners).

[2] "*Missing millions*" refers to the gap between the estimated incident and notified cases of TB in a year. These are suspected cases of TB that are not diagnosed or diagnosed but not reported (notified) to the national TB program by the private sector [16].

apprehension that such services would be hugely affected following the lockdowns [11].

Common symptoms like fever, cough and breathlessness are applicable to both COVID-19 and pulmonary TB. Both diseases mainly affect the lungs with close contact being the main mode of transmission. However, the incubation period from exposure to disease is typically longer in case of TB compared to COVID-19, often with slower onset [27]. Currently available body of literature on COVID-19 infection among TB patients is limited, however it is feared that TB-COVID-19 coinfections will result in adverse treatment outcomes. People reported having experienced increased stigma owing to overlapping symptoms of both diseases. Results from a global community-based survey revealed that 56% of people with TB in Kenya in a sample of 159 experienced feelings of shame. As an advocate from Kenya puts it, "*From conversations with community health workers, there is a lot of stigma around COVID. 'You cannot cough in peace,' one said, as any cough is mistaken as COVID, and when you become a suspected case, you are forcefully quarantined. There is a need for more education on human rights around COVID*". 50% faced trouble accessing TB care owing to restricted access to public transportation. 62% of TB patients from India in a sample of 58 reported hesitation in visiting a regular health facility out of fear of contracting COVID-19. While most TB patients were given additional medicines for ensuring uninterrupted home-based treatment, urgent need was expressed for non-medical support including economic, nutritional and psychosocial support [23].

In India, significant restrictions were imposed on "face to face" assessments and mobility of the population as part of COVID-19 infection control strategies and the national lockdown. It has been estimated that between 2020 and 2025 there could be 6.3 million TB cases with 1.4 million deaths additionally from TB implying a heavy setback of at least 5 years in the fight to counter TB, owing to the pandemic. In the said period, India would be expected to witness 40,685 additional deaths for every month taken to return to normal services related to TB care [25]. Thus, the ongoing pandemic imposed challenges in accessing healthcare service with respect to other morbidities like TB that present similar emergencies in terms of their contagiousness and therefore require early diagnosis and treatment commencement in order to break the chain of transmission of infection. Unilateral focus on battling COVID-19 bears clear indication that the pandemic period potentially led to a spike in underdiagnosed and underreported cases of TB which already present major hindrances in TB control. While the impact of COVID-19 on TB notifications at the national level has been researched by few [1, 17, 19], none has explored the TB notification status by state and healthcare sector in India.

Notification of TB may be defined as the process of reporting TB cases diagnosed, to relevant authorities in the health system, which would in turn report the same to the World Health Organization (WHO) through their respective national TB control programs or equivalent programs dedicated to TB management. TB cases that are screened in public health facilities are usually notified through routine systems. The concern, however, lies with the large chunk of cases that are identified and eventually treated outside the domain of the public sector (specifically the national TB program) in the private sector and fail to get notified in several settings. The problem automatically assumes greater significance in high TB burden settings like India where the dominance of the private sector is also formidable.

20.1.1 Objectives

The current study draws motivation and hopes to build on the gap in existing literature. Given this context, the objectives of the current study are to *compare monthly TB notifications at all-India level during the pandemic years of 2020 and 2021 with monthly TB notifications in the normal (reference) year of 2019; to explore the sectoral shares (private vs public) in total TB notification in India during the pandemic years; to compare select Indian states in terms of their public TB notification status during the pandemic years, with their own pre-COVID performance and all-India level performance during COVID.*

20.2 Review of Literature

Despite a public TB program being operational in full swing in the country for years now, the TB situation continues to be appalling with India in the lead of thirty countries recording the highest TB incidences across the globe. In fact, India contributes to more than a quarter of annual global TB incidence and almost a third of annual global deaths from TB [26]. A recent study has shown that prevalence of pulmonary TB which is contagious in nature is 295.9 (95% confidence interval; 201.1–390.6) per 1,00,000 population in India. Males had higher TB prevalence than females. Also, the rural population accounted for higher prevalence of the disease than urban population in India. Whereas most of the developed countries regard TB as a disease of the past owing to decline in TB prevalence following socio-economic development and implementation of effective strategies in TB management and control, many middle and low-income countries are yet to achieve any target relating to eradication of the disease [20]. A community survey in 2011 revealed that almost half of the TB patients in India sought treatment outside the government healthcare system, i.e. the private sector, and therefore could not be covered by the national notification system. Although efforts have been extended by the TB program to reach out to the non-government health facilities with the help of innovative strategies including public private mix strategies, such efforts turn out to be inadequate given the country's healthcare structure [21]. May 2012 onwards, notification of TB by private healthcare providers was made mandatory by the Government of India. The national TB program designed a case and web-based mechanism for online reporting of TB cases, called NIKSHAY.

20.2.1 Status of TB Notifications in India

Several studies reveal that TB notification rates in India are typically low, largely owing to an unregulated and heterogeneous private sector catering to a substantial chunk of patients with symptoms of TB, in India. A study from South India finds that the proportion of TB patients notified by a private hospital was substantially low (23.2%). Sputum smear positive cases were significantly more likely to be notified, while notifications of children were less likely. Qualitative analysis reveals that a crucial barrier to TB reporting related to misconceptions regarding notification and its procedure among healthcare workers [22]. Another study finds that most (75%) pharmacies were dispensing anti-TB medicines even without valid prescriptions despite half of the sales-persons expressing their awareness about the TB notification process. The study suggests improvement in notification from such pharmacies through systematic sensitization of sales-persons and addressing confidentiality-related concerns among patients that in fact encourage over the counter purchase of medicines without seeking institutional medical intervention [18]. In a cross-sectional study of private practitioners from the city of Mysore in Karnataka, India, 91% of study participants interviewed were aware that TB is notifiable. However, 82% were unaware of the notification process and the NIKSHAY portal of the government. Only one in six practitioners was NIKSHAY registered whereas one in three practitioners who notified, was without a NIKSHAY registration. The practitioners expected some incentives from the public programme in the line of free drugs, training, etc., to notify in NIKSHAY. 47% were of the opinion that mandate regarding notification should be supported by legal punitive steps [3].

20.2.2 Status of TB Notifications Globally During the Pandemic

Quite a few studies are testimony to the fact that the COVID-19 pandemic resulted in at least a standstill situation if not a worsened one with respect to progress in terms of other fatal morbidities like TB. The shortfall in TB notifications during the pandemic years of 2020 and 2021 were more glaring in countries with relatively high burdens of TB. COVID-19 adversely affected TB diagnosis and TB care in the Western Pacific Region [15]. Until COVID-19, TB was the leading cause of mortality from a single infectious agent. COVID-19 impaired access to essential services related to TB care. In sharp contrast to the global rise in TB detection rates from 2017 to 2019, TB notifications plummeted substantially in 2020 compared with 2019 by 18% globally with only partial recovery in 2021. The epidemiology of TB worsened in the pandemic times. The study suggests that focused attempts at TB management complemented by proportionate funds are an urgent requirement to negate the adverse impacts of COVID-19 on TB globally [4]. In the latter year of the pandemic, COVID-19 epicentres shifted to middle and low-income countries of the globe. South Asia and South America were particularly adversely affected in 2021. By the month of May, Latin America and the Caribbean crossed a million deaths from COVID-19, the chunk occurring in Brazil. India's daily count of emerging COVID-19 infections soared to an all-time high with 414,188 cases on May 6, 2021, with her neighbours like Nepal, Sri Lanka, Myanmar, Bangladesh and Indonesia experiencing similar hikes [30]. With TB as an already thriving health crisis in India, COVID-19 clearly was another blow.

20.3 Methods

The study uses monthly data from the NIKSHAY portal of the Government of India, dedicated to TB-related parameters like notification, treatment and cure rates under the national TB program. NI-KSHAY ("NI" meaning End and "KSHAY" meaning TB) constitutes a system for patient management for TB control in the National Tuberculosis Elimination Program. It has been developed and maintained by the Central TB Division of the Ministry of Health and Family Welfare, Government of India, in collaboration with the National Informatics Centre and the Country office of the World Health Organization, for India. It is utilized by health functionaries in the private and public sectors across the country for registering of cases in their domain; for ordering different types of laboratory based tests across India; for recording details related to treatment of the patient; for monitoring of compliance with treatment; and for transferring of cases between and among healthcare providers. It also operates as the National TB Surveillance System and facilitates reporting of surveillance data to the Government of India.

Data on TB notification at all-India level spanning from January 2019 to December 2021 has been used for analysis in the present study. The non-COVID-19 year 2019 has been taken as a normal year for reference. The first quarter of 2020 represents the time when the COVID-19 pandemic hit India and continued for some months as the first wave. Mid of the year 2021 marks the second wave of the pandemic. The exploratory study employs simple descriptive statistics and tools for graphical representation.

20.4 Results and Discussion

Figure 20.1 compares the month wise notification of TB cases at all-India level in 2020 which witnessed COVID-19 outbreak in the first quarter, with 2019, which was a regular year. It reveals that notifications started falling after February 2020. Meanwhile, a national lockdown of 21 days was declared to be effective from March 25, 2020, to April 14, 2020, during which phase, TB notifications were at major lows. The lockdown was then extended in three phases to continue till May 31, 2020. Notifications revived to some extent in the following months while remaining

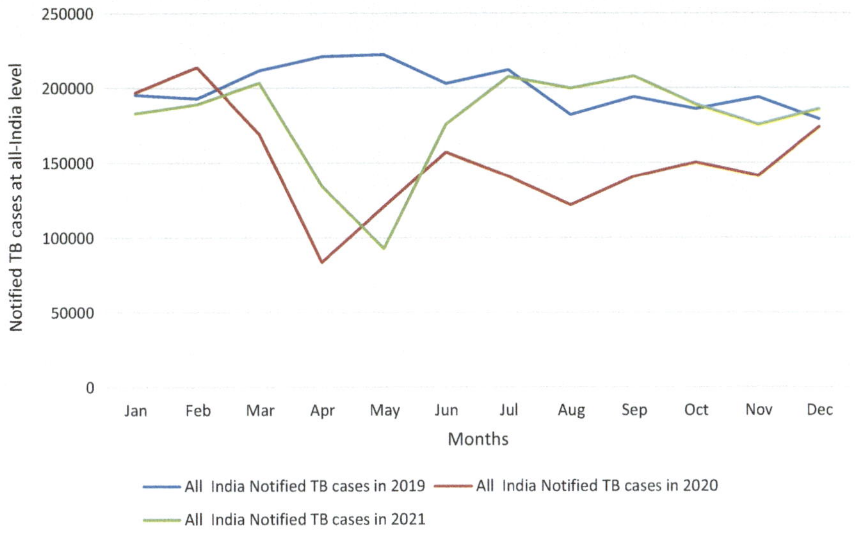

Fig. 20.1 Month wise TB notification at all-India level between 2019 and 2021. *Source* Data from NIKSHAY Portal of NTEP, Government of India

consistently lower than their 2019 counterparts. In 2021, notifications started to decline steeply from April when the second wave of COVID-19 hit the country and took a very sharp dip in May with signs of recovery from around July.

Table 20.1 shows percentage departures of TB notification figures during the COVID-19 years of 2020 and 2021, from the non-COVID-19 reference year 2019. The months of January and February in fact show increases in notification in 2020 and declines in notification in 2021, compared to 2019. March recorded a substantially steeper decline in notifications in 2020 than 2021, compared to 2019. The same trend holds in April with a reversal in May. The months of June and July went back to witnessing steeper shortfalls in TB notification in the first wave of the pandemic than the second, compared to the pre-pandemic year. In the months of August, September and October, TB notifications fell by around 20% or more in 2020 compared to 2019 while improving over 2019 figures in 2021. In November, percentage fall in TB notifications was three times more in 2020 than 2021, in comparison with 2019. The last month of the year witnessed a marginal dip in notifications in 2020 compared to 2019 while showing an improvement over 2019 figures in 2021.

Notifications had the sharpest dip in April 2020 with 62% lower figures compared to April 2019 followed by 46% lower notifications in May 2020 compared to May 2019. The ***mean percentage departure in the first year of COVID, i.e. 2020 is 23.5% (lower) than 2019*** with standard deviation of 19.21. ***In 2021, during the second wave of COVID, the mean departure of TB notifications from 2019 turns out to be 9.3% (lower)***, with 19.08 as standard deviation. ***Notifications declined by 39% in April 2021 compared with April 2019 and further by 58% in May 2021, compared with May 2019.***

Table 20.1 Percentage departures of monthly TB notifications in India during the COVID-19 years 2020 and 2021 from monthly TB notifications of the reference (normal) year 2019

Months	Percentage departure in 2020 compared to 2019	Percentage departure in 2021 compared to 2019
January	0.8	(−) 6.3
February	10.8	(−) 2.0
March	(−) 20.1	(−) 3.9
April	**(−) 62.1**	**(−) 39.0**
May	**(−) 45.7**	**(−) 58.3**
June	(−) 22.5	(−) 13.3
July	(−) 33.6	(−) 2.1
August	(−) 33.1	9.7
September	(−) 27.4	7.1
October	(−) 19.2	1.7
November	(−) 27.1	(−) 9.4
December	(−) 2.8	3.9

Source Authors' own calculations from data from NIKSHAY portal of NTEP, Government of India

Figure 20.2 shows that ***during the pandemic years of 2020 and 2021, share of TB notifications from the public healthcare sector in total TB notifications at all-India level declined in comparison to 2019.*** From 72% in 2019, the share in public notification fell to 69% and 68% in 2020 and 2021, respectively. Although the declines are marginal, the shortfalls in public notification are potentially of some significance given their coincidence with the pandemic.

An analysis of state wise performance reveals that in 2020, Jharkhand, Uttar Pradesh, Haryana, Telengana, Gujarat, Maharashtra and Bihar had lower than all-India share of publicly notified TB. In 2021, the poor performing states with respect to this share were Jharkhand, Madhya Pradesh, Haryana, Delhi, Gujarat, Maharashtra and Bihar. Thus, ***Haryana, Gujarat, Maharashtra, Jharkhand and Bihar consistently recorded lower than all-India share of publicly notified TB through both pandemic years***. Table 20.2 summarizes and compares the performance of these select states (in terms of their shares of publicly notified TB) in the COVID-19 period with national level figures during the pandemic and with their respective performances prior to the pandemic.

With respect to percentage departure in share of publicly notified TB from the corresponding all-India share during COVID-19, it can be seen that Haryana falls short by 4.35% in 2020 and 2.94% in 2021; Maharashtra lags by 13.04% in 2020 and 19.12% in 2021; Gujarat records a shortfall by 7.25% in 2020 and 5.88% in 2021; Jharkhand falls short by a much lower 2.90% in 2020 and 1.47% in 2021; on the contrary, Bihar slides down by as high as 24.64% in 2020 and by an even higher 30.88% in 2021.

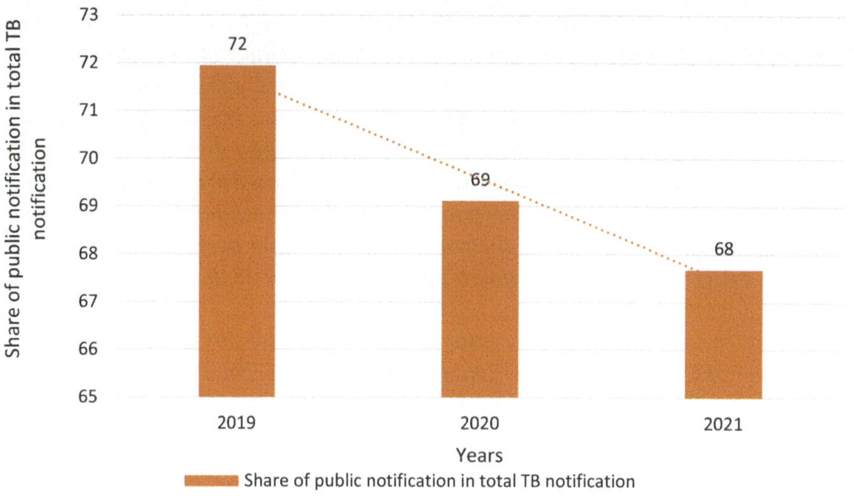

Fig. 20.2 Percentage shares of public notification in total TB notification in India in the pandemic years, in comparison with the reference (normal) year 2019. *Source* Authors' own calculations from data from NIKSHAY portal of NTEP, Government of India.

Table 20.2 Performance comparison in terms of public TB notification status of select Indian states during the pandemic, with respect to their pre-pandemic performance and all-India level performance during the pandemic

States	Percentage departure in share of publicly notified TB from all-India share during COVID-19		Percentage departure in share of publicly notified TB from respective state shares pre-COVID-19 (2019)	
	2020	2021	2020	2021
Haryana	(−) 4.35	(−) 2.94	(−) 5.71	(−) 5.71
Maharashtra	**(−) 13.04**	**(−) 19.12**	**(−) 6.25**	**(−) 14.06**
Gujarat	(−) 7.25	(−) 5.88	(−) 3.03	(−) 3.03
Jharkhand	(−) 2.90	(−) 1.47	**(−) 14.10**	**(−) 14.10**
Bihar	**(−) 24.64**	**(−) 30.88**	**(−) 18.75**	**(−) 26.56**

Source Authors' own calculations from data from NIKSHAY portal of NTEP, Government of India

In the context of percentage departure in share of publicly notified TB from respective state level shares pre-COVID-19 in 2019, it can be seen that the states of Haryana, Gujarat and Jharkhand record similar departures in both years of the pandemic; Maharashtra slides down by 6.25% in the first year and by an even higher 14.06% in the second year of COVID-19; a similar trend is shown by Bihar with a 18.75% decline in TB notifications in 2020 and a 26.56% decline in 2021.

To summarize, ***both in 2020 and 2021, Bihar records the highest percentage decline in the share of publicly notified TB compared to the all India-share in the***

respective years, with Maharashtra recording second highest decline. Also, both Maharashtra and Bihar record greater percentage declines in 2021 compared to 2020, while percentage declines are lower for Haryana, Gujarat and Jharkhand in the second year of pandemic compared to the first. Further, analysis reveals that ***the eastern states of Jharkhand and Bihar suffered the largest declines in the public shares of TB notification during the COVID years, compared to their respective pre-COVID shares in 2019. While Haryana, Gujarat and Jharkhand record the same extents of setback compared to their own pre-pandemic performances through both the COVID years, the situation tends to worsen for Maharashtra and Bihar in 2021.***

Analysis reveals that ***mean percentage departure of monthly TB notifications in the first year of COVID was much higher than that in the second year at the national level.*** However, lack of and in some critical COVID-19 phases, even complete shutdown of out-patient settings, community-based TB service, use of contact tracing as a tool for active screening of TB, etc., severely affected TB care delivery in both years [14]. The pandemic exposed India's already precarious healthcare delivery system and raised concerns over the impact of the COVID-19 pandemic on TB, a much older public health crisis. ***Substantially low TB notifications suggest increase in instances of both underdiagnosis and underreporting of TB, implying that the COVID crisis aggravated the already existing "missing millions" problem with respect to TB patients in India.*** Share of public notification in total TB notification may be interpreted as a proxy for general acceptability and inclusiveness of the government program to control TB, particularly in the context of a country like India where the healthcare system is fragmented and a substantially large and unregulated private sector dominates healthcare delivery. ***According to our finding, the pandemic years witnessed a decline in the share of public notification, which may be indicative of general hesitation to visit crowded public health institutions for fear of contracting COVID infection, among other factors. A few identified states like Haryana, Maharashtra, Gujarat, Jharkhand and Bihar seem to bear double the brunt as our analysis reveals that they not only recorded lower shares of public TB notification compared to all-India level shares in both the pandemic years, but also recorded substantial declines in this indicative share during the pandemic when compared with their respective pre-COVID performances.*** Findings with respect to the states of Bihar and Jharkhand should be particularly intriguing, given the revelation of the National Family Health Survey reports [6–8] regarding concentration (rates per 1,00,000 population higher than national average) of self-reported and medically treated TB in the eastern and northeastern parts of India.

Although people-centred delivery of TB prevention, diagnosis and treatment services were supposed to be maintained by health systems in tandem with the COVID-19 response [28] the COVID-19 waves resulted in a prolonged period of disruption in TB services in India with variations across states. Closure of health facilities from fear of contracting COVID-19 resulted in reduced access to TB services for diagnosis. There were reduced initiatives for implementation of active TB case screening campaigns among high-risk population coupled with supply chain management difficulties. NTEP program officers and health workers from the field were deployed to COVID-19 activities which led to minimal focus on TB service provision. Patients stranded in different parts of the country due to lockdown defaulted on treatment as they struggled to get medicines through health system. Non-availability of private transport and courier services led to problems in sputum sample collection and dispatch of the same to molecular diagnostic laboratory and culture drug susceptibility laboratory. Use of NTEP labs for testing of COVID-19 reduced the TB testing capacity and diagnosis of new cases, fuelled drug resistant TB and created constraints in following up with existing cases [29]. To catch up with previously achieved gains in TB control, revival strategies in NTEP services proposed for the post-pandemic scenario, constituted a boost to case finding from both public and private sectors, revival of sample collection through smoother transport services, modified DOTS modalities like supply of monthly medicine including home delivery to patients and family DOTS, patient and provider support services, surveillance and monitoring [10, 29]. A survey on community perspective of the impact of the pandemic on TB reveals that with COVID-19 and consequent social isolation, inequities and human rights-related barriers to TB services increased. As a healthcare worker from Morocco quotes, *"If we had used a quarter of the resources allocated to COVID… we would have eliminated TB a long time ago"*. According to an advocate from India, "*TB and COVID are both diseases. Ignoring one to tackle the other is shocking*". As a frontline healthcare worker from the US quotes, *"The additional fear and stigma generated by COVID-19 is contributing to the hesitancy among symptomatic TB patients to present to care"*. Similar observation is reiterated in the quote of a TB patient from Kenya—*"I am sometimes rejected by family members because they think I have the same signs of covid. I'm even thinking of moving from the family"* [23].

Although TB, another deadly infection, remains a public health priority for India even in the era of the newfound pandemic, there has unfortunately been minimal concern and media attention to generate large scale public awareness about the impact of COVID-19 on TB. ***The dilemma in public health management is glaringly visible in the unequal and rather biased allocation of economic and political resources in favour of COVID in the face of an already existing, high burden of TB.***

20.5 Conclusion

Two crucial aspects that might have directly affected the epidemiology of TB in the COVID-19 era include the worsening of poverty and undernutrition which are major trigger factors for TB and underdiagnosis of active TB cases due to difficulty in accessing healthcare services coupled with underreporting of diagnosed cases. This is apprehended to result in a steep rise TB incidence while also broadening the gap between TB cases estimated and notified. Additionally, effective implementation of WHO recommended digital technologies like therapy supported by video, electronic medication monitors, etc., to help TB patients complete treatment in a high burden setting like India, given the dominance of COVID-19 in the information and media space, remains questionable. **The current study reveals two vital points of concern**

with respect to the impact of the pandemic on TB control in India—*first,* a fall in TB notification during the COVID-19 years and *second*, a fall in the share of publicly notified TB in total TB notification, with identification of the relatively poor performing states. The overall as well as state-specific insights provided by the study are expected to help policymakers adopt a more focused approach to TB surveillance in general and utilization of the national TB program in particular by the Indian masses, particularly post-COVID-19 with greater focus on the relatively poorer performing states. Where COVID-19 and TB are likely to coexist for years or even decades from now, funding for the former should not be eating into the existing meagre public health infrastructure of India, rather it should add to the capacity to enhance efforts at prevention and control of both public health maladies.

The COVID-19 pandemic has significantly affected the TB care trajectory, limiting TB detection and reporting. However, an enquiry into the intersectionalities between COVID-19 and TB, might highlight dual threat from the diseases and also help determine shared solutions in addressing the two infectious diseases simultaneously. The present enquiry on the effects of the recent pandemic on notifications relating to another public health crisis from ancient times puts across a few obvious yet pertinent grounds of comparison between the two as well as discrepancies in the ways of addressal at least in the context of a developing country like India. As COVID-19 jolted both the health infrastructure and economy of India over a short span of time, the nation faced a "*life versus livelihood trade-off*". Much older health crises like TB also present a similar trade-off as a typically prolonged treatment tenure in case of TB hinders employment and income. Further, the COVID-19 crisis laid bare the socio-economic vulnerabilities of millions in a country like India yet again, whereas one of the major reasons behind the persistence of high TB incidence in the country till date has been lack of preventive measures through proper recognition of the socio-economic determinants of the disease and unidirectional focus on curative measures. Although the private healthcare sector worked in tandem with the public sector in efficient management of COVID-19 cases, the need for an overhauling of the Indian public health infrastructure itself has been widely felt and expressed both through scholarly contributions and policy prescriptions at the level of the government. On the contrary, a fatal morbidity as TB has been thriving for decades with substantially high disease burden coupled with mismanagement of cases by the private sector which is also well documented. Government's addressal of TB only through attempts to meet "detection" and "cure" targets have been more of a compliance than willingness to eradicate the disease. Last, there has been a glaring lacuna in promoting public education and awareness regarding TB, when compared with COVID-19 which is much recent but has garnered attention of Indian media and policymakers at large, an outcome of which is minimal or no awareness about the impact of COVID-19 on the TB epidemiology. *The novelty of the current piece of research lies in the possibility that it brings under the scanner one of the crucial TB surveillance parameters of the government by offering a glimpse of the status of TB notifications in the era of COVID-19. Insights into comparison of national level as well as state level TB notification rates pre and post pandemic are expected to invite policymakers to delve deeper into the reasons behind the failure of the Indian health care system in its current form, to combat the burden of TB in general and during the COVID crisis in particular.*

Stronger surveillance and political willingness to eradicate TB, supported by campaigns to improve general public awareness about TB, especially in times of another public health crisis like COVID-19, is the call of the hour. Consolidated effort from governments, both at the level of states and the centre, NGOs and policymakers is needed to curb the burden of TB by urgently ensuring continuity in TB-related essential service delivery to restore pace in identification and treatment of TB cases.

References

1. Aggarwal, A. N., Agarwal, R., Dhooria, S., Prasad, K. T., Sehgal, I. S., & Muthu, V. (2021). Impact of COVID-19 pandemic on tuberculosis notifications in India. *Lung India, 39*(1), 89–91. https://doi.org/10.4103/lungindia.lungindia_604_21
2. Behera, D. (2021). TB control in India in the COVID era. *Indian Journal of Tuberculosis, 68*(1), 128–133. https://doi.org/10.1016/j.ijtb.2020.08.019
3. Chadha, S. S., Nagaraja, S. B., Trivedi, A., Satapathy, S., Devendrappa N. M., & Devi Sagili, K. (2017). Mandatory TB notification in Mysore city, India: Have we heard the private practitioner's plea? *BMC Health Services Research, 17*(1). https://doi.org/10.1186/s12913-016-1943-z
4. Falzon, D., Zignol, M., Bastard, M., Floyd, K., & Kasaeva, T. (2023). The impact of the COVID-19 pandemic on the global tuberculosis epidemic. *Frontiers in Immunology, 14*. https://doi.org/10.3389/fimmu.2023.1234785
5. Ghosh, A. (2020). *A silent casualty: Drop in TB reporting, states admit lockdown issues*. https://indianexpresscom/article/india/tuberculosis-cases-drop-amid-lockdown-coronavirus-covid-19-6368736/
6. GOI. (2005–2006). *National Family Health Survey*. https://rchiips.org/nfhs/index.shtml
7. GOI. (2015–2016). *National Family Health Survey*. https://rchiips.org/nfhs/index.shtml
8. GOI. (2019–2021). *National Family Health Survey*. https://rchiips.org/nfhs/index.shtml
9. GOI. (2020). *India TB report*. https://tbcindia.mohfw.gov.in/annual-reports/
10. GOI. (2020). *Rapid Response Plan to mitigate impact of COVID-19 Pandemic on TB Epidemic and National TB Elimination Program (NTEP) activities in India*. https://tbcindia.gov.in
11. GOI. (2020). *West Bengal Cares for persons with TB during Covid-19 pandemic*. https://tbcindia.mohfw.gov.in/wp-content/uploads/2023/04/WestBengalCaresforTBpatientsinCOVIDtimes.pdf
12. Husain, A. A., Monaghan, T. M., & Kashyap, R. S. (2021). Impact of COVID-19 pandemic on tuberculosis care in India. *Clinical Microbiology and Infection, 27*(2), 293–294. https://doi.org/10.1016/j.cmi.2020.08.014
13. Jain, V. K., Iyengar, K. P., Samy, D. A., & Vaishya, R. (2020). Tuberculosis in the era of COVID-19 in India. *Diabetes & Metabolic Syndrome: Clinical Research & Reviews, 14*(5), 1439–1443. https://doi.org/10.1016/j.dsx.2020.07.034
14. Khetrapal, S., & Bhatia, R. (2020). Impact of COVID-19 pandemic on health system and Sustainable development goal 3. *Indian Journal of Medical Research, 151*(5), 395. https://doi.org/10.4103/ijmr.ijmr_1920_20
15. Oh, K. H., Yanagawa, M., Morishita, F., Glaziou, P., Rahevar, K., & Yadav, R. P. (2024). Changing epidemic of tuberculosis amidst the COVID-19 pandemic in the western Pacific region: Analysis of tuberculosis case notifications and treatment outcomes from 2015 to 2022. *The Lancet Regional Health—Western Pacific, 47*, Article 101104. https://doi.org/10.1016/j.lanwpc.2024.101104

16. Pai, M., & Dewan, P. (2015). Testing and treating the missing millions with tuberculosis. *PLOS Medicine, 12*(3), Article e1001805. https://doi.org/10.1371/journal.pmed.1001805
17. Rai, S., & Unisa, S. (2023). Assessing the impact of the COVID-19 pandemic on tuberculosis case. *Demography India, 52*(2), 108–118. https://iasp.ac.in/uploads/journal/009-1708491926.pdf
18. Revathy, R., Sahoo, S. K., & Rehman, T. (2021). Status of mandatory tuberculosis case notification among pharmacy outlets in Urban Puducherry: A cross-sectional descriptive study. *Indian Journal of Community Medicine, 46*(2), 344–345. https://doi.org/10.4103/ijcm.IJCM_499_20
19. Sahu, S., & Nagtode, N. (2023). Impact on tuberculosis notification during COVID-19 pandemic in India: A narrative review. *Cureus*. https://doi.org/10.7759/cureus.44087
20. Sathiyamoorthy, R., Kalaivani, M., Aggarwal, P., & Gupta, S. K. (2020). Prevalence of pulmonary tuberculosis in India: A systematic review and meta-analysis. *Lung India, 37*(1), 45. https://doi.org/10.4103/lungindia.lungindia_181_19
21. Satyanarayana, S., Nair, S. A., Chadha, S. S., Shivashankar, R., Sharma, G., Yadav, S., Mohanty, S., Kamineni, V., Wilson, N. C., Harries, A. D., & Dewan, P. K. (2011). From where are tuberculosis patients accessing treatment in India? Results from a cross-sectional community based survey of 30 districts. *PLoS ONE, 6*(9), Article e24160. https://doi.org/10.1371/journal.pone.0024160
22. Siddaiah, A., Ahmed, M. N., Kumar, A. M., D'Souza, G., Wilkinson, E., Maung, T. M., & Rodrigues, R. (2019). Tuberculosis notification in a private tertiary care teaching hospital in South India: A mixed-methods study. *British Medical Journal Open, 9*(2), Article e023910. https://doi.org/10.1136/bmjopen-2018-023910
23. STOP-TB-PARTNERSHIP. (2020). *The impact of COVID-19 on the TB pandemic: A Community Perspective*. https://www.stoptb.org/sites/default/files/imported/document/BoardDocs/33/33-06/33-6.2_The_Impact_of_COVID_on_TB_Response_A_Community_Perspective_0.pdf
24. STOP-TB-PARTNERSHIP. (2020). *The potential impact of the covid-19 response on tuberculosis in high-burden countries: A modelling analysis*. https://stoptb.org/assets/documents/news/Modeling%20Report_1%20May%202020_FINAL.pdf
25. STOP-TB-PARTNERSHIP. (2021). *One year on, new data show global impact of COVID-19 on TB epidemic is worse than expected*. https://www.stoptb.org/sites/default/files/imported/document/20210316_TB_and_COVID_One_Year_on_Media_Brief_FINAL.pdf
26. WHO. (2019). *Global tuberculosis report*. https://www.who.int/publications/i/item/global-tuberculosis-report-2019
27. WHO. (2020). *Covid 19: considerations for tuberculosis (TB) care*. https://www.who.int/docs/default-source/documents/tuberculosis/infonote-tb-covid-19.pdf
28. WHO. (2020). *World Health Organisation (WHO) Information Note: Tuberculosis and Covid 19*. https://www.who.int/docs/default-source/documents/tuberculosis/infonote-tb-covid-19.pdf
29. WHO. (2021). *Health and Care Worker Deaths during COVID-19*. https://www.who.int/news/item/20-10-2021-health-and-care-worker-deaths-during-covid-19
30. Zimmer, A. J., Klinton, J. S., Oga-Omenka, C., Heitkamp, P., Nawina Nyirenda, C., Furin, J., & Pai, M. (2021). Tuberculosis in times of COVID-19. *Journal of Epidemiology and Community Health, 76*(3), 310–316. https://doi.org/10.1136/jech-2021-217529

Paramita Barman Sen is Assistant Professor of Economics under the Faculty of Commerce and Management at St. Xavier's University, Kolkata. She is an M.A. in Economics from Jawaharlal Nehru University (JNU), New Delhi and M.Phil in Economics from University of Calcutta. She has worked in the capacity of research assistant in a National Rural Health Mission (NRHM) funded project of the Department of Health and Family Welfare, Government of West Bengal. She has more than ten years of experience in teaching and has published in peer-reviewed and Scopus-indexed journals. Her research interest lies in Economics of Social Sector.

Arijita Dutta is Professor at Department of Economics, University of Calcutta. Her research interests are Health Economics, Quantitative Development Economics, Labour Economics, Gender Economics and Economics of Entrepreneurship. She has published widely in National and International Journals. She has completed research projects funded by Indian and international agencies. She has been closely associated with health policy making of the state of West Bengal. She is currently associated with the global project of Empowering Women Globally as knowledge partner. She has been a visiting professor at University of Le Havre, France; UNU Merit, Maastricht, the Netherlands to name a few.

Chapter 21
In the Winds of the Pandemic: The Socio-Psychological Impact and Transformation

Ipseeta Satpathy, Debashree Chakravarty, Md. Zahir Uddin Arif, and B. C. M. Patnaik

Abstract The COVID-19 epidemic altered the interactions, structure, and psychology of global society. Economic recovery was not possible either, struck with lockdowns, isolation, and constant fear of COVID-19 on people's minds. Widespread anxiety, depression, and psychoses became a new social order indeed! But evolution prevails, people society and norms around the globe had to shift and learn how to navigate this new environment.

The pandemic changed people's everyday life including how they educated themselves, in countries like India, the pandemic pushed the use of digital educational technologies with or without a valid need because of the shutdown of schools and universities. When millions of students benefited from online education, it revealed the existing inequalities in technology, usage, and connection to the online space. The most affected after the transition were not the students but the regions which are already disadvantaged making it even more difficult for the students to access education by sitting at a computer. Furthermore, the rapid shift to virtual education posed psychological challenges for both students and educators, as they navigated the new pressures of remote learning. Psychological stability is crucial for managing stress and adapting to change. In education, especially with the move to online learning, it supports concentration, motivation, and a positive learning environment. Maintaining psychological well-being enhances individual quality of life and strengthens social connections, productivity, and overall societal functioning.

I. Satpathy · D. Chakravarty (✉) · B. C. M. Patnaik
School of Management, KIIT University, Bhubaneswar, India
e-mail: 2381104@ksom.ac.in

I. Satpathy
e-mail: ipseeta@ksom.ac.in

B. C. M. Patnaik
e-mail: bcmpatnaik@ksom.ac.in

Md. Z. U. Arif
Department of Marketing, Faculty of Business Studies, Jagannath University, Dhaka, Bangladesh
e-mail: arif@mkt.jnu.ac.bd

S. Gangopadhyay et al. (eds.), *Pandemic Diaries*,
https://doi.org/10.1007/978-981-96-5415-4_21

This paper tends to explore the socio-psychological impact of the Coronavirus, pandemic, and the transformation. In the context of change, education with a particular focus on India's shift to online learning turns the highlights to the urgent need for addressing digital shift and providing psychological support to ensure a more inclusive and sustainable education system in the post-pandemic era.

Keywords Pandemic · COVID-19 · Socio-psychological impact · Online learning · Education system

21.1 Introduction

The COVID-19 pandemic swept across the globe like an unpredictable storm, altering every aspect of life. Beyond the physical health crisis, the pandemic has unleashed a profound socio-psychological impact, reshaping how individuals and communities helm their daily lives. This period of uncertainty and disruption brought about an unprecedented wave of anxiety, stress, and emotional strain, challenging our mental and emotional resilience in ways we could never have imagined [29, 30].

At the heart of this upheaval was the widespread adoption of social distancing measures, leading to prolonged social isolation for many. This forced separation from family, friends, and colleagues intensified feelings of loneliness, fear, and uncertainty. Coupled with the fear of infection and economic instability, the pandemic disrupted not just our routines, but also our emotional equilibrium [7, 24].

As society faces these emotional winds, it is essential to explore the psychological toll and the transformative social changes brought by this global crisis. This report aims to investigate the social and psychological impact of the disease outbreak while suggesting ways in which people, communities, and institutions can come to terms with this ongoing change as well as safeguard mental health and promote resilience.

The pandemic has emerged as the most significant crisis globally since the Second World War, profoundly challenging human society [2, 4]. As both a health emergency and a widespread societal threat, pandemic has dramatically impacted every facet of life, affecting economies, social structures, and disproportionately harming populations that is vulnerable [14]. With its rapid spread across more than hundred countries and millions of reported deaths, the pandemic has made history. By 2020 spring, nearly a billion individuals were under quarantine or lockdown [35]. By 2022 November, around six hundred thirty million cases that are confirmed and deaths more than 6.5 million had been globally reported. The regions most affected by confirmed cases—Europe, Americas and Southeast Asia—continue to bear the heaviest burdens [36].

The pandemic's far-reaching consequences have disrupted nearly every aspect of human life. It has imposed immense health, economic, environmental, and social challenges on a global scale [4]. Prolonged lockdowns and restrictions, while effective in controlling the virus's spread, have amplified many of these challenges [3, 19]. Nowhere have the impacts been felt more intensely than in the social domain [20].

Social life so as livelihoods have been thrown into disarray [1, 22] and fundamental aspects of social interaction, such as mobility, social capital [6], education [5, 25, 26], and equality [13, 28] have deteriorated. The epidemic has exacerbated social inequalities and increased vulnerabilities, leaving lasting scars on social cohesion and human coexistence [16, 34].

Psychological stability is crucial for maintaining overall well-being and navigating life's challenges effectively. It allows individuals to manage stress, adapt to change, and make sound decisions under pressure. Stable mental health promotes emotional resilience, helping people cope with uncertainty and setbacks, which was especially important during the pandemic. Without psychological stability, individuals are more vulnerable to anxiety, depression, and burnout, which can impair personal, professional, and academic performance. In contexts such as education, particularly with the sudden shift to online learning, psychological stability is vital for fostering concentration, motivation, and a positive learning environment. Maintaining psychological stability not only improves individual quality of life but also strengthens social connections, productivity, and overall societal functioning.

The COVID-19 pandemic has prompted a widespread transformation in institutions and behaviors across various sectors, "shock effects" across various sectors, including education. The impact on students has been immense. By April 9, 2020, more than 1.5 billion learners worldwide, ranging from primary to tertiary education levels, were unable to attend school. The sudden and large-scale closure of educational institutions forced nations and communities to quickly adapt by turning to various digital learning platforms [11]. In this urgent shift from traditional classrooms to online teaching, deeper discussions about national education policies and theoretical frameworks were largely overlooked.

The current state of formal education systems can be interpreted through Philip Strong's 1990 model of "epidemic psychology," which describes societal reactions to an epidemiological crisis through three overlapping stages: fear, explanation, and action. In this context, the "epidemic of fear" raises a critical question that, how educational systems can and individual learners effectively navigate this extraordinary situation? Strong's model highlights collective psychological responses, using "epidemic" metaphorically to capture the emotional and cognitive responses to the pandemic.

21.2 Ambivalence in Psychological Stability During the Pandemic

The COVID-19 pandemic brought about a state of psychological ambivalence for many individuals, as they experienced conflicting emotions and mental states. On the one hand, there was anxiety, fear, and uncertainty driven by health risks, social isolation, and economic instability. On the other hand, periods of lockdown and social distancing created moments of reflection, stronger family bonds, and new coping mechanisms. This ambivalence in psychological stability caused fluctuations in emotional well-being, with individuals often oscillating between feelings of vulnerability and resilience. The unpredictability of the pandemic contributed to this emotional conflict, making it difficult for people to maintain consistent mental health, as they adapted to an ever-changing and uncertain reality. Stable mental health promotes emotional resilience, helping people cope with uncertainty and setbacks, which was especially important during the COVID-19 pandemic. Without psychological stability, individuals are more vulnerable to anxiety, depression, and burnout, which can impair personal, professional, and academic performance.

Mental strength alludes to a singular's ability to oversee difficulties and adjust to outer tensions, developing through self-improvement. It has been broadly explored as a significant asset for adapting to outer pressure. A review led on Middle Easterner youth uncovered that uneasiness related with Coronavirus adversely affects mental strength, adding to expanded psychological wellness issues. Mental flexibility has remarkably declined since the beginning of the pandemic contrasted with pre-pandemic times. In any case, people areas of strength for with exhibit better profound guideline, more prominent life fulfillment, and worked on generally speaking emotional wellness and prosperity [8, 9, 23]. Mental versatility mitigates the negative mental impacts of the pandemic, keeping up with emotional wellness. Research proposes that strength is a critical indicator of positive feelings, [31] with more significant levels of versatility being related with additional consistent good profound encounters [38]. Exceptionally versatile people will quite often encounter more steady feelings, though those with lower flexibility are more helpless to close to home problems. Zhang [39] accentuated the need to work on profound guideline, especially for weak gatherings, for example, left-behind kids, to reinforce versatility [27]. These outcomes highlight the cozy connection among strength and close to home steadiness, attesting the underlying speculation that mental versatility emphatically predicts profound dependability in secondary school understudies.

Mood states, which refer to enduring yet mild emotional conditions, can be either positive or negative. The "broaden-and-build" hypothesis of positive feelings recommends that a good state of mind can extend a singular's scope of contemplations and activities, supporting the improvement of individual assets to more readily deal with difficulties [33]. Research by Zautra [40] found that while positive and negative moods operate independently, a positive mood can significantly reduce negative feelings during stressful events. By disrupting negative emotional experiences caused by tension, positive moods enhance resilience and adaptability. Kuang [15] also noted that emotional contagion enhances positive moods while reducing negative ones. Positive emotional states foster calmness and emotional balance, which in turn contribute to the growth of psychological resilience [38]. Individuals who maintain a positive mood tend to have higher resilience and experience more positive emotions. Those with greater psychological resilience are better at managing emotions and maintaining mental balance under stress. Based on this, the second hypothesis proposes that a positive mood mediates the relationship between resilience and emotional stability.

Although social distancing has played a crucial role in controlling physical health during the pandemic, the prolonged absence of social interaction and physical activity has negatively affected people's moods [39]. For example, the United Kingdom's lockdown led to an increase in negative moods among its citizens [10]. Negative moods have been associated with poorer performance, while emotional stability weakens the link between negative moods and poor outcomes. Emotional stability is a key factor in predicting negative mood states. Individuals with negative moods often focus on sad and distressing emotions and are more likely to recall painful memories [21]. Individuals with lower mental strength will generally encounter pessimistic mind-sets while confronting distressing life altering situations, outrageous profound responses are frequently battling to control [17]. Studies demonstrate major areas of strength for a connection among versatility and negative mind-set states. In this way, the third speculation recommends that pessimistic state of mind states go about as a middle person in the association among flexibility and close to home solidness. [18]

There is areas of strength for a between mind-set states, mental versatility and profound dependability. In any case, the intervening impacts of positive and pessimistic mind-set states on the connection among strength and profound dependability have not been entirely analyzed. As close to home strength is a critical sign of psychological wellness, the Coronavirus pandemic affects the mental prosperity of secondary school understudies. Gaining a deeper understanding of their emotional stability during this period offers valuable insights and contributes to psychological research. The main objective of this study is to examine how emotional instability is affected by both having a positive or negative mood and at the same time assessing the relevance of the resilience factor among high school students.

21.3 Enhancing Health Outcomes Through Preventive Measures, Addressing Misinformation, and Reducing Socio-Psychological Stigma and Inequalities During the Pandemic

The pandemic has shown us how important it is to be prepared for a global health crisis. To truly improve health outcomes, we need to think about the bigger picture—taking steps not only to prevent the spread of illness but also to address the deeper impacts on our lives, communities, and mental well-being.

Simple preventive actions like getting vaccinated, keeping a safe distance, and practicing good hygiene are essential for keeping the virus in check. But these steps alone can fall short if false information spreads unchecked. Misinformation can lead to confusion, fear, and even resistance to following health guidelines that are meant to protect us all.

That's why fighting misinformation is just as important as the measures themselves. It's not just about sharing facts—it's about listening to people, answering their concerns, and building trust within communities. Misunderstandings thrive where there's a lack of connection, so honest conversations can make all the difference.

The pandemic has also exposed deeper issues, like stigma, prejudice, and inequalities, that make it harder for some people to get the help they need. These challenges are often overlooked but have a real impact on our collective health. By addressing these social barriers, creating a more inclusive environment, and supporting each other, we can ensure that everyone has the chance to stay healthy and safe during these tough times.

It's not just about policies or protocols—it's about people coming together to care for one another.

A COVID-19 diagnosis often brings a wave of panic and emotional distress. Many people find themselves grappling with sadness or depression long after recovering from the illness. At the same time, some discover a renewed appreciation for life, recognizing the value of their health and the moments they might have previously taken for granted.

The nationwide lockdowns implemented to contain the virus brought profound changes to daily life. From self-isolation to enforced quarantine, these measures, while necessary, have taken a heavy psychological toll. Studies published in The Lancet reveal that quarantine is linked to symptoms of post-traumatic stress disorder (PTSD), confusion, and anger—effects that can linger for years.

For some, quarantine has led to devastating consequences, such as increased rates of suicide and family disputes. Those who already struggle with mental health conditions like anxiety or depression are at even greater risk, and new cases of these conditions have surfaced in alarming numbers. People experiencing worsening symptoms, including thoughts of self-harm, need professional mental health support immediately.

The uncertainty surrounding the virus has amplified feelings of anxiety. Resource shortages and a world seemingly paused in time added to the overwhelming sense of unease. Busy cities became ghost towns, their empty streets and silent transport systems reflecting the disruption of normal life. For many, even stepping outside to buy necessities felt terrifying.

This shift in reality created a ripple effect. Research from the SARS outbreak showed that even after quarantine ended, people avoided crowds and public spaces for weeks. Hospital staff who endured the stress of quarantine still reported symptoms of depression years later. The lesson is clear: while quarantine helps contain outbreaks, it also comes with deep psychological costs.

Public health officials must carefully balance the need for restrictions with their emotional and social impact. Transparency is crucial—people need to understand why measures are necessary and how long they will last. Simple efforts, like ensuring access to essentials, clear communication, and ways to stay engaged during isolation, can make a big difference.

Yet, during the COVID-19 pandemic, mental health often took a backseat to discussions about infection and death rates. People feared losing their jobs due to travel restrictions or isolation, felt the sting of reduced social contact, and endured stigma if they showed flu-like symptoms. Meanwhile, endless media coverage and panic buying created an environment of relentless stress. Frontline workers, from doctors to volunteers, bore the brunt of these challenges. They faced exhaustion, burnout, and the fear of spreading the virus to loved ones. The pandemic was more than a health crisis; it was a mental health crisis, too. Beyond the physical measures, we need to recognize the human toll, offering compassion, understanding, and support to those still carrying its weight.

21.3.1 Current Difficulties and Reactions to Coronavirus in India

India has taken bold steps to tackle the COVID-19 pandemic, shutting down schools, malls, cinemas, and public spaces while enforcing strict social distancing and home quarantine rules. While these measures have helped control the spread of the virus, they haven't come without challenges. Protests have emerged from people who feel these restrictions are too harsh, highlighting the tension between public health needs and personal freedoms.

The situation feels almost like a wartime response, requiring the kind of drastic actions usually reserved for natural disasters or conflicts. With no vaccine or definitive treatment available, the pandemic has tested governments and healthcare systems around the world like never before.

As doctors and scientists race to develop effective treatments, people have a critical role to play. Messages about social distancing and good hygiene are everywhere, and many states have imposed strict quarantine rules to prevent the virus from spreading within communities. But government mandates and collective efforts alone can't solve this crisis—it also takes empathy, understanding, and support from society as a whole.

The economic toll has been devastating, particularly for small businesses struggling to survive. Urgent government intervention is needed to support these businesses and protect livelihoods. Vulnerable groups—such as those in long-term isolation or from lower-income backgrounds—require special attention. They need clear information about the help available to them and tailored support to get through this tough time.

While COVID-19 hotlines are available, they tend to focus on physical health and preventing infections, often overlooking the mental health struggles of those affected. Institutions like AIIMS and NIMHANS have stepped up to offer mental health support, but more comprehensive efforts are essential. Lessons from China's later pandemic response showed how important online counseling and regular communication can be. Strategies like digital check-ins, group counseling, and maintaining daily routines helped ease feelings of isolation and distress. Without adequate emotional and psychological support, people may resist quarantine measures, putting themselves and others at greater risk. It's vital to approach this challenge with care and compassion, balancing enforcement with understanding to ensure everyone's well-being. This pandemic isn't just a health crisis—it's a human crisis. While policies and protocols are essential, so is kindness, patience, and the willingness to help each other through this unprecedented time.

21.3.2 Managing Mental Well-Being During Pandemic

The pandemic has brought challenges that can feel overwhelming at times. Constant news updates, social media posts, and conversations with friends and family can stir up fear and anxiety, making us react in ways that don't feel like ourselves. But understanding the reasons behind measures like lockdowns can help us manage these emotions and protect our mental well-being.

Lockdowns are in place to slow the spread of the virus, limiting trips outside the home to essentials like groceries or medicines. When stepping out is unavoidable, it's best for one healthy family member to take on the task. It's also a good idea to familiarize yourself with nearby medical facilities, just in case someone in your household falls ill.

While some people might enjoy the slower pace of social isolation, it can also feel restrictive or monotonous after a while. To stay positive, try setting up a daily routine filled with meaningful activities. If negative emotions start to creep in, channel them into something you enjoy—maybe painting, gardening, reading, or even cooking. Don't forget to take care of your body by eating well, drinking plenty of water, and staying active with some simple indoor exercises.

Supporting those around you can also bring a sense of purpose. Share resources or lend a helping hand to elderly family members or neighbors who might struggle with daily tasks. If you have children at home, involve them in household chores. It's a great way for them to feel responsible, pick up new skills, and stay engaged.

Sticking to a structured schedule and staying connected with loved ones—whether through calls, video chats, or even writing letters—can help ease feelings of isolation. Remember, this is a tough time for everyone, but finding small ways to stay grounded and support each other can make a big difference.

21.3.3 Embrace Reliable Information, Avoid Misinformation

Getting the right information can go a long way in easing fear and anxiety. To protect yourself and those around you, stick to updates and advice from trusted sources. Avoid rumors or unverified claims, and take the time to cross-check facts before sharing them. Avoid sensational news or social media rumors, as they can distort reality and worsen your mental state. Instead of fixating on stories of illness, focus on positive outcomes and recovery stories. Adhere to established guidelines such as maintaining good hand hygiene and practicing physical distancing to protect yourself and others.

It's important to remember that symptoms of a common cold differ from those of COVID-19, which have been clearly outlined by health authorities. Practicing proper etiquette, like covering sneezes and coughs, and avoiding spitting in public, is essential. Most people with COVID-19 experience mild symptoms and only need to isolate until they are no longer contagious, typically about two weeks. Hospitalization is generally only necessary for those experiencing severe symptoms, such as difficulty breathing. Most individuals recover without needing hospital care.

21.3.3.1 Managing Emotional Challenges

During times of anxiety, take a few moments to practice deep, slow breathing to calm your mind. Try to distance yourself from the thoughts causing your anxiety and focus on something peaceful. When you feel anger or irritation, calming techniques such as counting down from 10 to 1 or engaging in a distraction can be helpful. Addressing fear involves asking yourself key questions: What aspects are within my control? Am I worrying excessively about the worst possible outcome? How have I managed stress in the past? What actions can I take to stay positive?

Loneliness and sadness are common emotions during challenging times. Stay connected with friends and family through calls or messages, discussing joyful topics, sharing interests, and exchanging ideas. If these feelings persist for several days despite your efforts to improve your mood, reach out to someone you trust. If your emotions become overwhelming, and you start feeling hopeless or as if life is not worth living, contact a mental health helpline for support.

21.3.4 Maneuvering Stress and Support During COVID-19

As the challenges developed of the COVID-19 pandemic, managing stress and supporting each other is crucial. To keep yourself and those around you healthy, try to avoid certain behaviors. Firstly, resist the urge to turn to tobacco, alcohol, or other substances to deal with emotions or boredom. These can worsen physical and mental health and weaken your immune system. If you have a history of substance use, seeking professional help can make a big difference, especially during tough times. Additionally, it's important not to stigmatize or judge those who contract the virus. While it's essential to keep your distance for safety, showing empathy and support can go a long way. If you ever feel overwhelmed, hopeless, or like life is too much to handle, please reach out to a mental health helpline. There's always someone ready to listen and help you through tough times—you don't have to face it alone.

21.3.4.1 Emotional Challenges Post-COVID-19 Recovery

Recovering from COVID-19 is a relief, but it can come with emotional challenges as you return to daily life. You might worry about your loved ones getting sick or feel hurt and frustrated if people keep their distance due to misunderstandings about the illness. These reactions can feel isolating and overwhelming.

It's also natural to feel guilt if your illness meant you couldn't work or care for others, leaving you with frustration or even sadness.

To navigate these feelings, focus on small steps to help yourself heal emotionally. Try practicing mindfulness, picking up hobbies you enjoy, or sharing your recovery story to inspire hope in others. Remind yourself—and those around you—that recovery is possible, and be a source of support for anyone dealing with similar emotions. You've come through a tough experience, and your strength can make a difference for both you and others.

21.3.4.2 Recognizing Mental Health Struggles in Others

While taking care of your own mental health, keep an eye on those around you who might be struggling emotionally. Watch for signs like changes in sleep patterns, difficulty concentrating, worsening physical health, or an increased reliance on alcohol, tobacco, or other substances. These can be signals that someone is finding it hard to cope. Offer a listening ear, express your support, and if needed, gently encourage them to reach out to a mental health professional or a helpline for help.

21.3.4.3 Supporting Those with Pre-existing Mental Health Conditions

People with pre-existing mental health challenges have faced added difficulties during the pandemic. The stress and uncertainty of this time can make conditions worse, while social isolation might leave them feeling more withdrawn, irritable, or anxious. For some, access to therapy or medications has become more difficult, creating additional hurdles.

Family members and caregivers can make a big difference. Providing reassurance, sticking to regular routines, and encouraging positive engagement can help. Whether it's through small activities, kind words, or simply being there, your support matters. Don't hesitate to reach out to mental health professionals or helplines—they can offer valuable guidance and assistance during this challenging time.

21.4 Adapting to Online Learning: Psychological Effects and Adaptation

The shift to online learning has been a major change for students, teachers, and parents alike, especially during the pandemic. While it brings benefits like flexibility and convenience, it also comes with its share of challenges. Students may feel isolated without the energy of a classroom, and teachers might struggle to adapt to digital teaching methods. For parents, supporting their children's education at home can be a source of stress.

The reliance on technology can add another layer of difficulty, with technical glitches and the need for self-discipline making it hard for everyone to stay focused and motivated.

To cope with these changes, it's important to acknowledge the emotional toll and find strategies to manage it. Creating a structured routine, staying connected through virtual interactions, and taking regular breaks can make the experience more manageable. Addressing mental well-being alongside academic progress is crucial to creating a positive and productive learning environment.

The pandemic has disrupted education globally, impacting almost every student. While the challenges are undeniable, with the right support and strategies, it's possible to adapt and even thrive in this new way of learning. The abrupt beginning of the pandemic and ensuing school terminations set off an earnest and boundless endeavors to adjust and enhance in training. As the situation evolved rapidly, many schools and education systems transitioned to remote learning almost overnight [12, 32]. This shift involved using various mediums such as television, radio, the internet, and even traditional mail to continue teaching in new ways. Remote learning quickly became the standard approach to education during this period, and educators demonstrated remarkable adaptability and support for this new mode of instruction. Indeed, it is widely recognized that "this crisis has stimulated innovation within the education sector" [37].

Nonetheless, the immediate adaptations made in response to the pandemic may not align with the long-term needs of education in a post-pandemic world. The rapid changes were primarily focused on maintaining continuity in schooling and finding immediate solutions for remote teaching rather than reimagining the future of education. While these short-term adaptations were crucial, they may be viewed as insufficient for addressing the broader, more enduring challenges that lie ahead.

21.4.1 Adapting to COVID-19: The Challenges and Transformations in Indian Education

The Indian government's decision to promulgate a lockdown and close educational institutions was a crucial measure to enforce social distancing during the COVID-19 pandemic. This action has profoundly impacted the education system, particularly affecting students in rural areas where classroom-based learning is prevalent. The pandemic has disrupted traditional educational practices by halting schools, universities, and various exams, including admissions, entrance tests, and competitive examinations. With no clear end to the pandemic in sight, educational institutions remain closed, leading to widespread interruptions in learning. The shift to online education has emerged as a primary response; however, its implementation has been uneven. While some schools and universities have transitioned successfully to virtual learning, numerous low-income private and government schools encounter substantial difficulties, leading to widespread disruptions in education. The study focuses on analyze the impact on the Indian education system of COVID-19 by examining how the pandemic has affected educational practices, assessing the benefits of online learning for students, parents, and educators, and identifying strategies to mitigate the negative effects on students to ensure a more effective educational experience in the future.

21.4.2 Digital Infrastructure in India: Adapting to the New Norm

Prior to the COVID-19 lockdown, few could have anticipated the profound shift in India's educational landscape. Schools that once prohibited electronic gadgets are now navigating the complexities of online learning. This transition has been particularly challenging for educators, who must adapt to new teaching methods and design effective virtual lessons. Addressing these challenges through targeted workshops and training can help teachers adjust to this new mode of instruction.

In the year 2017–2018 the report of NSSO on Household Social Consumption on Education in India reveals less than fifteen percent of rural households have internet access, whereas forty-two percent of urban households do. This disparity underscores the considerable challenges in providing online education to all students nationwide (Table 21.1).

Access to having electricity is a major obstacle to effective online learning in India. According to a survey of 2017–2018 conducted by the Ministry of Rural Development, only forty-seven percent of Indian households received electricity for 12 h of daily, and about more than thirty-six percent of schools are completely without having electricity. The pandemic has shone a light on the struggles faced by students from lower-income families as they adapt to remote learning. Many lack access to essential tools like computers or smartphones, while their parents,

Table 21.1 Statistics of computer literacy and Internet access across Indian states

Sr No	State	Rural	Rural	Urban	Urban
		Operate Computer (%)	Access Internet (%)	Operate Computer (%)	Access Internet (%)
1.	Andhra Pradesh	1.5	10.4	11.6	29.5
2.	Assam	3.7	12.1	30.8	46.9
3.	Bihar	2.7	12.5	20.0	38.6
4.	Chhattisgarh	3.2	10.6	22.0	34.6
5.	Delhi	NA	NA	34.7	55.8
6.	Gujrat	4.4	21.1	20.1	49.1
7.	Haryana	5.9	37.1	29.5	55.5
8.	Himachal Pradesh	10.5	48.6	28.3	70.6
9.	Jammu & Kashmir	3.5	28.7	16.0	57.7
10.	Jharkhand	1.3	11.9	15.6	40.2
11.	Karnataka	2.0	8.3	22.9	33.5
12.	Kerala	20.1	46.9	27.5	27.5
13.	Madhya Pradesh	2.3	9.7	17.2	35.4
14.	Maharashtra	3.3	18.5	27.4	52.0
15.	Odisha	1.8	5.8	17.2	31.2
16.	Punjab	9.4	39.4	26.7	57.1
17.	Rajasthan	6.4	18.5	26.6	49.9
18.	Tamil Nadu	11.6	14.4	24.7	24.8
19.	Telangana	1.6	9.9	17.6	41.9
20.	Utter Pradesh	7.0	35.2	32.5	64.3
21.	Uttarakhand	4.0	11.6	22.3	41.0
22.	West Bengal	3.3	7.9	23.0	36.0

Source 75th round of National Sample survey conducted between July 2017 and June 2018

unfamiliar with digital technology, often find it hard to offer support. Inconsistent internet connectivity and inadequate infrastructure make the situation even more difficult for students and teachers across the country.

21.4.3 Challenges and Opportunities for Teachers and Students

The sudden shift to online education has been a tough adjustment for teachers and students alike. Teachers often face challenges like unsuitable home environments, interruptions, and distractions. Schools and colleges, constrained by tight budgets, struggle to invest in technology, provide sufficient training, or offer clear guidance for online learning. Technical difficulties, such as unreliable internet, unfamiliarity with digital platforms, and concerns about online security, add to the frustration. Teachers also have to navigate personal hurdles, including limited technical skills and the challenge of integrating technology into their teaching methods.

For students, the transition has been equally daunting. Many lack quiet spaces to study or access to devices, particularly in rural or economically disadvantaged areas. The absence of in-person interaction and the constant demands of navigating online platforms can leave students feeling isolated and overwhelmed.

Despite these challenges, there have been silver linings. The shift to digital education has encouraged the use of new tools, fostering creativity and innovation in teaching and learning. Students and educators are now more comfortable with digital platforms, gaining skills that will benefit them long-term. Online learning has opened doors for global connectivity, offering opportunities to interact with peers worldwide. It has also encouraged better time management among students and sparked interest in flexible, open-distance learning models.

However, the transition has also deepened existing inequalities. Rural and underprivileged students face significant barriers, including limited access to devices and the internet, while many urban students enjoy better resources and support. This digital divide has highlighted stark disparities, making it clear that more needs to be done to level the playing field.

21.5 Conclusion

In conclusion, the pandemic has changed the world in profound ways, reshaping healthcare, economies, and education. In India, the transition to online learning has brought to light critical gaps in access to technology and digital skills, particularly between urban and rural students. While online education has kept learning alive, it has also introduced new challenges for teachers and students, emphasizing the importance of supporting their mental well-being. To ensure the benefits of online education reach everyone, we must tackle these inequalities head-on. By continuing to innovate, adapt, and invest in inclusive solutions, we can build a more equitable education system where no student is left behind.

References

1. Agarwal, R., & Gopinath, M. G. (2021). A proposal to end the COVID-19 pandemic. *International Monetary Fund.*
2. Al-Omoush, K. S., Simón-Moya, V., & Sendra-García, J. (2020). The impact of social capital and collaborative knowledge creation on e-business proactiveness and organizational agility in responding to the COVID-19 crisis. *Journal of Innovation and Knowledge, 5*(4), 279–288.
3. Brady, O., & Wilder-Smith, A. (2021). What is the impact of lockdowns on dengue? *Current Infectious Disease Reports, 23*, 1–8.
4. Chakraborty, I., & Maity, P. (2020). COVID-19 outbreak: Migration, effects on society, global environment, and prevention. *Science of the Total Environment, 728*, Article 138882.
5. Chaturvedi, R., Lui, B., Aaronson, J. A., White, R. S., & Samuels, J. D. (2022). COVID-19 complications in males and females: Recent developments. *Journal of Comparative Effectiveness Research, 11*(9), 689–698.
6. Elgar, F. J., Pickett, W., Pförtner, T. K., Gariépy, G., Gordon, D., Georgiades, K., Davison, C., Hammami, N., MacNeil, A. H., Da Silva, M. A., & Melgar-Quiñonez, H. R. (2021). Relative food insecurity, mental health and well-being in 160 countries. *Social Science & Medicine, 268*, 113556.
7. Fofana, N. K., Latif, F., Sarfraz, S., Bashir, M. F., & Komal, B. (2020). Fear and agony of the pandemic leading to stress and mental illness: An emerging crisis in the novel coronavirus (COVID-19) outbreak. *Psychiatry Research, 291*, Article 113230.
8. Gundogan, S. (2021). The mediator role of the fear of COVID-19 in the relationship between psychological resilience and life satisfaction. *Current Psychology, 40*(12), 6291–6299.
9. Hatun, O., & Kurtça, T. T. (2023). Self-compassion, resilience, fear of COVID-19, psychological distress, and psychological well-being among Turkish adults. *Current Psychology, 42*(23), 20052–20062.
10. Ingram, D. J., Coad, L., Milner-Gulland, E. J., Parry, L., Wilkie, D., Bakarr, M. I., Wilkie, D., Bakarr, M. I., Benítez-López, A., Bennett, E. L., Bodmer, R., Cowlishaw, G., El Bizri, H. R., Eves, H. E., Fa, J. E., Golde, C. D., Iponga, D. M., Minh, N. V., Morcatty, T. Q., Mwinyihali, R., ... & Abernethy, K. (2021). Wild meat is still on the menu: Progress in wild meat research, policy, and practice from 2002 to 2020. *Annual Review of Environment and Resources, 46*(1), 221–254.
11. Jandrić, P. (2020). Postdigital research in the time of Covid-19. *Postdigital Science and Education, 2*, 233–238.
12. Kamenets, E. A., Gusarova, E. A., Milovanova, N. V., Itkis, Y. S., Strokova, T. V., Melikyan, M. A., Garyaeva, I. V., Rybkina, I. G., Nikitina, N. V., & Zakharova, E. Y. (2020). Hepatic glycogen synthase (GYS2) deficiency: Seven novel patients and seven novel variants. *JIMD Reports, 53*(1), 39–44.
13. Kim, S. J., & Bostwick, W. (2020). Social vulnerability and racial inequality in COVID-19 deaths in Chicago. *Health Education & Behavior, 47*(4), 509–513.
14. Kithiia, J., Wanyonyi, I., Maina, J., Jefwa, T., & Gamoyo, M. (2020). The socio-economic impacts of Covid-19 restrictions: Data from the coastal city of Mombasa, Kenya. *Data in Brief, 33*, Article 106317.
15. Kuang, Y., Xu, M. X., Yang, S. W., Ding, Y., Zheng, R., Taplin, J., & Li, S. (2021). Which information frame is best for reporting news on the COVID-19 pandemic? An online questionnaire study in China. *Psychology Research and Behavior Management*, 563–574.
16. Laurencin, C. T., & McClinton, A. (2020). The COVID-19 pandemic: A call to action to identify and address racial and ethnic disparities. *Journal of Racial and Ethnic Health Disparities, 7*, 398–402.
17. Li, C., Yang, Y., & Ren, L. (2020). Genetic evolution analysis of 2019 novel coronavirus and coronavirus from other species. *Infection, Genetics and Evolution, 82*, Article 104285.
18. Mitchell, K., & Štulhofer, A. (2021). Online sexual harassment and negative mood in Croatian female adolescents. *European Child & Adolescent Psychiatry, 30*(2), 225–231.

19. Mofijur, M., Fattah, I. R., Alam, M. A., Islam, A. S., Ong, H. C., Rahman, S. A., Najafi, G., Ahmed, S. F., Alhaz Uddin, Md., & Mahlia, T. M. I. (2021). Impact of COVID-19 on the social, economic, environmental, and energy domains: Lessons learnt from a global pandemic. *Sustainable Production and Consumption, 26*, 343–359.
20. Motamed-Jahromi, M., & Kaveh, M. H. (2021). The social consequences of the Novel Coronavirus Disease (COVID-19) outbreak in Iran: Is social capital at risk? A qualitative study. *Interdisciplinary Perspectives on Infectious Diseases, 2021*(1), 5553859.
21. Niedenthal, P. M., & Setterlund, M. B. (1994). Emotion congruence in perception. *Personality and Social Psychology Bulletin, 20*(4), 401–411.
22. Paul, E., Steptoe, A., & Fancourt, D. (2021). Attitudes towards vaccines and intention to vaccinate against COVID-19: Implications for public health communications. *The Lancet Regional Health–Europe, 1*.
23. Peker, A., Cengiz, S., & Nebioğlu Yıldız, M. (2021). The mediation relationship between life satisfaction and subjective vitality fear of COVID-19 and problematic internet use. *Turkish Journal of Clinical Psychiatry, 24*(2), 199–206.
24. Pietrabissa, G., & Simpson, S. G. (2020). Psychological consequences of social isolation during COVID-19 outbreak. *Frontiers in Psychology, 11*, 2201.
25. Pitas, N., & Ehmer, C. (2020). Social capital in the response to COVID-19. *American Journal of Health Promotion, 34*(8), 942–944.
26. Pokhrel, S., & Chhetri, R. (2021). A literature review on the impact of COVID-19 pandemic on teaching and learning. *Higher Education for the Future, 8*(1), 133–141.
27. Poole, D. C., Behnke, B. J., & Musch, T. I. (2021). The role of vascular function on exercise capacity in health and disease. *The Journal of Physiology, 599*(3), 889–910.
28. Robinson, E., Sutin, A. R., Daly, M., & Jones, A. (2022). A systematic review and meta-analysis of longitudinal cohort studies comparing mental health before versus during the COVID-19 pandemic in 2020. *Journal of Affective Disorders, 296*, 567–576.
29. Saladino, V., Algeri, D., & Auriemma, V. (2020). The psychological and social impact of COVID-19: New perspectives of well-being. *Frontiers in Psychology, 11*, Article 577684.
30. Sanderson, W. C., Arunagiri, V., Funk, A. P., Ginsburg, K. L., Krychiw, J. K., Limowski, A. R., Olesnycky, O. S., & Stout, Z. (2020). The nature and treatment of pandemic-related psychological distress. *Journal of Contemporary Psychotherapy, 50*, 251–263
31. Sugawara, N., Yasui-Furukori, N., Fukushima, A., & Shimoda, K. (2021). Attitudes of medical students toward COVID-19 vaccination: Who is willing to receive a third dose of the vaccine? *Vaccines, 9*(11), 1295.
32. Sun, P., Lu, X., Xu, C., Sun, W., & Pan, B. (2020). Understanding of COVID-19 based on current evidence. *Journal of Medical Virology, 92*(6), 548–551.
33. Tugade, M. M., Devlin, H. C., & Fredrickson, B. L. (2021). Positive emotions. *The Oxford Handbook of Positive Psychology*, 18–32.
34. Van Dorn, A., Cooney, R. E., & Sabin, M. L. (2020). COVID-19 exacerbating inequalities in the US. *Lancet (London, England), 395*(10232), 1243.
35. Wassler, P., & Talarico, C. (2021). Sociocultural impacts of COVID-19: A social representations perspective. *Tourism Management Perspectives, 38*, Article 100813.
36. World Health Organization, & United Nations Children's Fund. (2020). Levels and trends in child malnutrition: Key findings of the 2020 edition. *UNICEF/WHO/World Bank Group Joint Child Malnutrition Estimates*. World Health Organization.
37. World Health Organization. (2021). COVID-19 clinical management: Living guidance, 25 January 2021 (No. WHO/2019-nCoV/clinical/2021.1). *World Health Organization.*
38. Yang, Y., Peng, F., Wang, R., Guan, K., Jiang, T., Xu, G., Sun, J. & Chang, C. (2020). The deadly coronaviruses: The 2003 SARS pandemic and the 2020 novel coronavirus epidemic in China. *Journal of Autoimmunity, 109*, 102434.
39. Zhang, Y., Zeng, G., Pan, H., Li, C., Hu, Y., Chu, K., Han, W., Chen, Z., Tang, R., Yin, W., Chen, X., Hu, Y., Liu, X., Jiang, C., Li, J., Yang, M., Song, Y., Wang, X., Gao, Q. & Zhu, F. (2021). Safety, tolerability, and immunogenicity of an inactivated SARS-CoV-2 vaccine in healthy adults aged 18–59 years: A randomised, double-blind, placebo-controlled, phase 1/2 clinical trial. *The Lancet Infectious Diseases, 21*(2), 181–192.

40. Zautra, A. (2020). Zautra, Alex. In *Encyclopedia of Personality and Individual Differences* (pp. 5837–5839). Cham: Springer International Publishing.

Ipseeta Satpathy Senior Professor at the School of Management, KIIT University, Bhubaneswar, India, with over 22 years of experience in teaching, research, and administration. she has authored six textbooks, including Environment Management, Business Communication and Personality Development, and Transnational Marketing Strategy. She has published more than 200 research papers in national and international journals, including Scopus-indexed journals, and have presented at global conferences, and won several Best Paper Awards. She has been a visiting professor at Lynchburg College, USA, and have contributed extensively to the field of management education.

Debashree Chakravarty PhD Scholar at the School of Management, KIIT University, Bhubaneswar, India. Her research interests include organizational behavior, human resource management, and business strategy. She has worked on various research papers. She has aimed to contribute to the development of sustainable business practices. As a scholar, she is engaged in interdisciplinary studies to explore business ethics and corporate governance issues.

Md. Zahir Uddin Arif Professor in the Department of Marketing, Faculty of Business Studies, Jagannath University, Dhaka, Bangladesh. His research interests include Marketing, Brand Marketing, Pharmaceutical Marketing, Consumer Behavior, Human Resource Management, Career Development, Tourism, Green Economy, Development Issues, Women & Gender Issues, Ethics & Social Responsibility, and International Business. He has authored numerous research papers and contributed to interdisciplinary studies. He is actively engaged in teaching, research, and consulting in those areas, making significant contributions to both academia and industry. His work also focuses on sustainable business practices and social responsibility.

B. C. M. Patnaik Professor at the School of Management, KIIT University, Bhubaneswar, India, with over 20 years of experience in educational research, content development, and creative initiatives. He has specialized in Accounts & Finance, organizational behavior, and staff training and development. He has published more than 150 research papers, including 34 in Scopus-indexed journals. He is committed to student development and have received the Best Faculty Award four times. He has excelled in managing both academic and administrative responsibilities.

Chapter 22
Sustainable Consumption of Food During Covid-19: A Conceptual Framework Through Literature Review with Focus on Awe

Arundhati Das Chatterjee and Soma Sur

Abstract Ongoing environmental degradation and continuous research on its impact on human lives have made it clear that one of the ill-effects of ecological imbalance is emergence of life-threatening diseases that can create devastation. Even before mankind fathomed its true implication the world witnessed Covid-19 (2019–2022) that wreaked havoc globally. Among the different areas of life which got affected in the pandemic it is noted that consumers' choice behavior with regard to consumption during Covid-19 has become very prominent. One plausible reason behind it is that consumers' behavior now have started shifting towards sustainable consumption. It therefore becomes pertinent to study whether pandemic experience have brought this change.

This study utilizes the theoretical knowledge of Awe which is a powerful emotion conveying perceptual vastness. Within the realm of academic studies, it has been found to create pro-environmental behavior. This research considers it as a construct and further identifies other relevant variables to develop a model of sustainable consumption during Covid-19. Six key elements along with subconstructs are used in the framework. It helps to explain the factors of motivation behind consumers' shift in consumption behavioral pattern during the pandemic.

Keywords Sustainable consumption · Covid-19 · Pandemic · Awe · Consumer behavior

---

A. Das Chatterjee (✉)
St. Xavier's University, Kolkata, India
e-mail: arundhati.dchatterjee@gmail.com

S. Sur
Xavier Business School, St. Xavier's University, Kolkata, India
e-mail: soma.sur@sxuk.edu.in

S. Gangopadhyay et al. (eds.), *Pandemic Diaries*,
https://doi.org/10.1007/978-981-96-5415-4_22

22.1 Introduction

According to *Sustainable Development Report* sustainable consumption occurs when ecological footprint is less or has a smaller impact with respect to its biocapacity. This indicates that consumers should limit their usage of resources to those that are naturally produced in the given area and should refrain from using more than the boundary of resources [15]. The main objective of sustainable consumption is to reduce the impact of goods and services on environment through consumption [53]. Companies gradually have started introducing policies that is aligned with the goal of sustainability. However, it is seen that even if the concept of sustainability is making way into the society it is not always translated into sustainable consumer behavior [40]. Sustainable consumer behavior is likely to occur out of awareness linked to consequences over deterioration in natural or social environment [37]. The factors contributing to sustainable consumer behavior are myriad and complex. Special times like the pandemic have altered the pattern of consumer behavior like never before. Unexpected outbreak of Covid-19 has led to changes in consumers' dietary habits and consumers behavioral pattern changed progressively during the pandemic's evolutionary phases [6]. Researches have shown that consumers' buying pattern became moderate with more healthier choices thereby shifting towards sustainable consumption [7]. Researches have tried to understand consumers response with regard to food consumption during the pandemic with the help of different theories in consumer behavior. Prior research has looked into the relevance of Awe in making a choice decisions regarding utilitarian products [58]. However, psychological impact during Covid-19 has been profuse which have altered consumers' behavior and there are elements of abstractness that cannot be explained merely through utilitarian or hedonic consumer behavioral models.

This study with the help of literature review identifies the commonalities within the publications to build a framework that can rationalize consumers' shift towards a more sustainable food consumption.

The research at first briefly discusses the concept of sustainable consumption particularly with regard to food and beverages as found in existing literatures before moving in to the research objectives, methodology and final analysis. It takes the help of *Awe* as a theoretical construct and then reviews selected papers to identify the key factors of motivation behind sustainable consumption during Covid-19. These factors helped in developing a framework for the study.

22.2 Survey of Literature

22.2.1 Sustainable Consumption

The term "*sustainable consumption*" has assumed varied definitions in existing researches. [43] proposes 27 definitions by reviewing literatures between 1997 and 2020. [56] defines it as consumption of an environmentally friendly product or purchasing local [49]. [19] mentions certain agreed-upon features of sustainable consumption some of which are satisfying basic requirements, consuming moderately, minimizing waste and pollution and making ecologically friendly consumption choices. However, a working definition of sustainable consumption was proposed in the Oslo Symposium, 1994 as "the use of goods and services that respond to basic needs and bring a better quality of life, while minimising the use of natural resources, toxic materials and emissions of waste and pollutants over the life cycle, so as not to jeopardise the needs of future generations" (OECD, 2008). Influenced by the definition a common vision has been developed by researchers across the world and that is one of the key element of achieving sustainability in consumption is for the consumers to adopt a new attitude towards lifestyle. Consumer behavior depends on a wide array of internal and external stimuli [17] like external environment. The outbreak of the coronavirus during 2019 provided a drastic change in external environment due to which consumers started to respond and change their behavioral pattern.

22.2.2 The Pandemic and Consumer Behavior

From the beginning of the year 2020 reports of individuals being infected by SARS-Cov2 caused a stir in most of the countries of the world. By March 2020, major countries enforced stringent lockdown measures. Lockdown affected individuals' lives like never before. It was one of the special occasions in history when the entire world was unanimous in its responds against a certain event. Since lives came to a standstill the challenges of meeting the basic requirements of life started to affect individuals' behavior. Researchers found that consumers' behavior was oriented towards online procurement [35] preference towards ordering from virtual stores through increased uses of mobile apps increased [2] in order to maintain isolation protocol during the pandemic. Lockdown initially triggered a sense of disbelief followed by stress and apprehension which boosted stockpiling of food [28] and overconsumption [3]. Literature finds that with gradual flattening of the curve the overconsumption pattern gradually became moderate with consumers opting for sustainable healthy food choices [48]. Studies find that environmental awareness and concerns [27], an element embedded in the concept of sustainable consumption often helps in consuming green [10]. The consensus opinion among researchers and marketers is that Covid-19 radically altered the functioning of modern society by creating significant changes in behavioral patterns of consumers [18, 24]. Pandemic was the perfect context through which a rise in green behavior among consumers was seen [45].

22.2.3 Reviewing the Concept of Awe

Awe is an emotional response to a perception of a large stimulus faced by individuals that overwhelm their knowledge or mental state. It is a complex emotion associated with a negative feeling of shock, fear and anxiety and also a positive feeling of beauty and virtue. Awe initiating stimulus are characterized by two essential features [26]. One of it is "Perceived vastness" which in this case means the worldwide impact and negative consequences of Covid-19 should be considered as a threat with perceived vastness. The other notion is "Need for accommodation" which under this context means adapting through cognitive and behavioral changes. Both the features qualify the context of Covid-19. Though application of awe in the field of social science research is limited awe has been linked to green marketing and sustainable consumption. Studies find experience of awe leads to pro-social behavior [58]. The "need for accommodation" as conveyed in the theory is applied consumers' expansion of conventional frame of reference and responding to stimuli like pandemic through a new cognitive way in this case by purchasing green [26]. Existing literature also finds that awe can impact consumer behavior through re-examining attitude towards consumption and correct unhealthy lifestyle [11]. The value of self-transcendence as is embedded behind the concept of awe evoking theory act as an impetus to work towards self-less, pro-social and green behavior thereby motivating sustainable consumption. Hence, Awe can be used as a relevant construct in this contextual study.

22.2.4 Gap in the Study

1. Studies of sustainable consumption as a part of green consumer behavior have shown that there exists a complex relationship between its antecedents and the behavioral effect which requires more understanding.
2. While consumption pattern during Covid-19 reflects a shift towards green sustainability, a traditional green consumer behavior model may not be sufficient to explain the factors behind this transition.
3. The conceptual characteristics of Awe namely "Perceived vastness" and "Need for accommodation" are relevant for the analysis in Covid-19. However, studies related to sustainable consumption pattern in the pandemic do not utilize the concept to a great extent.
4. With rising environment deterioration there are looming threats of more such pandemic situations. A conceptual framework on the factors that triggers a shift

in behavioral pattern of consumers during pandemic situation is not prevalent in existing studies.

22.3 Methods

The research design used in this study is to gain an insight on the factors that motivated consumers towards sustainable food consumption during the Covid-19 period. The first step includes understanding the concept of "*Awe*" which is a relevant and important idea in this context. For the purpose relevant publications were selected from database Dimensions.AI. The keywords used for the purpose was "*Sustainable Consumption* (or) *consumer behaviour during Covid 19…*" The filter applied in the search technique was publication year between 2020 and 2024. 399 papers came as a search result. The data set was downloaded in an excel format and filter was applied with "food." Based on the criteria 23 papers were chosen and reviewed. The initial selection process is shown in a schematic Fig. 22.1.

The descriptive studies relating to the publications as provided by Dimensions.AI are first provided in Figs. 22.2 and 22.3. *Next* the selected articles are reviewed and their themes along with the country of research are listed and tabulated in Table 22.1. Once the papers are analyzed a further analysis is done to see number of times a

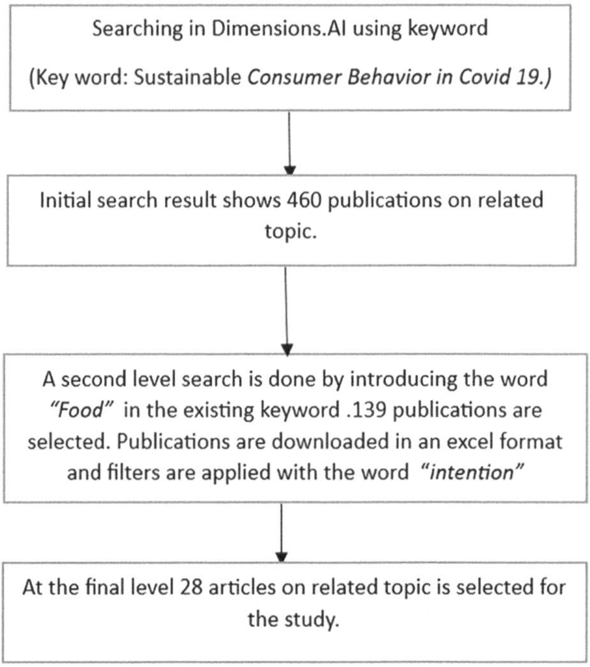

Fig. 22.1 Schematic flow chart on the selection process

factor has been chosen for studying sustainable consumer behavior. Publications with similar variables are grouped and analyzed for better understanding. It is shown in Table 22.2. After the common factors of motivation behind the sustainable behavior is identified and analyzed a conceptual framework is developed using them as a construct for this study.

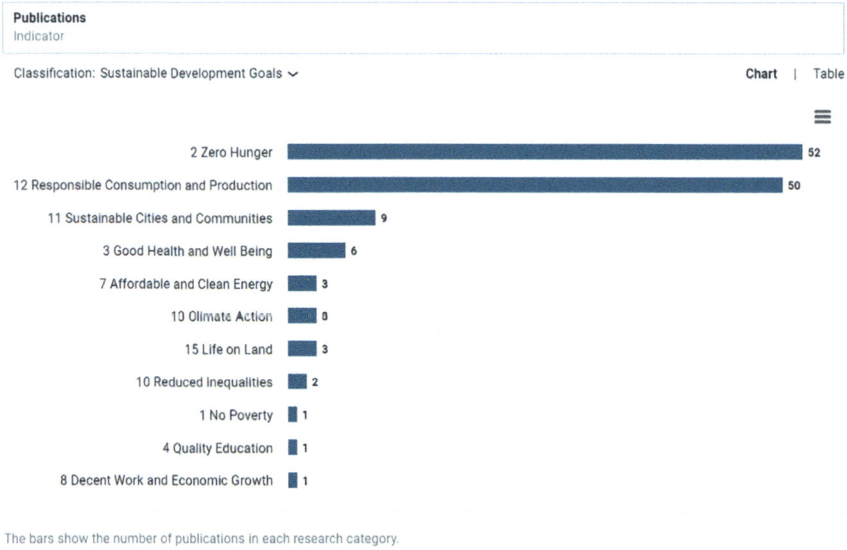

Fig. 22.2 Bar chart showing grouping of publications according to SDG. *Source* Dimensions.AI

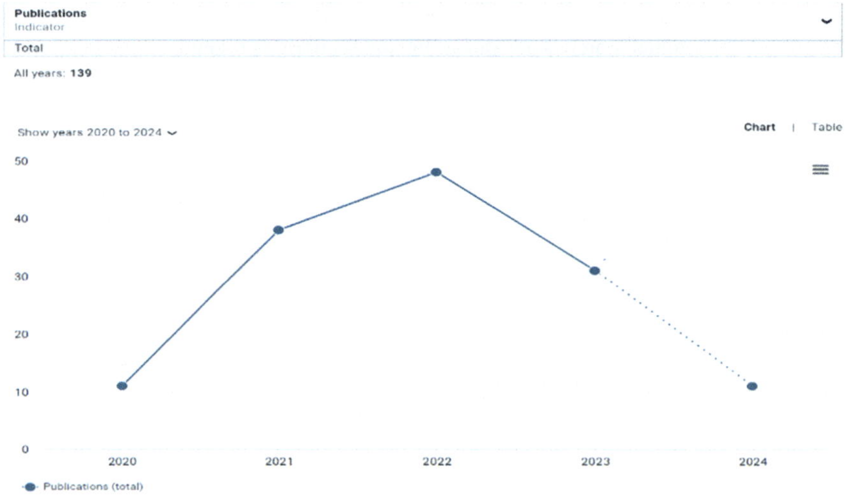

Fig. 22.3 Publication trend in the year (2020–2024). *Source* Dimensions.AI

Table 22.1 Reviewed articles on the basis of themes, authors and country of study

Authors	Country	Themes
1. [48]	Italy	Impact of lockdown on self-perceived food choice, impulse buying and food waste production. Lockdown altered consumers' behavior with respect to food leading to a more positive and sustainable mode of consumption
2. [29]	China	Impact of lockdown on sustainable food consumption. Gender, age, household size and attitude towards risk contribute in shift in consumers' sustainable food choice behavior
3. [13]	Zimbabwe	Nexus between organic food adoption behavior and perceived consumer longevity after pandemic experience. Attitude towards green, subjective norms and perceived behavioral control contributes in developing green purchase intent
4. [34]	France	How lockdown changed food choice motives with respect to nutritional quality. Although there was a decrease in nutritional quality on an average the lockdown has altered food choice motive for some with increased consumption of sustainable food choices
5. [55]	Italy	Recovery during post-Covid-19 period with regard to consumer purchase and consumption behavior. Mass media, optimism of small things influences food, health and environment leading towards a more sustainable society
6. [57]	Poland	How food traceability has influenced consumers' choices in the pandemic period. Purchasing locally produced food and consumers' awareness regarding cleanliness of shops before purchase increased. Accuracy of information regarding food labels also became a priority
7. [42]	China	Extending TPB model by adding constructs like moral attitude, health awareness in the study of consumers' food choice behavior during the pandemic era. The extended variables have a positive impact on consumers food choices leading to a sustainable consumption but relationship between subjective norm and green purchase intent was not so significant
8. [5]	Romania	How sustainable food consumption pattern concerning the two broad strategies of sustainable consumption, namely sustainable product choice and sustainable dietary composition of food, has been altered during pandemic. Purchase of local food increased and fresh vegetables increased while food waste and organic purchase decreased
9. [44]	Italy	Awareness of Sapienza University, Italy regarding sustainable food consumption during pandemic. The findings help in developing a model on sustainable food consumption during Covid-19
10. [7]	Italy	Over view of changes in consumption pattern during Covid-19 lockdown. How evolution of consumption pattern relies to a great extent on evolution of supply chain. Sustainable food consumption during Covid-19 required a sustainable food system existing in the country that will include infrastructure to move goods rather than consumers

(continued)

Table 22.1 (continued)

Authors	Country	Themes
11. [51]	Greece	Impact of Covid-19 on consumers' motive to consume quality beverage particularly local Samos' Greek wine. Wine consumption decreased during Covid-19 with consumers' preferring to purchase online or from local stores rather than from supermarket. Sustainable and good quality wine are the choices of consumers but the consumption became moderate during pandemic
12. [30]	China	Impact of the pandemic on food consumption pattern and food diversity with focus on income challenges affecting consumption. Individual food consumption decreased and consumers obtaining self-production method like kitchen made or kitchen gardening is likely to get less affected during pandemic times
13. [23]	Germany	Whether society will try to maintain sustainable consumption even after Covid-19 period. Environmental and societal concern followed by a voluntary simplistic consciousness deteriorated among consumers
14. [41]	Germany	From the perspective of sustainable food consumption how Covid-19 has started a different diet trend among different population groups. More ready-made meals were consumed and canned product with more shelf life were consumed. Consumption of alcohol and sweets increased indicating a downward trend in sustainable consumption
15. [52]	Poland	Examine the reduction in food waste by students at the University of Zielona Gora, Poland during the lockdown phase. The study considers two phases one during Covid-19 lockdown and the other after the lockdown was withdrawn but the threat of pandemic still persisted. Results suggest no remarkable change in consumption pattern with regard to food among the students. However, consumer preferred frequent purchases while in second phase it was not so
16. [12]	Romania	A bibliometric analysis of publications on sustainable consumption of organic food. Publication trend suggest although pattern of consumers' food consumption have shifted more towards a sustainable path interest in research have been moderate
17. [37]	Romania	Analysis of consumers' sustainable food behavior during the pandemic. Consumers response indicates a positive attitude towards sustainable food behavior which is mainly influenced by factors like age and education level
18. [1]	Iran	Changing trend in sustainable consumer behavior in chain stores during Covid-19 times in Iran. The pandemic has got a significant role in developing attitude towards sustainable consumption
19. [10]	Malaysia	Consumers' green product purchase attitude and intentions during Covid-19. Study suggests fear plays a key role in developing the intentions
20. [38]	UK	The consumer behavioral pattern in sourcing meals and accessing foods before and during the pandemic. Health concern is predominant in the changing pattern of behavioral intentions that motivate the process
21. [38]	Pakistan	Factors responsible for green purchase behavior during Covid-19. Fear factor played a positive role in shifting towards sustainable purchase alongside with psychological distress and mortality salience

(continued)

Table 22.1 (continued)

Authors	Country	Themes
22. [16]	Review	Explains the effect of pandemic on green purchase behavior with the help of an integrated model
23. [22]	Malaysia	The impact of the fear of Covid-19 on consumer behavioral intention to purchase green with special focus on the impact of Covid-19 on health, intolerance of uncertainty and social media information
24. [9]	Malaysia	Exploring attitude-behavior gap in organic food consumption during Covid-19. The studies find that motives and barriers coexist in creating positive attitude towards sustainable consumption during Covid-19
25. [36]	Turkey	Exploring impact of consumer behavior during and after lockdown. Anxiety and fear have led to adoption of consumption practices some of which were temporary and did not persist after lockdown
26. [33]	Pakistan	Exploring the sustainable food consumption from the purchase intention to the purchase competition phase during Covid-19
27. [32]	China	Understanding organic consumption behavior based on Stimulus-Organism-Response model. Associations are found between anxiety, uncertainty and sustainable consumption with perceived values
28. [46]	India	Understanding how consumers' choice decisions have shifted towards sustainability due to the Covid-19 and the marketing opportunities that it had opened

Table 22.2 Variables used in majority of the studies that is used as a construct

1. Attitude	
Towards sustainability	[9, 12, 13, 16, 38, 55]
Towards risk	[29]
2. Health Concern	[12, 44, 30, 42]
3. Perceptions	
About longevity	[13]
About sustainable product	[12, 37]
About risk	[29]
4. Stress/Fear	
Regarding Covid-19 infection	[1, 41, 45]
Regarding Jobs	[48]
5. Awareness/Information	
About environment	[44, 23]
About the pandemic	[29]
About food and beverages	[51]

According to the themes related to sustainable consumption of food and consumer behavior during the pandemic period it is found that Covid-19 protocols particularly the lockdown has impacted consumer behavior in both positive and negative ways. The major studies have tried to find out how isolation, apprehensions and uncertainties have impacted consumers' purchase and pattern of consumption. While standardized constructs of consumer behavioral models have been used in many articles' others have tried to incorporate variables to address it contextually. The studies are also highly skewed towards European countries as can be seen from Table 22.1.

22.4 Results and Discussion

Initial search results on sustainable consumer behavior suggest that the publication topic is relevant. Figure 22.2 points out that the topic is important from the perspective of *Sustainable Development Goals.* Particularly publications related to SDG*2 (Zero Hunger)* and *SDG12 (Responsible Consumption and production)* are high in the bar diagram. On the other hand, trend study in Fig. 22.3 highlights that post-Covid-19 publications on the related topic have been reducing. However, publications reached its peak during 2022. These results show that while much needs to be done with regard to SDG commitments customizing the targets at a contextual level like sustainability at times of pandemic is also required. As such there is a scope for further research in this area.

22.4.1 Factors Motivating Sustainable Consumption

The analyses have identified 5 major variables with some subtopics related to the variables in the publications. The variables are *Attitude, Health Concern, Perceptions, Stress and Awareness.* Each of the constructs has been used contextually with a focus on pattern of food consumption during the pandemic. Table 22.2 states how the variables are treated in each of the literatures. It is to be mentioned that based on the frequency of occurrence in the publications these 5 variables are selected while other variables on the related topic are not considered in this study. The explanation of the factors as found in the literatures reviewed is provided below which in the process helped in developing the constructs for the conceptual framework.

22.4.2 Attitude

Attitude has been treated differently in the articles on the related topic of food consumption during Covid-19. [30] analyses impact on attitude of consumer depending on "access-to-food" mode namely cooking at home, ordering from home, eating at restaurants. While consumer experiences are the main motivation in the last two cases during the Covid-19 phase the study finds consumers' procuring mainly through cooking which is healthier and sustainable. [12] investigates the Covid-19 impact on consumers' attitude towards organic consumption. Organic food is still a niche market although in recent times awareness about sustainable consumption is causing it to expand rapidly. The pandemic helped raise awareness about health issues due to which there is a consumer orientation towards organic food consumption. [13] finds that attitude is the main predictor of behavioral intent to purchase organic in the post-Covid-19 period. The study stresses on the fact that pandemic experience acted as a stimulus for the consumers to have a green orientation but organic food marketers should help further in creating positive attitudes towards organic food for those consumers who are still skeptic. Again [55] noted that during Covid-19 people have altered their attitude towards consumption and resumed a traditional approach of buying local and eating home-made food. However, according to the findings consumers' behavioral responses were the results of financial constraints. As such it is highlighted in the study that spending capacity is one of the barriers of maintaining sustainable consumption in food.

Covid-19 experience altered consumers' dietary habits due to associated stress. Attitude towards risk exerts an influence on behavioral intentions and in this study green purchase intent for consumption. [29] defines two essential components linked with risk; one which is uncertainty and the other is consequences. According to that study risk is defined as "uncertainty that matters" because "uncertainty without any consequence will not pose any risk." Results of the study highlights that organic food is still relatively an unfamiliar notion with less information and so in times of pandemic consumers who have neutral or aversive attitude towards risk purchased less organic food.

So, this review finds that developing attitude towards sustainable consumption during Covid-19 is mediated by the role of risk.

22.4.3 Health Concern

According to [22] consumer's risk perception has influenced their purchase intention for organic food during pandemic [21] and because consumers see organic items to be healthier, [47] out of health concern they opt for sustainable consumption. [59] which states organic food consumption during pandemic is linked to health concerns and immunity boosting. [42] states that studies have shown health concern have often acted as a motivator for consumers' environmentally friendly choices [25]. The results of the study by [42] highlights a positive relationship between health consciousness and green purchase intent during Covid-19. However, health concerns during Covid-19 did not always result in green consumption [30] reports that a reduction in consumption pattern of green vegetables was noted in parts of China which was probably a result of reduction in household income during the pandemic. Thus, income can be a moderating element in sustainable consumption of food during Covid-19.

22.4.4 Perceptions

The usage of consumers' perception as a construct in the context of sustainability during Covid-19 is varied. [37] observes that often awareness of environmental issues creates a perception about sustainable consumption as the key factor that will solve the problem [31] [37] further states that consumers' perception of sustainable consumption is multi layered [8] and may be associated with voluntary restraint of consumption out of ethical consideration[50] or buying green [14]or creating minimum wastes. However, it was observed in the study [37] that although there exists a positive perception of sustainable consumption during Covid-19 but there exists an attitude behavioral gap. The study explains price to be one of the reason causing barriers in sustainable consumption like buying from locals during pandemic.

[13] observes that consumer longevity remains the ultimate goal of green food adoption as it is considered healthy. Perception about longevity act as one of the key factors that influences green behavior and often organic consumption. The study finds that organic food consumption during Covid-19 is positively related to consumer's perception about physical wellness and longevity. On the other hand, [29] mentions in the study that risk perception can exert influence on behavioral intention. Perceptions of risk involves consumers' evaluation of problems to which they might be exposed in the future. [29] states that health risk is one of the major perceived risks during Covid-19 period which could have motivated consumers to opt for sustainable food.

22.4.5 Stress/Fear

The pandemic wreaked havoc and affected consumers' behavior in multiple ways. Its psychological impact is manifested through behavioral changes during Covid-19 like increasing consumption to cope with stress. Alongside with health concerns population was affected with the uncertainty of a possible job loss or unemployment [41]. Studies find that increased consumption of food and beverages particularly alcohol was common which hints at a negative impact on sustainable consumption. Also consuming processed "comfort food" act as a coping strategy [4]. Studies associate emotional anxiety [48] with working during lockdown as well thereby altering consumption pattern.

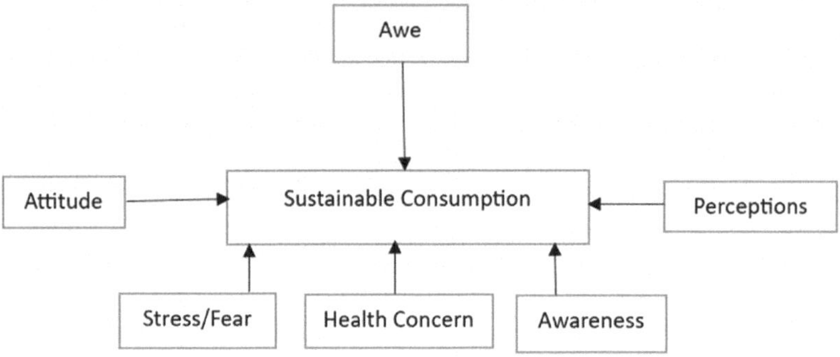

Fig. 22.4 Framework of sustainable consumption during Covid-19

22.4.6 Knowledge/Awareness

Existing literatures point out that consumers with enhanced knowledge or awareness tend to make sustainable food choices [39]. Knowledges concerning eco-label and environment often has positive correlation towards pro-environmental consumer behavior [54]. During pandemic research points out that consumers with more information on Covid-19 are likely to make sustainable choices as more knowledge means more perceive severity about disease [29]. Study relating to Greek wine suggests that consumers' purchase of wine entailed gathering in information on health attributes and quality [51]. Again, mentions that Covid-19 is an opportunity to shift consumption pattern to a more sustainable path. However, contrary to the above results [23] states that the notion of sustainable consumption is susceptible to change under volatile situation like the pandemic.

Based on the findings it can be stated that all the above variables are of interest in the context of pandemic. This study therefore proposes a conceptual model on the basis of the review literature (Fig. 22.4).

22.5 Conclusion

Research finds that although there has been a massive influx of research publications during Covid-19 [20] limited studies pertaining to impact of Covid-19 on consumer behavior exist. Sustainable consumption during this period was studied with regard to different sectors but understanding the consumption behavior with regard to food is relevant as it is the most basic need of existence. With the advent of research on sustainable consumption and food being one of the most essential items, it becomes pertinent to study the impact of the pandemic on consumers' green behavioral choices with respect to food and beverages.

The concept of sustainable consumption has many intricacies in its definition itself which are often not distinctly addressed. As such it becomes more explicit when the study is conducted contextually at a more micro-level. So sustainable consumption at the backdrop of the pandemic becomes more intriguing as a study topic. Since the study of consumer behavior is largely psychological in nature capturing the abstractness of the emotional elements becomes important. Literatures have addressed the topic through different theoretical approaches with *Theory of Planned Behavior* being applied most frequently. However, it has been observed that the overwhelming magnitude with which Covid-19 affected lives of people can be best studied with Awe whose application in the field of Covid-19 studies alongside with consumer behavior is limited.

Previous researches have highlighted that constant environmental degradation and as a consequence the looming threat of more pandemics in future. The study develops a framework that is not only relevant for Covid-19 but leaves a scope for application and research in years to come, if the world is gripped by similar situation. From the marketers point of view one needs to understand the triggers which causes consumers to respond sustainably in a post-Covid-19 world so that there may be better communication on the benefits of sustainable products. It will also help to bring out customized offerings for the consumers.

The study is limited since it has included literatures only from 2 databases namely *Dimension.AI and Semantic Scholar.* More studies would have helped in understanding the topic better. Also, the literatures are largely contextual on the basis of countries in which research was held as well as on the basis of sociodemographic characteristics. Since consumers' response to Covid-19 was constrained by several government-imposed protocols each study had its own limitations based on the economic vulnerability of the consumer data set. The aspect of income which remains one of the essential determinants of consumer behavior during Covid-19 was not considered in most of the articles studied in this research.

References

1. Aghaei, M., Sahebi, A. G., & Kordheydari, R. (2021). The effect of Covid-19 on sustainable consumption behavior in chain stores. *Asia-Pacific Journal of Management and Technology, 2*(02), 35–51. https://doi.org/10.46977/apjmt.2021v02i02.005
2. Andronie, M., Lăzăroiu, G., Ştefănescu, R., Ionescu, L., & Cocoşatu, M. (2021). Neuromanagement decision-making and cognitive algorithmic processes in the technological adoption of mobile commerce apps. *Oeconomia Copernicana, 12*(4), 1033–1062. https://doi.org/10.24136/oc.2021.034
3. Babbitt, C. W., Babbitt, G. A., & Oehman, J. M. (2021). Behavioral impacts on residential food provisioning, use, and waste during the COVID-19 pandemic. *Sustainable Production and Consumption, 28*, 315–325. https://doi.org/10.1016/j.spc.2021.04.012
4. Bracale, R., & Vaccaro, C. M. (2020). Changes in food choice following restrictive measures due to Covid-19. *Nutrition, Metabolism, and Cardiovascular Diseases, 30*(9), 1423–1426.
5. Brumă, I.-S., Ulman, S.-R., Tanasă, L., & Cautisanu, C. (2022). Implications of COVID-19 pandemic on sustainable consumption patterns. Evidence from Iasi County, Romania. *Frontiers in Sustainable Food Systems, 6*. https://doi.org/10.3389/fsufs.2022.1050977
6. Bustos, S. (2020). *COVID-19 and the food phenomena*. Food and Agriculture Organization of the United Nations (FAO).

7. Cavallo, C., Sacchi, G., & Carfora, V. (2020). Resilience effects in food consumption behaviour at the time of Covid-19: Perspectives from Italy. *Heliyon, 6*(12), Article e05676. https://doi.org/10.1016/j.heliyon.2020.e05676
8. Chan, R. Y. K. (2001). Determinants of Chinese consumers' green purchase behavior. *Psychology and Marketing, 18*(4), 389–413. https://doi.org/10.1002/mar.1013
9. Cheah, W. K. A., & Aigbogun, O. (2022). Exploring attitude-behaviour inconsistencies in organic food consumption during the COVID-19 pandemic in the Klang Valley, Malaysia. *Cleaner and Responsible Consumption, 7*, Article 100077. https://doi.org/10.1016/j.clrc.2022.100077
10. Chen, X., Rahman, M. K., Rana, Md. S., Gazi, Md. A. I., Rahaman, Md. A., & Nawi, N. C. (2022). Predicting consumer green product purchase attitudes and behavioral intention during COVID-19 pandemic. *Frontiers in Psychology, 12*. https://doi.org/10.3389/fpsyg.2021.760051
11. Cheung, W., Luke, M. A., & Maio, G. R. (2014). On attitudes towards humanity and climate change: The effects of humanity esteem and self-transcendence values on environmental concerns. *European Journal of Social Psychology, 44*(5), 496–506. https://doi.org/10.1002/ejsp.2037
12. Cramarenco, R. E., Burca-Voicu, M. I., & Dabija, D.-C. (2023). Organic food consumption during the COVID-19 pandemic. A bibliometric analysis and systematic review. *Amfiteatru Economic, 25*(Special 17), 1042. https://doi.org/10.24818/EA/2023/S17/1042
13. Dangaiso, P. (2023). Extending the theory of planned behavior to predict organic food adoption behavior and perceived consumer longevity in subsistence markets: A post-peak COVID-19 perspective. *Cogent Psychology, 10*(1). https://doi.org/10.1080/23311908.2023.2258677
14. Finisterra do Paço, A. M., & Raposo, M. L. B. (2008). Determining the characteristics to profile the "green" consumer: an exploratory approach. *International Review on Public and Nonprofit Marketing, 5*(2), 129–140. https://doi.org/10.1007/s12208-008-0010-9
15. Fischer, M., Foord, D., Frecè, J., Hillebrand, K., Kissling-Näf, I., Meili, R., Peskova, M., Risi, D., Schmidpeter, R., & Stucki, T. (2023). *Sustainable Consumption* (pp. 105–116). https://doi.org/10.1007/978-3-031-25397-3_7
16. George, A., & Nair, A. S. (2022). Reflections on green purchase behaviour in the era of COVID-19: A conceptual framework. *Vision: The Journal of Business Perspective*, 097226292210873. https://doi.org/10.1177/09722629221087363
17. Gorrgitano, M., & Sodano, V. (2014). Sustainable food consumption: Concept and policies. *Quality-Access to Success, 15*, 207–212.
18. Gąsiorowska-Mącznik, E. (2022). The impact of Covid-19 pandemic on consumer purchasing decisions. *Scientific Papers of Silesian University of Technology. Organization and Management Series, 2022*(162), 167–179. https://doi.org/10.29119/1641-3466.2022.162.9
19. Haider, M., Shannon, R., & Moschis, G. P. (2022). Sustainable consumption research and the role of marketing: A review of the literature (1976–2021). *Sustainability, 14*(7), 3999. https://doi.org/10.3390/su14073999
20. Harper, L., Kalfa, N., Beckers, G. M. A., Kaefer, M., Nieuwhof-Leppink, A. J., Fossum, M., Herbst, K. W., & Bagli, D. (2020). The impact of COVID-19 on research. *Journal of Pediatric Urology, 16*(5), 715–716. https://doi.org/10.1016/j.jpurol.2020.07.002
21. Hesham, F., Riadh, H., & Siham, N. K. (2021). What have we learned about the effects of the COVID-19 pandemic on consumer behavior? *Sustainability, 13*(08).
22. Hu, P., Bhuiyan, M. A., Rahman, M. K., Hossain, M. M., & Akter, S. (2022). Impact of COVID-19 pandemic on consumer behavioural intention to purchase green products. *PLoS ONE, 17*(10), Article e0275541. https://doi.org/10.1371/journal.pone.0275541
23. Hüttel, A., & Balderjahn, I. (2022). The coronavirus pandemic: A window of opportunity for sustainable consumption or a time of turning away? *Journal of Consumer Affairs, 56*(1), 68–96. https://doi.org/10.1111/joca.12419
24. Islam, T., Pitafi, A. H., Arya, V., Wang, Y., Akhtar, N., Mubarik, S., & Xiaobei, L. (2021). Panic buying in the COVID-19 pandemic: A multi-country examination. *Journal of Retailing and Consumer Services, 59*, Article 102357. https://doi.org/10.1016/j.jretconser.2020.102357

25. Katt, F., & Meixner, O. (2020). Is it all about the price? An analysis of the purchase intention for organic food in a discount setting by means of structural equation modeling. *Foods, 9*(4), 458. https://doi.org/10.3390/foods9040458
26. Keltner, D., & Haidt, J. (2003). Approaching awe, a moral, spiritual, and aesthetic emotion. *Cognition and Emotion, 17*(2), 297–314. https://doi.org/10.1080/02699930302297
27. Kennedy, E. H., & Hauslik, D. (2018). The practice of green consumption. In *Environment and Society* (pp. 187–206). Springer International Publishing. https://doi.org/10.1007/978-3-319-76415-3_9
28. Koch, J., Frommeyer, B., & Schewe, G. (2020). Online shopping motives during the COVID-19 pandemic—Lessons from the crisis. *Sustainability, 12*(24), 10247. https://doi.org/10.3390/su122410247
29. Li, S., Kallas, Z., & Rahmani, D. (2022). Did the COVID-19 lockdown affect consumers' sustainable behaviour in food purchasing and consumption in China? *Food Control, 132*, Article 108352. https://doi.org/10.1016/j.foodcont.2021.108352
30. Lian, T., Haq, S. ul, Shahbaz, P., Zhao, L., Nadeem, M., & Aziz, B. (2022). Changing food patterns during the pandemic: Exploring the role of household dynamics and income stabilization strategies. *Sustainability, 15*(1), 123. https://doi.org/10.3390/su15010123
31. Liu, Y., Qu, Y., Lei, Z., & Jia, H. (2017). Understanding the evolution of sustainable consumption research. *Sustainable Development, 25*(5), 414–430. https://doi.org/10.1002/sd.1671
32. Liu, C., Zheng, Y., & Cao, D. (2021). Similarity effect and purchase behavior of organic food under the mediating role of perceived values in the context of COVID-19. *Frontiers in Psychology, 12*. https://doi.org/10.3389/fpsyg.2021.628342
33. Mahmood, B., Mahmood, A., & Ali, Y. T. (2022). Sustainability in food purchase decision A qualitative study on factors influencing sustainable consumer behavior in Pakistan. *Global Journal for Management and Administrative Sciences, 3*(1), 101–118.
34. Marty, L., de Lauzon-Guillain, B., Labesse, M., & Nicklaus, S. (2021). Food choice motives and the nutritional quality of diet during the COVID-19 lockdown in France. *Appetite, 157*, Article 105005. https://doi.org/10.1016/j.appet.2020.105005
35. McKinsey. (2020). *Consumer sentiment and behavior continue to reflect the uncertainity of the COVID-19 crisis*. https://www.Mckinsey.Com/Capabilities/Growth-Marketing-and-Sales/Our-Insights/a-Global-View-of-How-Consumer-Behavior-Ischanging-amid-Covid-19
36. Mol, S., Karakaş Geyik, S., & Süren, Y. (2024). The impact of the COVID-19 pandemic on fish and food perception, consumption, and purchasing behaviors of Turkish consumers. *Aquatic Research, 7*(1), 15–29. https://doi.org/10.3153/AR24003
37. Muresan, I. C., Harun, R., Arion, F. H., Brata, A. M., Chereches, I. A., Chiciudean, G. O., Dumitras, D. E., Oroian, C. F., & Tirpe, O. P. (2021). Consumers' attitude towards sustainable food consumption during the COVID-19 pandemic in Romania. *Agriculture, 11*(11), 1050. https://doi.org/10.3390/agriculture11111050
38. Pan, Y., & Rizov, M. (2022). Consumer behaviour in sourcing meals during COVID-19: Implications for business and marketing. *Sustainability, 14*(21), 13837. https://doi.org/10.3390/su142113837
39. Peschel, A. O., Grebitus, C., Steiner, B., & Veeman, M. (2016). How does consumer knowledge affect environmentally sustainable choices? Evidence from a cross-country latent class analysis of food labels. *Appetite, 106*, 78–91. https://doi.org/10.1016/j.appet.2016.02.162
40. Phipps, M., Ozanne, L. K., Luchs, M. G., Subrahmanyan, S., Kapitan, S., Catlin, J. R., Gau, R., Naylor, R. W., Rose, R. L., Simpson, B., & Weaver, T. (2013). Understanding the inherent complexity of sustainable consumption: A social cognitive framework. *Journal of Business Research, 66*(8), 1227–1234. https://doi.org/10.1016/j.jbusres.2012.08.016
41. Profeta, A., Siddiqui, S. A., Smetana, S., Hossaini, S. M., Heinz, V., & Kircher, C. (2021). The impact of Corona pandemic on consumer's food consumption. *Journal of Consumer Protection and Food Safety, 16*(4), 305–314. https://doi.org/10.1007/s00003-021-01341-1
42. Qi, X., & Ploeger, A. (2021). Explaining Chinese consumers' green food purchase intentions during the COVID-19 pandemic: An extended theory of planned behaviour. *Foods, 10*(6), 1200. https://doi.org/10.3390/foods10061200

43. Quoquab, F., & Mohammad, J. (2020). A review of sustainable consumption (2000 to 2020): What we know and what we need to know. *Journal of Global Marketing, 33*(5), 305–334. https://doi.org/10.1080/08911762.2020.1811441
44. da Rocha Ramos, L. A., Zecca, F., & Del Regno, C. (2022). Needs of sustainable food consumption in the pandemic era: First results of case study. *Sustainability, 14*(16), 9852. https://doi.org/10.3390/su14169852
45. Sajid, K. S., Hussain, S., Hussain, R. I., & Mustafa, B. (2022). The effect of fear of COVID-19 on green purchase behavior in Pakistan: A multi-group analysis between infected and non-infected. *Frontiers in Psychology, 13*. https://doi.org/10.3389/fpsyg.2022.826870
46. Salah, A., Khaled, S. D. A., Alomari, K., & Tabash, M. (2023). COVID-19 pandemic roles on consumer behaviour towards sustainable transitions: A retail industry survey. *International Journal of Innovation and Sustainable Development, 17.*
47. Savarese, M., Wismer, W., & Graffigna, G. (2020). Conceptualizing "free-from" food consumption determinants a systematic integrative literature review focused on gluten and lactose. *Food Quality and Preference, 10410.*
48. Scacchi, A., Catozzi, D., Boietti, E., Bert, F., & Siliquini, R. (2021). COVID-19 lockdown and self-perceived changes of food choice, waste, impulse buying and their determinants in Italy: QuarantEat, a cross-sectional study. *Foods, 10*(2), 306. https://doi.org/10.3390/foods10020306
49. Schmitt, E., Keech, D., Maye, D., Barjolle, D., & Kirwan, J. (2016). Comparing the sustainability of local and global food chains: A case study of cheese products in Switzerland and the UK. *Sustainability, 8*(5), 419. https://doi.org/10.3390/su8050419
50. Shaw, D., & Newholm, T. (2002). Voluntary simplicity and the ethics of consumption. *Psychology and Marketing, 19*(2), 167–185. https://doi.org/10.1002/mar.10008
51. Skalkos, D., Roumeliotis, N., Kosma, I. S., Yiakoumettis, C., & Karantonis, H. C. (2022). The impact of COVID-19 on consumers' motives in purchasing and consuming quality Greek wine. *Sustainability, 14*(13), 7769. https://doi.org/10.3390/su14137769
52. Skawińska, E., Zalewski, R. I., & Wyrwa, J. (2023). Students' food consumption behavior during COVID-19 lockdown. *Sustainability, 15*(12), 9449. https://doi.org/10.3390/su15129449
53. Sreen, N., Yadav, R., Kumar, S., & Gleim, M. (2021). The impact of the institutional environment on green consumption in India. *Journal of Consumer Marketing, 38*(1), 47–57. https://doi.org/10.1108/JCM-12-2019-3536
54. Taufique, K. M. R., Vocino, A., & Polonsky, M. J. (2017). The influence of eco-label knowledge and trust on pro-environmental consumer behaviour in an emerging market. *Journal of Strategic Marketing, 25*(7), 511–529. https://doi.org/10.1080/0965254X.2016.1240219
55. Timpanaro, G., & Cascone, G. (2022). Food consumption and the Covid-19 pandemic: The role of sustainability in purchasing choices. *Journal of Agriculture and Food Research, 10*, Article 100385. https://doi.org/10.1016/j.jafr.2022.100385
56. Tomşa, M.-M., Romonţi-Maniu, A.-I., & Scridon, M.-A. (2021). Is sustainable consumption translated into ethical consumer behavior? *Sustainability, 13*(6), 3466. https://doi.org/10.3390/su13063466
57. Walaszczyk, A., Koszewska, M., & Staniec, I. (2022). Food Traceability as an element of sustainable consumption—pandemic-driven changes in consumer attitudes. *International Journal of Environmental Research and Public Health, 19*(9), 5259. https://doi.org/10.3390/ijerph19095259
58. Yang, Y., Li, O., Peng, X., & Wang, L. (2020). Consumption trends during the COVID-19 crisis: How awe, coping, and social norms drive utilitarian purchases. *Frontiers in Psychology, 11*. https://doi.org/10.3389/fpsyg.2020.588580
59. Śmiglak-Krajewska, M., & Wojciechowska-Solis, J. (2021). Consumer versus organic products in the COVID-19 pandemic: Opportunities and barriers to market development. *Energies, 14*(17), 5566. https://doi.org/10.3390/en14175566

Arundhati Das Chatterjee is a Research scholar in Management at St. Xavier's University, Kolkata and working as Research Assistant at Father Lafont Centre for Excellence in Research and Innovation in St. Xavier's University, Kolkata. She is M.Sc. in Economics from the University of Calcutta and a Diploma holder in Marketing management from the Narsee Monjee Institute of Management Studies. She has received a National Scholarship Merit Certificate (Secondary Level). She has a prior experience in Post Graduate Management teaching in B-schools. She has published papers related to sustainability. Her area of interest includes Sustainability, Behavioural economics and consumer behaviour.

Soma Sur is a Professor and Former Dean of Xavier Business School of St. Xavier's University, Kolkata and is presently the Hony. Director of Father Lafont Centre for Excellence in Research and Innovation at the University. She has published widely in national and international journals and book chapters in countries like USA, Malaysia, Philippines, Sri Lanka, Indonesia, and India. She has also presented papers in various national and international conferences. She has been widely appreciated in international forums for her lecture deliberations and invited talks in Loyola Marymount University, Los Angles, California, USA, Jiatong University, Bejing, China, and Keio University, Japan.

The manufacturer's authorised representative in the EU is Springer Nature Customer Service Centre GmbH, Europaplatz 3, 69115 Heidelberg, Germany. If you have any concerns regarding our products, please contact ProductSafety@springernature.com

Printed and bound by CPI Group (UK) Ltd, Croydon, CR0 4YY
01/10/2026
02225936-0014